# Origami⁷

The Proceedings from the
7th International Meeting on
Origami in Science, Mathematics, and Education

## Volume 4:

## Engineering Two

# Origami[7]

The Proceedings from the
7th International Meeting on
Origami in Science, Mathematics, and Education

## Volume 4:

## Engineering Two

## Editorial Board

Robert J. Lang, Mark Bolitho and Zhong You
Norma Boakes, Chris Budd, Yan Chen,
Mary Frecker, Simon Guest, Thomas Hull,
Yves Klett, Jun Mitani, Jorge Pardo,
Glaucio Paulino, Mark Schenk, Tomohiro Tachi,
Ryuhei Uehara, and Patsy Wang-Iverson

© 2018 OSME
All rights reserved

Printed and bound by CPI Group (UK) Ltd,
Croydon, CR0 4YY
ISBN (Vol 1): 978-1-911093-89-3
ISBN (Vol 2): 978-1-911093-90-9
ISBN (Vol 3): 978-1-911093-91-6
**ISBN (Vol 4): 978-1-911093-92-3**
ISBN (Set of 4): 978-1-911093-93-0

Published by Tarquin
Suite 74, 17 Holywell Hill
St Albans AL1 1DT
United Kingdom

info@tarquingroup.com

# Origami[7] Volume 4 Contents

# Origami[7] General Preface

Origami[7] is a collection of papers published for the **7th International Meeting on Origami in Science, Mathematics and Education (7OSME),** held at Oxford University in the United Kingdom from September 4–7, 2018. 7OSME is the seventh conference in a series dedicated to research in the applications of origami and folding in the conference title fields, as well as in technology, design and history.

The First International Meeting of Origami Science and Technology, December 6–7, 1989, was organised by Professor Humiaki Huzita at the Casa di Ludovico Ariosto, in Ferrara, Italy. This conference brought together people interested in the mathematics and application of origami. Five years later the Second International Meeting of Origami Science and Technology, November 29–December 2, 1994, was held at Seian University of Art and Design, in Shiga, Japan. Organised by Professor Koryo Miura, the meeting grew to include presentations in origami design, education and history.

The third meeting, March 19–11, 2001, was organised by Professor Thomas Hull, at Asilomar, Pacific Grove, California, USA. The meeting adopted the title, the Third International Meeting of Origami Science, Mathematics, and Education. The acronym OSME was adopted for the series and the meeting became known as 3OSME. Subsequent meetings have followed this convention.

The fourth meeting, 4OSME, September 6–10, 2006, was organised by Robert J. Lang, at the California Institute of Technology, Pasadena, California. The fifth, 5OSME, July 13–17, 2010, was organised by Eileen Tan, Benjamin Tan, and Patsy Wang-Iverson and was held at the Singapore Management University, Singapore. 6OSME, August 10–13, 2014, was organised by Ichiro Hagiwara, Koryo Miura and Toshikazu Kawasaki, and was held at the University of Tokyo, Japan.

7OSME, 4–7 September 2018, was held at Oxford University, Oxford, United Kingdom. The conference was organised by a team led by Professor Zhong You, Oxford University, that included Mark Bolitho, former Chairman of the British Origami Society; Professor Yan Chen, Tianjin University; Professor Simon Guest, Cambridge University; Professor Chris Budd, Bath University; and Mark Schenk, University of Bristol. 7OSME marked a return of the OSME conference to Europe, the first since the inaugural meeting in 1989.

Selected papers presented and published for 7OSME comprise the volume that you are now reading.

Both presentations and published papers for 7OSME were reviewed by a dedicated committee chaired by Robert J Lang. The committee comprised 17 members: Norma Boakes, Mark Bolitho, Chris Budd, Yan Chen, Mary Frecker, Simon Guest, Tom Hull, Yves Klett, Jun Mitani, Jorge Pardo, Glaucio Paulino, Mark Schenk, Tomohiro Tachi, Ryuhei Uehara, Patsy Wang-Iverson and Zhong You. The committee coordinated a peer review process through a group of more than 130 reviewers.

During the review process, 207 two-page abstracts were received for consideration. The authors of 166 abstracts were invited to submit full papers. Of these 110 full paper submissions were received, resulting in the selection of 96 papers that were selected for publication.

Papers published in Origami[7] were spread across six categories: 11 papers in art and design, 5 papers in education, 2 papers in history, 5 papers in other science applications, 23 papers in mathematics, and 50 papers in engineering.

In the nearly 30 years since the first OSME conference was held, the scientific and technical aspects of origami have exploded in interest across the fields represented in this work. The connections between folding and diverse fields have proven to be subtle, deep, and rich; origami research may now be found at mainstream technical conferences ranging from materials research to mechanical engineering, robotics, and computer science. Despite the expansion of venues where origami research may be found, the OSME series of conferences continues to grow, a testament to the vibrancy and potential in origami and its affiliated fields. Each OSME conference has grown in size as new applications of folding are explored and developed. Engineering has seen particular interest; half of the published papers in this work came from this area of research. The expansion of origami as an academic field of study has been driven in part by dedicated university research centres at major universities around the world, including postgraduate programmes in Oxford University in the UK, Tianjin University in China, and nationally funded programs in applications of origami at numerous universities in the USA.

Beyond engineering, other folding fields are still well represented. In this work, Volume One includes papers on art and design, education, and science. Papers published range from the exploration of folding in architecture, to presenting folding sculpture as contemporary art. There are also studies of new creative folding processes and methods. The papers categorised as Science include research into new and emerging research areas.

Volume Two includes papers on the mathematics of folding. Papers range from exploration of the mathematics inherent in folded structures, to the use of folding techniques to explore and illustrate mathematical concepts. Several papers look into the mathematics connected with specific folding structures, including the Miura-ori pattern and origami tessellations.

Volumes Three and Four include papers on engineering. Papers published include research into diverse folding applications and products, adding strength to defensive shields, reinforcing tube strength, development of space sails, and the applications of folding processes to robotics. Some papers also explore the mechanical properties of the Miura fold, curved folding, and tessellations.

Origami[7] contains a unique collection of papers. Thanks to the efforts of the review team and the authors, who wrote, re-wrote and responded to reviews, the collection represents the leading edge of applied origami research.

The OSME conference brings together practitioners from different disciplines, connected by their common interest in folding and its application. The sharing of these ideas through published papers will lead to further collaboration and inspiration. The papers presented give an idea of the current state of applied origami. They also give some indication of new and emerging trends that we will see explored in the years ahead.

We hope that you enjoy Origami[7].

Mark Bolitho, Zhong You, and Robert J. Lang,

On behalf of The Editors of Origami[7]

September 2018

# Origami[7] Volume Four Preface

Origami, the ancient art of creating unique patterns and shapes through paper folding, is inspiring engineers to design multi-functional materials and structures that bend, stretch and curve, overcoming various traditional design constraints. While mathematicians are interested in the geometric aspects of origami objects, such as the foldability of origami patterns, engineers, are more focused on developing novel application-oriented structures and materials with deployable features.

In the recent years, origami has been applied to various high-end engineering applications at varying scales. These range from shape-changing materials and surfaces with enhanced and interesting properties, to medical devices and the construction of space satellites.

The third and fourth volume of Origami[7] includes papers that relate to the application of origami and folding in Engineering. Volume three includes papers relating to, Deployable structures inspired by origami, Structural design using origami, Fabrication, and Modular origami and metamaterials.

- *Deployable structures inspired by origami*: Origami and folding mechanisms can be applied to structural systems to create transformable objects. The papers in this section explore the application of origami to specific designs. These include the paper the "Design of an Origami-Inspired Deployable Aerodynamic Locomotive Fairing" (authored by Tolman et al.), "Deployable Sub-module for Space Solar Power Station Based on Origami", (authored by Tang et al.). The construction of a "Rigidly Foldable Rotational Erection System (RES)" (authored by Miyamoto). The application of a light weight defensive shield is explored in the paper "Origami-Based Deployable Ballistic Barrier" (authored by Seymour et al.).

- *Structural design using origami*: Origami can be applied to an object's structural designs to develop new inherent mechanical properties and behaviours. "A Generative Shape Grammar for Piecewise Cylindrical Surfaces and Curved-Crease Origami" looks at a general design approach to curved folding (authored by Gattas) and the paper "Curved-Crease Origami with Multiple States" explores the structural designs made with curved folds as opposed to straight line creases (authored by Lee et. al.). "Programmable 3-D surfaces using origami tessellations" (authored by Yuan et. al.) and "Designing Self-Blocking Systems with Non-Flat-Foldable Degree-4 vertices" (authored by Foschi and Tachi), look at how folding can change structural properties of a surface. The paper "Multiphysics Origami: Achieving Tunable Frequency Selective Surfaces from Origami Principles", (authored by Novelino et al.), explores how folding can change the tuneable frequency of selective surfaces

- *Fabrication*: Advances in fabrication techniques such as 3D printing, has opened up the potential for industrial-scale production of origami-based objects. As a result, this area of research has is now receiving greater attention from the scientific community. However, the fabrication of origami-based materials and structures can be challenging. Specifically, the replication of complicated creases and facets required to create the desired physical properties and behaviour of an object. This section of Origami[7] covers various aspects of fabrication relating to origami objects such as "Folding Fabrication of Curved-Crease Origami Spindle Beams" (authored by Cui et al.), "Fold Printing: Using Digital Fabrication of Multi-Materials for Advanced Origami Prototyping", (authored by Gardiner et al.) and "Kirigami fabrication of shaped, flat-foldable cellular materials based on the Tachi-Miura polyhedron" (authored by Calisch and Gershenfeld).

- *Modular origami and metamaterials*: Modular origami is a technique of paper folding involving more than one sheet of paper to create larger and more complicated structure. With this technique, each of the sheets of papers is folded to form a specifically designed module. The modules are assembled in two or three dimensions. These modular constructions (and conventional origami tessellations) can be used to create metamaterials with unique properties. This section of Origami[7] comprises two interesting studies on the mechanical behaviour and interactive designs of such modular origami objects, "A Modular Origami-inspired Mechanical Metamaterial" (authored by Yang and You) and "An Interactive Design System for Deltahedron-based Modular Origami" (authored by Tsuruta).

Zhong You

September 2018

# Microstructures using a Single Cell Origami Technique

*Qian He, Takaharu Okajima, and Kaori Kuribayashi-Shigetomi*

**Abstract:** *This paper describes using only a single technique, cell origami, to form various shapes of three-dimensional (3D) cell co-culture microstructures. This technique combines cell traction forces with origami folding. Cells are cultured on microplates that are fabricated by standard lithography. Through their traction forces, the cells can fold the microplates to directly form 3D cell co-culture microstructures from the 2D surface. In previous research, numbers of 3D cell co-culture microstructures were created. In this research, we develop various shapes of these 3D microstructures using cell origami. Depending on the design of the microplates, different shaped 3D cell co-culture microstructures can be easily and rapidly formed using this technique. Therefore, forming (1) 3D structures with (2) co-culturing cells inside and (3) various shapes, 3 important factors for mimicking human tissues, can be achieved easily using a single technique, cell origami.*

## 1 Introduction

Fabrication of *in vitro* 3D structures to mimic tissues *in vivo* is a challenge for regenerative medicine. Tissues are 3D structures that are formed by many different cell types. To mimic different tissues, three factors — 3D, co-culturing of different types of cells and the various shapes required — should be considered. Recently, several techniques have been developed to create *in vitro* 3D cell-laden structures [1-5]. These involve micro-sized structures that have various shapes, including blocks [2], fibres [3-5] and spheroids [2, 6]. However, different techniques such as microfluidics or moulding are required to produce each shape. Therefore, a single technique that allows the creation of various shaped 3D microstructures that mimic *in vivo* tissues is desirable in terms of time efficiency and practicality.

Cell origami was developed by our group and is a technique that combines cell traction forces and origami [7]. Previously, we produced dodecahedron cell co-culture microstructures easily and rapidly using this technique that enabled us to co-culture different types of cells inside the structures [8].

In this research, we produce various shapes of 3D cell co-culture microstructures using a single technique of cell origami (Figure 1). Different cell types are sequentially seeded and cultured on engineered microplates. These microplates have various shapes and are fixed to a flat glass surface by a sacrificial

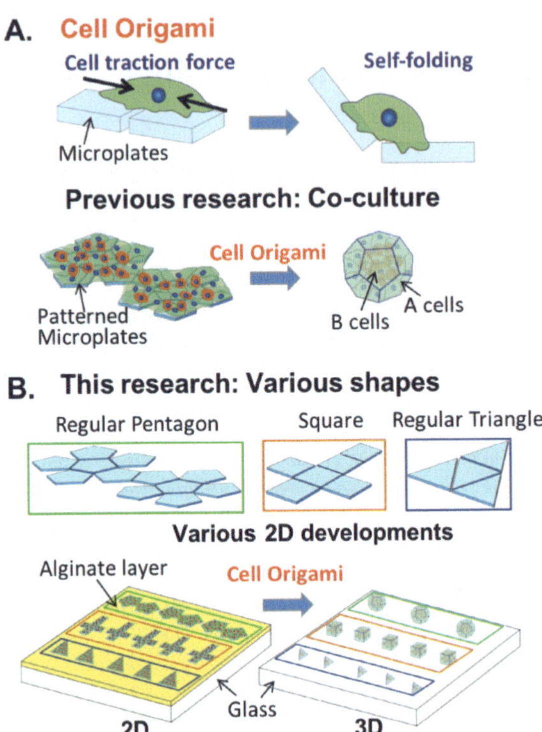



**Figure 1:** A, The cell origami technique. The theory behind the cell origami technique is that traction forces produced during cell growth can be used to self-fold the microplates and create 3D microstructures. In our previous research, dodecahedron microstructures containing co-cultured cells were created. B, In the present study, three different microplates were produced that allowed for three different types of 3D cell co-culture microstructures to be formed by this single technique, cell origami.

layer. The plates are detached from the surface as the sacrificial layer is degraded. This allows the cells to pull the plates by their traction forces, thereby forming 3D microstructures directly from the 2D microplates. During this process the cells wrap themselves inside the structure (Figure 1A). The shapes of the 3D co-culture microstructures can be easily changed by changing the shapes of the microplates (Figure 1B).

## 2  Materials and methods

### 2.1  Fabrication of the cell origami substrate

2.1.1 Production of the sacrificial layer

In total, 10 mg/mL sodium alginate (Wako, Japan) was spin-coated onto a $20 \times 20$ mm hydrophilic glass surface (Matsunami, Japan). By dipping the coated glass with $CaCl_2$ solution, a calcium alginate layer was formed on the glass surface. This microplates to be lifted off from the glass surface by cell traction forces.

## 2.1.2 Preparation of the microplates

Parylene C (Specialty Coating Systems, USA) was used to produce the microplates because it is biocompatible with cell cultures [9]. It is also transparent, which allows for visualization of the cells inside the formed microstructures. Parylene microplates were prepared using standard Micro Electro Mechanical Systems (MEMS) techniques as previously described [8]. Briefly, a 3-μm-thick parylene layer was deposited on the calcium alginate layer by chemical vapour deposition using a parylene deposition machine (LABCOTERPDS2010; Specialty Coating Systems, USA). This was followed by Aluminium (Al) vapour deposition (Figure 2I). Using a photolithographic technique, Al can be patterned through the use of a photoresist (PR) and a glass mask (Figure 2II). After the Al etching, the parylene regions that were not covered with Al were removed by $O_2$ plasma to expose the glass surface and define the shapes of the parylene microplates (Figure 2III). Next, a 2-methacryloyloxyethyl phosphorylcholine (MPC) polymer, that exhibits inhibitory properties against protein and cell adsorption [10], was applied to the exposed glass surface to prevent cells from attaching to the glass surface (Figure 2IV). Before culturing the cells, the Al with MPC polymer on the surface of the microplates was removed using an alkaline solution, NMD-3 developer, to expose the final parylene microplates surface (Figure 2V). Finally, 20 mg/mL fibronectin (FN) solution was applied to coat the microplates (Figure 2VI) to increase the attachment of cells and microplates.

Glass masks with various shapes were used to determine the shapes of the microplates. According to the different designs of masks, various shapes of microplates can be created.

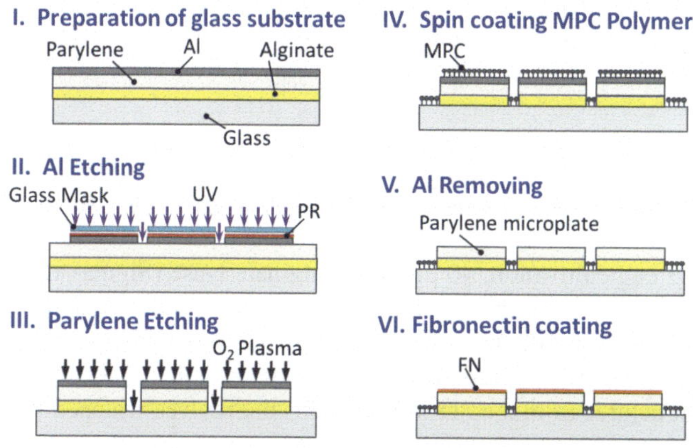


**Figure 2:** Steps in the production of microplates.

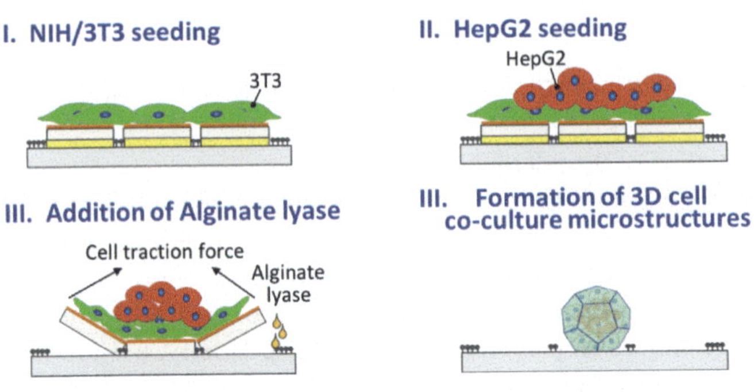


**Figure 3:** Co-culturing of NIH/3T3 and HepG2 cells on microplates and the self-folding process to form 3D microstructures.

## 2.2 Seeding and culturing cells on the origami substrate

NIH/3T3 cells (TKG299; RIKEN Cell Bank, Japan) and HepG2 cells (TKG0205; RIKEN Cell Bank, Japan) were used as the cell co-culture combination in this research. NIH/3T3 cells are fibroblast cells and HepG2 cells are liver cancer cells. The combination of these two types of cells is a classical model for cell co-culture research. These cells were cultured in High Glucose Dulbecco's modified Eagle's medium (HG-DMEM; Sigma-Aldrich, USA) with 10% foetal bovine serum (FBS; Sigma-Aldrich) and 1% penicillin-streptomycin solution (P/S; Sigma-Aldrich) at 37 °C with 5% $CO_2$. The cells were treated with 0.25% trypsin and collected by centrifugation to remove the trypsin.

Up to 1 mL of NIH/3T3 cell suspension ($5 \times 10^5$ cells/mL) was seeded onto the glass substrate with the microplates in a 40-mm non-adherent dish. After 4 h of cultivation, non-adherent cells were removed by changing the medium. NIH/3T3 cells were cultured for 24 h to allow them to form a confluent monolayer on the entire area of the microplates (Figure 3I). Next, a HepG2 cell suspension was seeded on top of the NIH/3T3 monolayer at the same seeding concentration and volume. Non-adherent HepG2 cells were also removed after 4 h of cultivation (Figure 3II).

## 2.3 Folding of microplates using cell traction forces to form 3D microstructures

After 4 h of HepG2 cell cultivation, 40 µg/mL of alginate lyase (Sigma-Aldrich) was added, as described previously [9]. The alginate lyase rapidly degrades the alginate sacrificial layer under the microplates to allow the self-folding process to commence. The cells then directly folded the microplates using their traction forces to form 3D microstructures from the 2D with the cells themselves wrapped inside (Figure 3III-IV).

NIH/3T3 and HepG2 cells were stained with CellTracker Green CMFDA Dye and CellTracker Red CMTPX Dye (Invitrogen, USA), respectively, before seeding onto the microplates. After the formation of the 3D microstructures, the cells inside were observed using a fluorescence microscope (DP72; Olympus, Japan) through the transparent parylene.

## 3   Results

### 3.1   Production of microplates with different shapes

Three different microplate shapes — a regular pentagon, a square and a regular triangle — were successfully produced. The side length of each microplate was 50 μm, and the distance between adjacent microplates was 5 μm (Figure 4). Our results show that by using a photolithographic technique, it is easy to design microplates with various shapes, and many microplates can be produced in parallel.

### 3.2   Self-folding processes

After the NIH/3T3 and HepG-2 cells were seeded and cultured on the microplates, self-folding of the microplates was activated by the addition of alginate lyase. The microplates started to peel off from the glass substrate, while at the same time the traction forces produced by the cells folded the detached microplates together (Figure 5). For the formation of the dodecahedron microstructure, the microplates at the edges started to be folded first by the cells (Figure 5I-II). Next, the cells at the centre of the shape folded the two central microplates (Figure 5III) and the final 3D microstructure was successfully formed. (Figure 5IV). This result shows that the cells can fold the microplates together automatically and easily via their cell traction forces to form 3D microstructures directly from 2D.

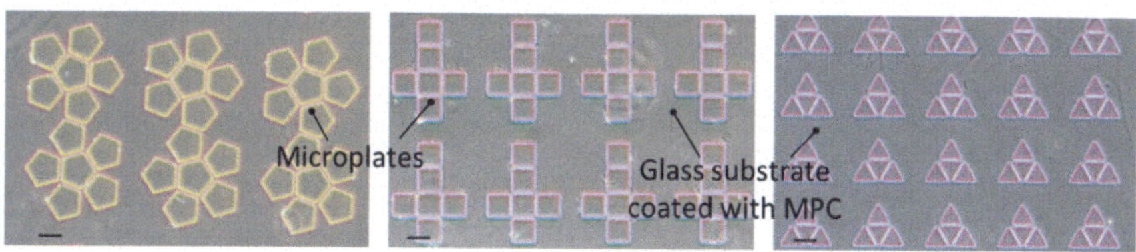


**Figure 4:** Microplate shapes. Three shapes — regular pentagon, square and triangle — were produced. Scale bar: 50 μm.

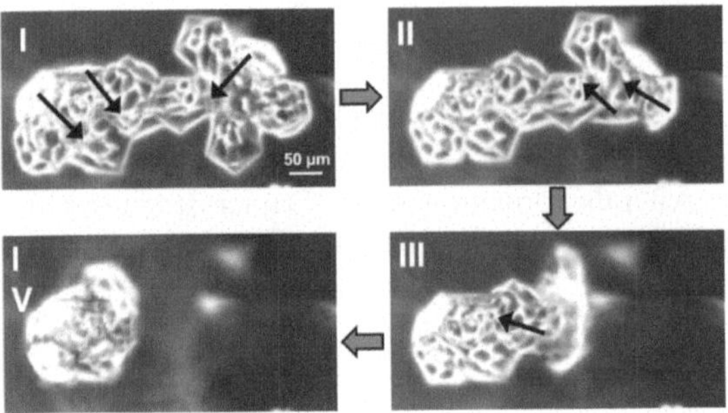

**Figure 5:** The cell origami self-folding process. After alginate lyase was added, the cells progressively folded the microplates together from the edges to the centre using cell traction forces to form a dodecahedron.

### 3.3  Formation of 3D cell co-culture microstructures with various shapes

Before folding commenced, the HepG2 (red) and NIH/3T3 (green) cells were cultured on the microplates for 28 h and 4 h, respectively (Figure 6). The NIH/3T3 cells occupied all areas of the microplates, and the HepG2 cells were on top of the NIH/3T3 cells. After the alginate sacrificial layer was degraded by the addition of alginate lyase, the 2D microplates were self-folded by the traction force of the NIH/3T3 cells to directly form 3D cell co-culture microstructures.

Depending on the different shapes of the microplates — pentagon, square and triangle — different shapes of 3D microstructures — dodecahedrons, cubes and tetrahedrons — were created. The NIH/3T3 cells covering the HepG2 cells were observed inside the 3D microstructures. These results demonstrate that 3D cell co-culture microstructures can be easily and rapidly formed from 2D directly using only a single technique, cell origami technique, without the need for any complicated processes or machinery.

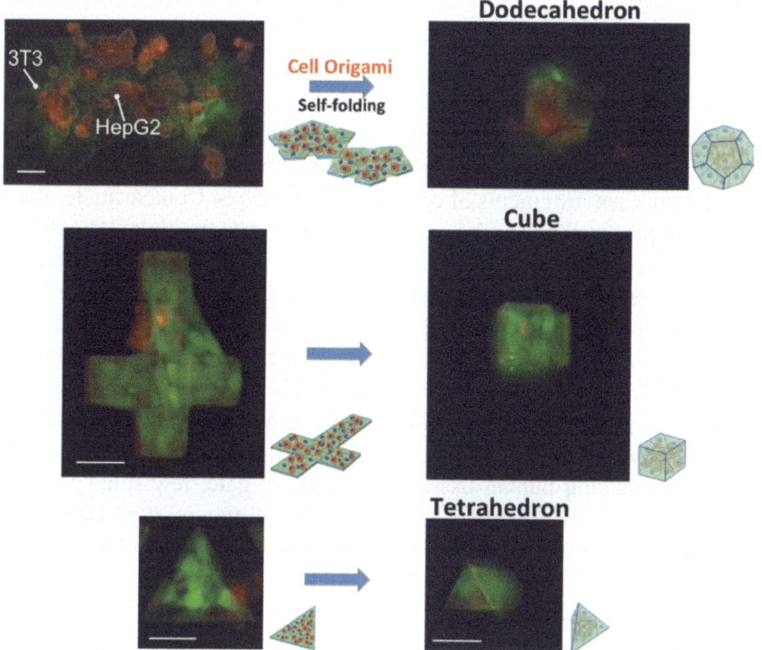


**Figure 6:** 3D cell co-culture microstructures with various shapes. According to the different shapes of the 2D microplates, various shapes of 3D cell co-culture microstructures were created using the cell origami technique. Scale bar: 50 μm.

## 4  Discussion

To generate human tissues *in vitro*, it is essential to be able to: (1) create 3D microstructures; (2) co-culture different types of cells inside these structures; and (3) create different shaped microstructures. Our research shows that cell origami is an efficient and practical method to achieve these three important factors. Without any complicated processes or extra machinery, and using cell traction forces, various shapes of 3D cell co-culture microstructures can be automatically formed directly from a 2D surface using cell origami techniques.

In comparison with our previous research in which only a single shape was created, three different shaped 3D cell co-culture microstructures — dodecahedron, cube and tetrahedron — were created using the cell origami technique in this study. Our results show that the different 3D microstructure shapes can be formed easily by using different 2D microplate shapes.

The results showed intact shapes of 3D cell co-culture microstructures (Figure 6). To create complete 3D cell co-culture microstructures without misfolding, it is important that the cells occupy all the microplates as a confluent monolayer such that they can form sufficient bridges and traction between adjacent microplates to allow folding. In our previous research, we found that the optimized to achieve the formation of 3D microstructures with various shapes.

There is enhanced maintenance of cell function in cells grown in 3D co-culture conditions in comparison with cells grown in 2D-culture conditions [11]. For drug screening, 3D cell co-culture microstructures might be more suitable for imitating the 3D microenvironments of cells than 2D surfaces. Comparing to other self-folding techniques, cell origami can be used to generate multiple 3D cell co-culture microstructures simply and rapidly without the need for any complicated processes or machinery [8], making it an ideal method for generating 3D cell co-culture microstructures for drug screening applications.

Recent researchers have found that the shapes of cells influence their functions [12-13] in 2D conditions. In the human body, tissues and organs are 3D structures which have specific shapes. These shapes might relate to the functions of the cells. Therefore, shape itself should be regarded as one of the factors which is important for mimicking human tissues. However, to date, few studies of 3D microstructure formation have focused on this aspect. Moreover, it is difficult to control the shapes of 3D microstructures using the other available techniques [2]. Cell origami has the advantage in that by simply changing the designs of microplates, 3D cell co-culture microstructures of various shapes can be easily created. Therefore, the 3D cell co-culture microstructures created by cell origami technique might be useful for regenerative medicine applications.


# 5  Conclusions

We have demonstrated that by using only a single technique, cell origami, 3D cell co-culture microstructures with various shapes can be easily and rapidly produced. Using different cell combinations, these 3D microstructures might be mimic human tissues, useful for drug screening and regenerative medicine applications.


# References


[1] Gurkan, U. A., Tasoglu, S., Kavaz, D., Demirel, M. C. & Demirci, U. Emerging technologies for assembly of microscale hydrogels. Adv. Healthc. Mater., 1, 149–158 (2012).

[2] Matsunaga, Y. T., Morimoto, Y. & Takeuchi, S. Molding cell beads for rapid construction of macroscopic 3D tissue architecture. Adv. Mater., 23, 90–94 (2011).

[3] Onoe H, Okitsu T, Itou A, Kato-Negishi M, Gojo R, Kiriya D, Sato K, Miura S, Iwanaga S, Kuribayashi-Shigetomi K, Matsunaga YT, Shimoyama Y, Takeuchi S. Metre-long cell-laden microfibres exhibit tissue morphologies and functions. Nat. Mater. 12, 584–90 (2013).

[4] Yamada, M., Sugaya, S., Naganuma, Y. & Seki, M. Microfluidic synthesis of chemically and physically anisotropic hydrogel microfibers for guided cell growth and networking. Soft Matter, 8, 3122 (2012).

687 (2013).

[6] Mironov V, Visconti RP, Kasyanov V, Forgacs G, Drake CJ, Markwald RR. Organ printing: tissue spheroids as building blocks. Nat. Biotechnol. 32, 773–785 (2014).

[7] Kuribayashi-Shigetomi, K., Onoe, H. & Takeuchi, S. Cell Origami: Self-Folding of Three-Dimensional Cell-Laden Microstructures Driven by Cell Traction Force. PLoS One, 7, e51085 (2012).

[8] He, Q., Okajima T., Onoe, H., Subagyo, A., Sueoka, K., Kuribayashi-Shigetomi, K. Origami-based self-folding of co-cultured NIH/3T3 and HepG2 cells into 3D microstructures. Sci. Rep., 8, article number: 4556, doi:10.1038/s41598-018-22598-x (2018).

[9] Chang TY, Yadav VG, De Leo S, Mohedas A, Rajalingam B, Chen CL, Selvarasah S, Dokmeci MR, Khademhosseini A. Cell and protein compatibility of parylene-C surfaces. Langmuir, 23, 11718–11725 (2007).

[10] Ishihara K, Iwasaki Y, Ebihara S, Shindo Y, Nakabayashi N. Photoinduced graft polymerization of 2-methacryloyloxyethyl phosphorylcholine on polyethylene membrane surface for obtaining blood cell adhesion resistance. Colloids Surf B Biointerfaces, 18: 325–335 (2000).

[11] Lu HF, Chua KN, Zhang PC, Lim WS, Ramakrishna S, Leong KW, Mao HQ. Three-dimensional co-culture of rat hepatocyte spheroids and NIH/3T3 fibroblasts enhances hepatocyte functional maintenance. Acta Biomater., 1, 399–410 (2005).

[12] Kristopher A. Kilian, Branimir Bugarija, Bruce T. Lahn, and Milan Mrksich Geometric cues for directing the differentiation of mesenchymal stem cells. PNAS, 107 (11) 4872-4877 (2010).

[13] von Erlach TC, Bertazzo S, Wozniak MA, Horejs CM, Maynard SA, Attwood S, Robinson BK, Autefage H, Kallepitis C, Del Río Hernández A, Chen CS, Goldoni S, Stevens MM. Cell-geometry-dependent changes in plasma membrane order direct stem cell signalling and fate. Nat. Mater., 17, 237–242 (2018).


## Acknowledgments

We thank Professor Shoji Takeuchi at The University of Tokyo for assisting with the glass mask production, and Professor Kazuhiko Ishihara at The University of Tokyo for providing the MPC polymer. We also thank Masaru Sakuma and Kazumi Nakahata for their technical assistance with the mask design and microplate production. This work was supported by a Grant-in-Aid for Young Scientists (B; JSPS KAKENHI Grant Number JP17K14094) from the Japan Society for the Promotion of Science, a Grant-in-Aid for Scientific Research (B; JSPS KAKENHI Grant Number JP26286030) and the Shiseido female research foundation. Part of this work was conducted at Hokkaido University and was supported by the Nanotechnology Platform Program of the Ministry of Education, Culture, Sports, Science and Technology (MEXT), Japan.

---

Qian He

Hokkaido University, Kita 8, Nishi 5, Kita-ku, Sapporo, Hokkaido, 060-0808

Japan E-mail: heather@ist.hokudai.ac.jp

Japan E-mail: okajima@ist.hokudai.ac.jp

Kaori Kuribayashi-Shigetomi  (corresponding author)
Hokkaido University, Kita 8, Nishi 5, Kita-ku, Sapporo, Hokkaido, 060-0808
Japan. E-mail: kaori@ist.hokudai.ac.jp

# Singular Behaviour on Folding Path Characterised by Rigid Foldability Analysis

*N. Watanabe*

***Abstract:*** *A formulation of the deformable condition in the singular flat state of a rigid origami model is presented, which incorporates the second-order term. The derivation is shown to correspond to a linkage problem. In addition, the singular behaviours of certain models on a folding path are characterised, considering their conformity to the rigid foldability condition.*

## 1  Introduction

A rigid origami model is useful for designing kinetic structures, and the characteristics of these unstable structures provide many interesting problems for structural engineers (e.g. [Filipov et al. 15]). The definition of a rigid origami model is that all flexure occurs along well-defined lines; thus, a rigid origami model can be considered to be a developed linkage model. Indeed, [Demaine and O'Rourke 07] presents comparisons between a linkage model and a rigid origami model. In previous research, an analysis method using a generalised inverse matrix was shown to be effective for stability classification and extraction of the deformable modes for truss structures with many hinges [Tanaka and Hangai 85, Hangai and Kawaguchi 91]. These problems about deformability of unstable truss structures correspond to the linkage rigidity problem. Further, [Connely and Whiteley 96] have proposed the concept of 'second-order rigidity' for linkages. The concept of a high-order mechanism or high-order rigidity was also introduced [Chen 11, Vassart et al. 00].

The rigid origami model was examined using a Gauss map [Huffman 76, Miura 91], and the condition that every facet angle and folding angle around a vertex must satisfied was investigated using a rotational matrix [Belcastro et al. 09, Kawasaki 97]. Further, a method of extracting a rigid foldable mode based on the condition of infinitesimal motion of the folding angles with every crease was developed [Tachi 09, Watanabe and Kawaguchi 09].

This paper refines the extraction of the rigid foldable mode from the perspective that the flat state is a singular state, focusing on the second-order term in the constraint equation presented by [Watanabe 15]. In addition, certain specific crease patterns are characterised. Recently, a simple method of mountain-valley (M-V) assignment that is rigid foldable for a single vertex crease pattern was proposed in [Abel et al. 16]. In comparison to that approach, the procedure proposed in this study also allows treatment of every possible folding motion angle that is rigid foldable, along with discussion of the rigid foldability of some multi-vertex models.

The remainder of this paper is organised as follows. In section 2, the formulation of the existing method using generalised inverse matrices to extract the non-extensional deformable modes for the linkage model is adjusted, focusing on the correspondence between the deformability of particular linkage structures with consideration of high-order terms in the constraint equation. In section 3, this formulation is applied to a rigid origami model and the characteristics of the flat-state singularity are demonstrated. To extract the rigidly deformable modes of the folding angles, both numerical and geometrical methods are shown. In section 4, based on examples of crease patterns, the validity of the proposed method is demonstrated. Finally, in section 5, specific behaviours of some models on the folding path are characterised by the rigid foldability condition.

## 2  Extraction of deformable mode for inextensible linkage

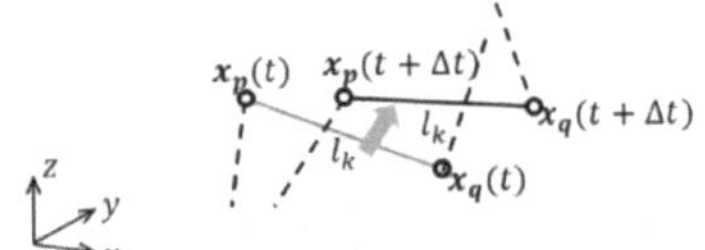


*Figure 1: Constraint condition of linkage model.*

In this section, the derivation of the homogeneous solution extracted from the constraint condition is presented, based on [Tanaka and Hangai 85, Hangai and Kawaguchi 91]. The constraint condition of the linkage model is shown in Figure 1. The length of member $k$ connecting nodes $p$ and $q$ with positions $x_p$, $x_q$, respectively, is represented by $l_k$; this condition can be expressed as follows:

$$g_k = \|x_p - x_q\| - l_k = 0 \tag{1}$$

Further, the constraint condition for the structure composed of $m$-members and $n$-nodes that satisfy Equation 1 can be described as follows:

$$(g_1, g_2, \cdots, g_m)^T = A(x_1, x_2, \cdots x_n) = 0 \tag{2}$$

After time $\Delta t$, the constraint equation given by Equation 2 remains sufficient; hence, the Maclaurin expansion of $A(0 + \Delta t)$ is obtained:

$$A(x_1, x_2, \cdots)|_{t=0} + A'\dot{x}|_{t=0}\Delta t + \frac{1}{2}(A'\ddot{x} + \dot{x}^T A''\dot{x})|_{t=0}\Delta t^2 + o(\Delta t^3) = 0 \tag{3}$$

Here, $x = (x_1, x_2, \cdots)^T = (x_{1x}, x_{1y}, x_{1z}, x_{2x}, x_{2y}, x_{2z} \cdots)^T$ and $\dot{\blacksquare}$ and $\blacksquare'$ represent differentiation by time and $x_i$, respectively.

For any $\Delta t$, the following relations must be satisfied:

$$A'\dot{x} = \sum_{i=1}^{3n} \frac{\partial g_k}{\partial x_i}\dot{x}_i = 0 \tag{4}$$

$$A''\dot{x} + \dot{x}^T A''\dot{x} = \sum_{i=1}^{3n} \frac{\partial g_k}{\partial x_i}\ddot{x}_i + \sum_{i=1}^{3n}\sum_{j=1}^{3n} \frac{\partial^2 g_k}{\partial x_i \partial x_j}\dot{x}_i\dot{x}_j = 0 \tag{5}$$

Here, $A'$, $A''$ are second- and third-order tensors, respectively. For example, for a member satisfying the condition expressed in Equation 1, Equation 4 becomes

$$\left[\frac{(x_p - x_q)}{l_k} \quad -\frac{(x_p - x_q)}{l_k}\right]\begin{bmatrix}\dot{x}_p \\ \dot{x}_q\end{bmatrix} = 0 \tag{6}$$

The homogeneous solution for Equation 4 is expressed as follows, using the generalised inverse matrix $A^+$:

$$\dot{x} = [I - A'^+ A']\alpha \tag{7}$$

In addition, the orthogonal basis for the null space, which represents the solution of the rigidly deformable mode in an infinitesimal range, is expressed as

$$H = [h_1 h_2 \cdots h_{3n}] = [I - A'^+ A'] \tag{8}$$

Next, considering the second-order term, Equation 5 can be modified to yield

$$A'\ddot{x} = -\dot{x}^T A''\dot{x} \tag{9}$$

The existence condition of the solution $x$ in the equation $Cx = b$ is expressed as $[I - CC^+]b = 0$. Thus, the existence condition of the value of $\ddot{x}$ in Equation 9 is given as

$$[I - A'A'^+][-\dot{x}^T A''\dot{x}] = 0 \tag{10}$$

Here, the term $(I - A'A'^+)$ is the orthogonal projection to Ker $A' = (\text{Im}A')^\perp$. When the deformation mode extracted by Equation 8 also satisfies Equation 10, the mode can be considered to be deformable on the second-order level, or in the finite range. Based on [Tanaka and Hangai 85], Figure 2 shows an example of the deformable mode in the infinitesimal/finite range.

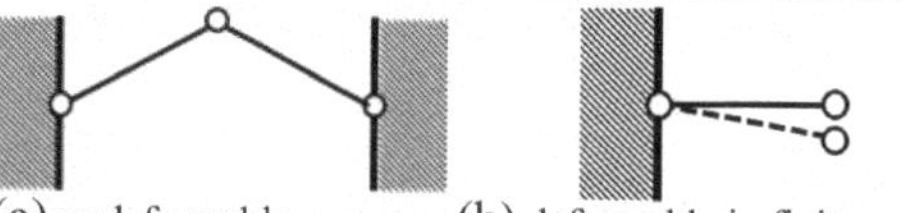

(a) undeformable          (b) deformable in finite range

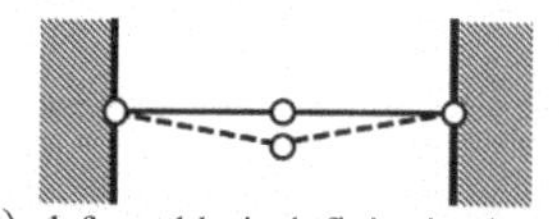

(c) deformable in infinitesimal range and undeformable in finite range

**Figure 2:** *Deformability in infinitesimal/finite range.*

# 3  Extension to rigid origami model: Extraction of deformable mode

As noted above, the rigid origami model can be considered to be a developed linkage model. As regards description of the deformation mode in the development diagram, the M-V information of each crease line is more recognisable than the deformation information of each node. Therefore, the folding angles are taken as variables instead of the nodal point displacements in this section.

## 3.1  Derivation of constraint condition

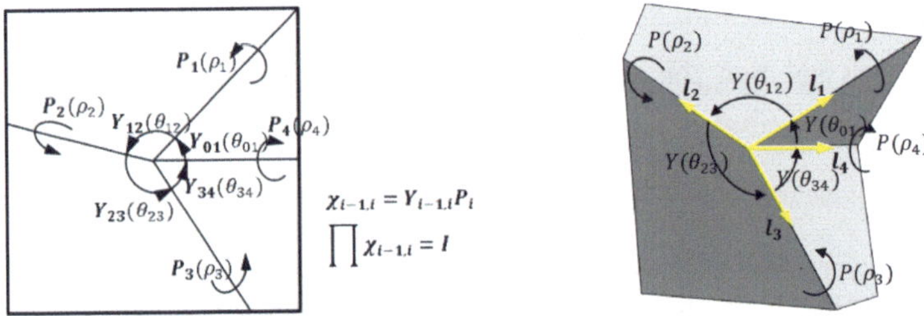

**Figure 3:** *Constraint condition for angles in a rigid origami model.*

Consider the condition for angles around a single vertex. The condition of a circuit around a vertex with $n$ crease lines, a series of facet angles $\theta_1, \theta_2, \cdots \theta_n$, and a series of folding angles $\rho_1, \rho_2, \cdots \rho_n$ ([Belcastro et al. 09, Kawasaki 97]), as shown in Figure 3, is expressed as

$$R(\rho) = \chi_{0,1}\chi_{1,2}\cdots\chi_{n-1,n} = I \tag{11}$$

$$\chi_{i-1,i} = YP = \begin{bmatrix} \cos\theta_{i-1,i} & -\sin\theta_{i-1,i} & 0 \\ \sin\theta_{i-1,i} & \cos\theta_{i-1,i} & 0 \\ 0 & 0 & 1 \end{bmatrix}\begin{bmatrix} 1 & 0 & 0 \\ 0 & \cos\rho_i & -\sin\rho_i \\ 0 & \sin\rho_i & \cos\rho_i \end{bmatrix} \tag{12}$$

After time $\Delta t$, the constraint equation given in Equation 12 remains sufficient. Therefore, the Maclaurin expansion of $A(0 + \Delta t)$ similarly applies to linkage models, where

$$R(\rho_1, \rho_2, \cdots)|_{t=0} + R'\dot{\rho}|_{t=0}\Delta t + \frac{1}{2}(R'\ddot{\rho} + \dot{\rho}^T R''\dot{\rho})|_{t=0}\Delta t^2 + o(\Delta t^3) = I \tag{13}$$

Here, $\blacksquare$ represents time differentiation and $\blacksquare'$ represents differentiation by $\rho_i$. The first-order term of Equation 13 must satisfy the expression

$$\boldsymbol{R}'\dot{\boldsymbol{\rho}} = \left( \sum_{i=1}^{n} \frac{\partial R_{kl}}{\partial \rho_i} \dot{\rho}_i \right) = \frac{\partial \boldsymbol{R}}{\partial \rho_1} \dot{\rho}_1 + \frac{\partial \boldsymbol{R}}{\partial \rho_2} \dot{\rho}_2 + \cdots \frac{\partial \boldsymbol{R}}{\partial \rho_n} \dot{\rho}_n = \boldsymbol{0} \qquad (14)$$

Note that $\boldsymbol{R}'\dot{\boldsymbol{\rho}}$ is a $3 \times 3$ matrix; hence, nine equations for the folding motion are obtained, which can be represented using a $9 \times n$ matrix based on the description given in Equation 4. Then,

$$\boldsymbol{R}'\dot{\boldsymbol{\rho}} = \begin{bmatrix} \dfrac{\partial R_{11}}{\partial \rho_1} & \dfrac{\partial R_{11}}{\partial \rho_2} & \cdots & \dfrac{\partial R_{11}}{\partial \rho_n} \\ \dfrac{\partial R_{12}}{\partial \rho_1} & \dfrac{\partial R_{12}}{\partial \rho_2} & \cdots & \dfrac{\partial R_{12}}{\partial \rho_n} \\ \vdots & \vdots & & \vdots \\ \dfrac{\partial R_{33}}{\partial \rho_1} & \dfrac{\partial R_{33}}{\partial \rho_2} & \cdots & \dfrac{\partial R_{33}}{\partial \rho_n} \end{bmatrix} \begin{bmatrix} \dot{\rho}_1 \\ \dot{\rho}_2 \\ \vdots \\ \dot{\rho}_n \end{bmatrix} = \boldsymbol{0} \qquad (15)$$

Here, $R_{ij}$ represents the $i,j$-component of $\boldsymbol{R}'$. As shown in [Tachi 09], by extracting three independent components, the condition given in Equation 15 can be expressed as

$$\boldsymbol{C}\dot{\boldsymbol{\rho}} = [\boldsymbol{l}_1 \boldsymbol{l}_2 \cdots \boldsymbol{l}_n] \begin{bmatrix} \dot{\rho}_1 \\ \dot{\rho}_2 \\ \vdots \\ \dot{\rho}_n \end{bmatrix} = \sum_{i=1}^{n} \dot{\rho}_i \, \boldsymbol{l}_i = \boldsymbol{0} \qquad (16)$$

Here, $\boldsymbol{l}_i$ is the direction cosine of each crease line. Similar to Equation 8, the rigid deformation mode can be obtained as

$$\boldsymbol{H} = [\dot{\rho}_1 \dot{\rho}_2 \cdots \dot{\rho}_n] = [\boldsymbol{I} - \boldsymbol{C}^+ \boldsymbol{C}] \qquad (17)$$

On the other hand, considering the second-order term of Equation 14, it is necessary to satisfy the relation

$$\boldsymbol{R}'\ddot{\boldsymbol{\rho}} + \dot{\boldsymbol{\rho}}^T \boldsymbol{R}'' \dot{\boldsymbol{\rho}} = \sum_{i=1}^{n} \frac{\partial R_{kl}}{\partial \rho_i} \ddot{\rho}_i + \sum_{i}^{n} \sum_{j}^{n} \frac{\partial^2 R_{kl}}{\partial \rho_i \partial \rho_j} \dot{\rho}_i \dot{\rho}_j = \boldsymbol{0} \qquad (18)$$

Although the left side of Equation 18 is composed of nine equations for the constraint condition of each component of the $3 \times 3$ matrix, the number of constraint conditions can be reduced by considering the matrix symmetry. The condition that vector $\dot{\boldsymbol{\rho}}$ is the deformable mode in a finite range is expressed in Equation 19, which is the existence condition when vector $\ddot{\boldsymbol{\rho}}$ is satisfied:

$$[\boldsymbol{I} - \boldsymbol{R}'\boldsymbol{R}'^+][-\dot{\boldsymbol{\rho}}^T \boldsymbol{R}'' \dot{\boldsymbol{\rho}}] = \boldsymbol{0} \qquad (19)$$

The given in Equations 18 and 19 correspond to those of Equations 5 and 10.

## 3.2  Numerical procedure

The problem of obtaining the vector that satisfies Equations 15 and 19 can be treated as a problem of simultaneous equations, where

$$R'\dot{\rho} = 0 \tag{20}$$

$$S(\dot{\rho})_{kl} = ([I - R'R'^{+}][-\dot{\rho}^{T}R''\dot{\rho}])_{kl} = 0 \tag{21}$$

First, when any vector $\dot{\rho}_0$ is substituted for $\dot{\rho}$ on the left side of Equation 20, the result of $S_{kl}(\dot{\rho}_0)$ becomes the specific value $r_{kl}$ instead of zero, as in the usual case. To bring the residual value $r_{kl}$ to zero, the Newton-Raphson method is applied as follows.

Let the substituted $\dot{\rho}$ be $\dot{\rho}^*$ and the obtained residual value be $r_{kl}{}^*$. The expression of the Newton equation corresponding to Equations 20 and 21 then becomes Equation 22a. Next, a series of $S_{kl}$ and $r_{kl}{}^*$ are expressed, as shown in equation 22b. The vector can be obtained using the generalised inverse, as in Equation 22c. Hence, the value of $\dot{\rho}^*$ is updated, as in Equation 22d. By repeating this operation until the residual $r^*$ converges to almost zero, the mode satisfying Equation 21 can be obtained.

$$\begin{bmatrix} R' \\ (\nabla S_{11}(\dot{\rho}^*))^T \\ (\nabla S_{12}(\dot{\rho}^*))^T \\ \vdots \\ (\nabla S_{33}(\dot{\rho}^*))^T \end{bmatrix} [\Delta\dot{\rho}] = \begin{bmatrix} 0 \\ r_{11}{}^* \\ r_{12}{}^* \\ \vdots \\ r_{33}{}^* \end{bmatrix} \tag{22a}$$

$$\begin{bmatrix} R' \\ (\nabla S(\dot{\rho}^*))^T \end{bmatrix} [\Delta\dot{\rho}] = \begin{bmatrix} 0 \\ r^* \end{bmatrix} \tag{22b}$$

$$[\Delta\dot{\rho}] = \begin{bmatrix} R' \\ (\nabla S(\dot{\rho}^*))^T \end{bmatrix}^{+} \begin{bmatrix} 0 \\ r^* \end{bmatrix} \tag{22c}$$

$$\dot{\rho}^* \leftarrow \dot{\rho}^* + \Delta\dot{\rho} \tag{22d}$$

## 3.3  Geometrical interpretation

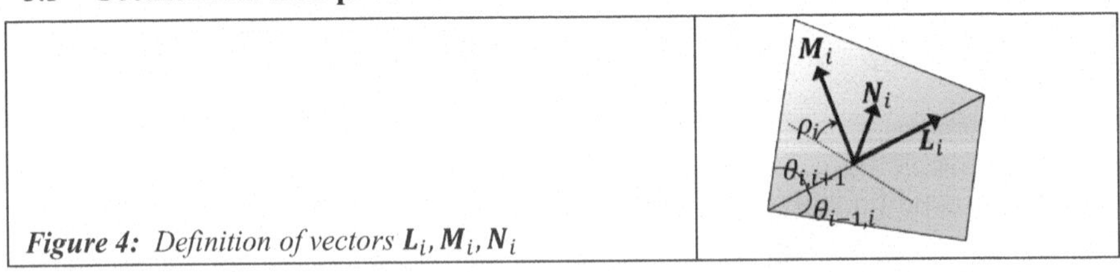


**Figure 4:** *Definition of vectors $L_i, M_i, N_i$*

In this section, the geometrical interpretation of the condition of vector $\dot{\rho}$ for the terms of each order in Equation 13 is discussed. The product from $\chi_{i,i+1}$ to $\chi_{j-1,j}$ as $T_{i\_j}$ is expressed, as indicated in Equation 23. Then, each component of $T_{0\_i}$ has the form given in Equation 24.

$$T_{i\_j} = X_{i,i+1} X_{i+1,i+2} \cdots X_{j-1,j} \tag{23}$$

$$T_{0\_i} = \begin{bmatrix} L_i^x & M_i^x & N_i^x \\ L_i^y & M_i^y & N_i^y \\ L_i^z & M_i^z & N_i^z \end{bmatrix} \tag{24}$$

Here, vector $L_i$ is the direction cosine of crease $i$ and vector $M_i$ is the unit vector on the plane surface created by creases $i$ and $i+1$, which is normal to vector $L_i$. Further, vector $N_i$ is equivalent to $L_i \times M_i$ (Figure 4). Matrix $P_i$ defined in Equation 12 is implemented in accordance with Equation 25; then, the first and second derivatives of matrix $R$ are expressed as detailed in Equations 26 and 27, respectively.

$$\frac{\partial X_{i-1,i}}{\partial \rho_i} = X_{i-1,i} P_i^{-1} \frac{\partial P_i}{\partial \rho_i} = X_{i-1,i} \begin{bmatrix} 0 & 0 & 0 \\ 0 & 0 & -1 \\ 0 & 1 & 0 \end{bmatrix} = X_{i-1,i} Q \tag{25}$$

$$\tag{26}$$

$$\frac{\partial R}{\partial \rho_i} = X_{0,i} \cdots \frac{\partial X_{i-1,i}}{\partial \rho_i} \cdots X_{n-1,0} = T_{0\_i} Q T_{i\_0}$$

$$\tag{27}$$

$$\frac{\partial^2 R}{\partial \rho_i \partial \rho_j} = X_{0,i} \cdots \frac{\partial X_{i-1,i}}{\partial \rho_i} \cdots \frac{\partial X_{j-1,j}}{\partial \rho_j} \cdots X_{n-1,0} = T_{0\_i} Q T_{i\_j} Q T_{j\_0} \quad (i < j)$$

Considering Equation 28 below, Equations 29 and 30 are obtained.

$$[T_{0\_i}][T_{0\_i}]^T = I, \qquad [T_{0\_i}][T_{i\_j}][T_{j\_0}] = I \tag{28}$$

$$\tag{29}$$

$$\frac{\partial R}{\partial \rho_i} = T_{0\_i} Q T_{i\_0} = T_{0\_i} Q T_{0\_i}^T = \begin{bmatrix} 0 & -L_i^z & L_i^y \\ L_i^z & 0 & -L_i^x \\ -L_i^y & L_i^x & 0 \end{bmatrix}$$

$$\tag{30}$$

$$\frac{\partial^2 R}{\partial \rho_i \partial \rho_j} = T_{0\_i} Q T_{i\_j} Q T_{j\_0} = T_{0\_i} Q T_{0\_i}^T T_{0\_j} Q T_{0\_j}^T$$

$$= \begin{bmatrix} 0 & -L_i^z & L_i^y \\ L_i^z & 0 & -L_i^x \\ -L_i^y & L_i^x & 0 \end{bmatrix} \begin{bmatrix} 0 & -L_j^z & L_j^y \\ L_j^z & 0 & -L_j^x \\ -L_j^y & L_j^x & 0 \end{bmatrix}$$

$$= \begin{bmatrix} -L_i^z L_j^z - L_i^y L_j^y & L_i^y L_j^x & L_i^z L_j^x \\ L_i^x L_j^y & -L_i^z L_j^z - L_i^x L_j^x & L_i^z L_j^y \\ L_i^x L_j^z & L_i^y L_j^z & -L_i^y L_j^y - L_i^x L_j^x \end{bmatrix} \quad (; i < j)$$

By extracting the three independent components from Equation 29, which correspond to the first-derivative term, Equation 16 can be obtained, as $\left(L_i^x, L_i^y, L_i^z\right)^T = l_i$. As regards the second-order derivative term, the case of $i = j$ is expressed in Equation 31. Then, considering the fact that $T_{0\_i}$ is an orthogonal matrix, Equation 32 can be obtained. This expression is considered to be one case of Equation 30.

$$\frac{\partial^2 R}{\partial \rho_i^2} = X_{0,i} \cdots X_{i-1,i} P_i^{-1} \frac{\partial^2 P_i}{\partial \rho_i^2} \cdots X_{n-1,0} = T_{0\_i}\begin{bmatrix} 0 & 0 & 0 \\ 0 & -1 & 0 \\ 0 & 0 & -1 \end{bmatrix} T_{i\_0} \tag{31}$$

$$\frac{\partial^2 R}{\partial \rho_i^2} = \begin{bmatrix} (L_i^x)^2 - 1 & L_i^y L_i^x & L_i^z L_i^x \\ L_i^x L_i^y & (L_i^y)^2 - 1 & L_i^z L_i^y \\ L_i^x L_i^z & L_i^y L_i^z & (L_i^z)^2 - 1 \end{bmatrix} \tag{32}$$

For the case that $i > j$, the component of matrix $\dfrac{\partial^2 R}{\partial \rho_i \partial \rho_j}$ is obtained by replacing $i$ with $j$ in Equation 30, because $\dfrac{\partial^2 R}{\partial \rho_i \partial \rho_j} = \dfrac{\partial^2 R}{\partial \rho_j \partial \rho_i}$.

In the infinitesimal range considered only to the first-derivative term around the vertex, the product of the mode of the folding motion $\dot{\rho}$ and matrix $C$ which is composed of direction cosines of every crease line, must be vector $0$, as stated in Equation 16. This condition can be expressed as follows: 'the sum of fold-line unit vectors $l_i$, weighted by $\dot{\rho}_i$, should be equal to zero'. Figure 5 (left) shows an example of the non-flat state in which Equation 16 is applicable, and the rank $(C)$ of this case is equal to 3. On the other hand, in the flat state, rank $(C)$ becomes deficient, because every crease line is on the same plane. Thus, the number of constraint conditions in Equation 16 is insufficient. This state is considered to be the singular state obtained from flatness, as shown in Figure 2(c). Figure 5 (right) shows an example of two M-V modes satisfying Equation 16. Case (B) is not foldable, but case (A) is. This example shows that the condition given by Equation 16 is insufficient in the flat state; the condition in consideration to the second-derivative term is also required.

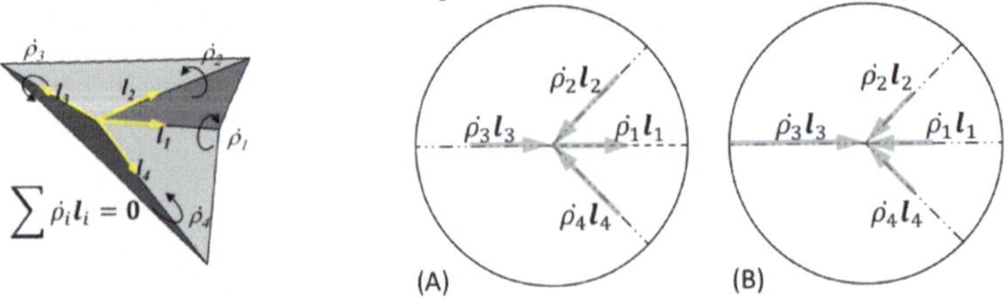


**Figure 5:**   *Satisfiability of condition (Equation 16) in singular state.*

Next, Equation 30 is considered in the flat state. Because $L_i^z = 0$, the following is obtained:

$$\frac{\partial^2 R}{\partial \rho_i \partial \rho_j} = \begin{bmatrix} -L_i^y L_j^y & L_i^y L_j^x & 0 \\ L_i^x L_j^y & -L_i^x L_j^x & 0 \\ 0 & 0 & -L_i^y L_j^y - L_i^x L_j^x \end{bmatrix} \tag{33}$$

The third term of Equation 13 is expressed as

$$R' \ddot{\rho} + \dot{\rho}^T R'' \dot{\rho} = \sum_{i=1}^{n} \begin{bmatrix} 0 & 0 & L_i^y \\ 0 & 0 & -L_i^x \\ -L_i^y & L_i^x & 0 \end{bmatrix} \ddot{\rho}_i$$

$$+ \sum_{\substack{i,j \\ i \leq j}}^{n} \begin{bmatrix} -L_i^y L_j^y & L_i^y L_j^x & 0 \\ L_i^x L_j^y & -L_i^x L_j^x & 0 \\ 0 & 0 & -L_i^y L_j^y - L_i^x L_j^x \end{bmatrix} \dot{\rho}_i \dot{\rho}_j$$

$$+ \sum_{\substack{i,j \\ i > j}}^{n} \begin{bmatrix} -L_j^y L_i^y & L_j^y L_i^x & 0 \\ L_j^x L_i^y & -L_i^x L_j^x & 0 \\ 0 & 0 & -L_j^y L_i^y - L_j^x L_i^x \end{bmatrix} \dot{\rho}_i \dot{\rho}_j = 0 \tag{34}$$

Considering the condition in the first-order derivative range given in Equation 35, the expression given in Equation 36 is obtained.

$$\sum_{i=1}^{n} \dot{\rho}_i \, l_i = \sum_{i=1}^{n} \dot{\rho}_i \begin{Bmatrix} L_i^x \\ L_i^y \end{Bmatrix} = 0 \tag{35}$$

$$\left( \sum_{i=1}^{n} L_i^\blacksquare \dot{\rho}_i \right) \left( \sum_{i=1}^{n} L_i^* \dot{\rho}_i \right) = \sum_{i=1}^{n} \sum_{j=1}^{n} L_i^\blacksquare L_j^* \, \dot{\rho}_i \dot{\rho}_j = 0 \tag{36}$$

In the above equations, $\blacksquare, *$ represent $x$ or $y$. From Equation 36, the diagonal components of the left side of Equation 34 becomes zero, accordingly. Further, component (1,2) of Equation 34 can be deformed to the following equivalent equation, considering Equation 36:

$$\sum_{j=1}^{n} \sum_{i=1}^{n} \frac{\partial^2 R_{12}}{\partial \rho_i \partial \rho_j} \dot{\rho}_i \, \dot{\rho}_j = \sum_{i \leq j}^{n} L_i^y L_j^x \dot{\rho}_i \dot{\rho}_j + \sum_{i > j}^{n} L_j^y L_i^x \dot{\rho}_i \dot{\rho}_j \tag{37}$$

$$= \sum_{\substack{i,j}} L_i^y L_j^x \dot{\rho}_i \dot{\rho}_j - \sum_{\substack{i,j \\ i>j}} L_i^y L_j^x \dot{\rho}_i \dot{\rho}_j + \sum_{\substack{i,j \\ i>j}} L_j^y L_i^x \dot{\rho}_i \dot{\rho}_j$$

$$= 0 - \sum_{\substack{i,j \\ i<j}} \left( L_i^x L_j^y - L_i^y L_j^x \right) \dot{\rho}_i \dot{\rho}_j = 0$$

From the deformation of Equation 37, the following is obtained:

$$\sum_{\substack{i,j \\ i<j}} \left( L_i^x L_j^y - L_i^y L_j^x \right) \dot{\rho}_i \dot{\rho}_j = \left( \sum_{\substack{i,j \\ i<j}} (\dot{\rho}_i l_i \times \dot{\rho}_j l_j) \right) \cdot e_z = \left( \sum_{j=2}^{n} \left( \sum_{i=1}^{j-1} \dot{\rho}_i l_i \right) \times \dot{\rho}_j l_j \right) \cdot e_z = 0 \qquad (38)$$

The geometrical interpretation of this expression (Figure 6) is as follows: 'while proceeding in anticlockwise order around the vertex, the operation of connecting each vector $\dot{\rho}_i l_i$ to the next one, head-to-tail, allows the realisation of a closed loop'.

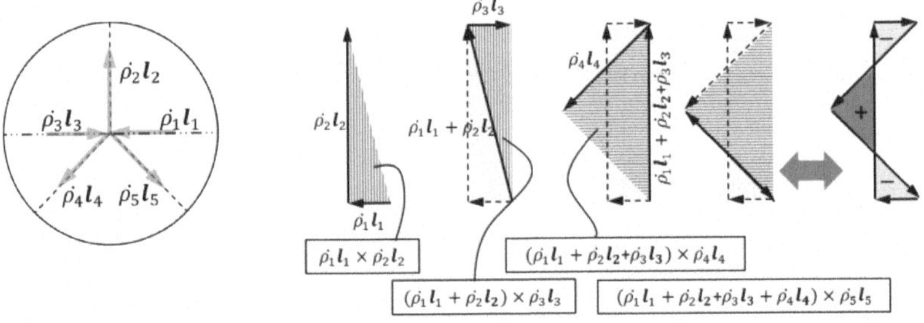

**Figure 6**: *Geometric interpretation of Equation 38.*

# 4. Examples

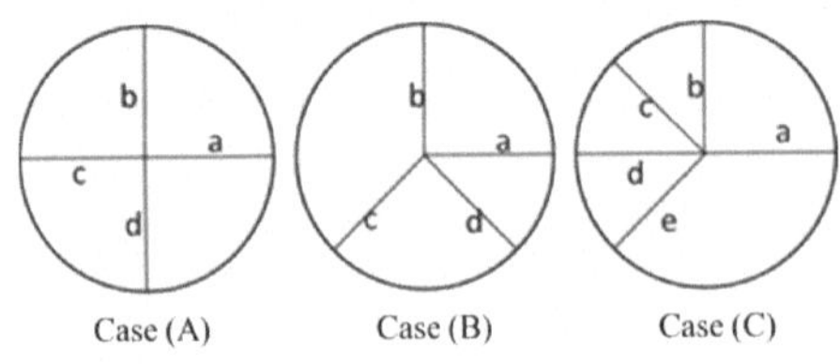

| | Flat Foldable | Rigid Foldable |
|---|---|---|
| Case (A) | ✓ | - |
| Case (B) | ✓ | ✓ |
| Case (C) | - | ✓ |

**Figure 7**: *Examples of crease patterns.*

The validity of the calculation method proposed in section 3 was examined as follows. For $n$ creases, the number of possible M-V assignments is $2^n$. For each of the $2^n$ M-V patterns in each case shown in Figure 7, the valid rigid foldable mode was extracted by performing a convergence calculation.

Here, cases (A) and (B) are flat foldable cases, whereas case (C) is not. On the other hand, cases (B) and (C) are rigid foldable cases, whereas case (A) is not. The calculation procedure is as follows: $\dot{\boldsymbol{\rho}}_0 = (\dot{\rho}_a, \dot{\rho}_b, \cdots) = (\pm 1, \pm 1, \cdots)$ are set as initial values. Then, the sign of each convergence value result $\dot{\boldsymbol{\rho}}^* = (\dot{\rho}_a^*, \dot{\rho}_b^*, \cdots)$ is compared with the initial value. If both signs of the modes about every crease exhibit agreement, the converged mode is considered to be the extracted deformable mode. Here, plus or minus signs are assigned to mountain or valley creases, respectively. Table 1 lists the $\dot{\boldsymbol{\rho}} = (\dot{\rho}_a, \dot{\rho}_b, \cdots)$ values obtained through the above procedure.

**Table 1**: *Extracted combinations of $\dot{\boldsymbol{\rho}}_i$.*

|          | a      | b      | c      | d      | e      |
|----------|--------|--------|--------|--------|--------|
| Case (A) | -      | -      | -      | -      |        |
| Case (B) | 1.207  | 0.500  | 1.207  | -0.500 |        |
|          | 0.500  | -1.207 | -0.500 | -1.207 |        |
|          | -0.500 | 1.207  | 0.500  | 1.207  |        |
|          | -1.207 | -0.500 | -1.207 | 0.500  |        |
| Case (C) | 0.666  | 0.482  | 0.783  | -0.923 | 1.464  |
|          | 0.770  | 1.229  | -1.023 | 0.986  | 0.716  |
|          | 0.491  | 1.147  | -0.300 | -0.231 | 1.322  |
|          | 1.455  | -0.534 | 1.032  | 0.530  | 0.277  |
|          | 0.969  | -0.272 | 1.289  | -0.582 | 0.905  |
|          | -0.969 | 0.272  | -1.289 | 0.582  | -0.905 |
|          | -1.455 | 0.534  | -1.032 | -0.530 | -0.277 |
|          | -0.491 | -1.147 | 0.300  | 0.231  | -1.322 |
|          | -0.770 | -1.229 | 1.023  | -0.986 | -0.716 |
|          | -0.666 | -0.482 | -0.783 | 0.923  | -1.464 |

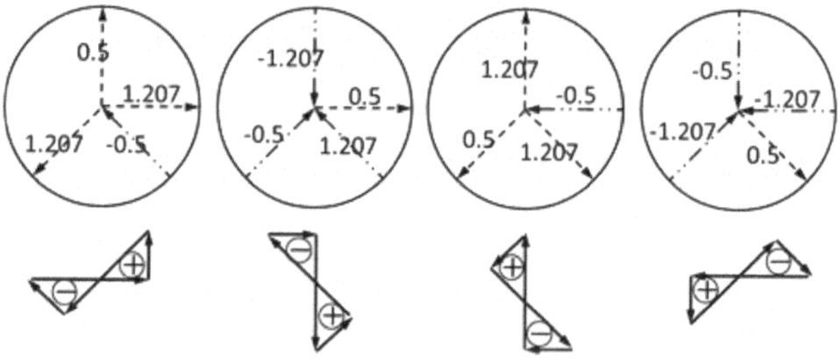

**Figure 8:** *Extracted M-V combinations for case (B).*

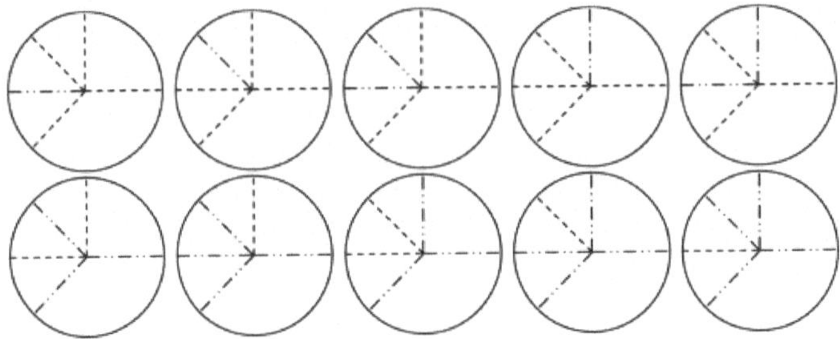

**Figure 9:** *Extracted M-V combinations for case (C).*

It can be observed that these results satisfy the condition proposed in [Abel et al. 16], for simple M-V assignment for creases around the single vertex. That is, all face angles $\theta$ made by adjoining creases which have the same sign should satisfy the condition $0 < \theta < \pi$, whereas all signs of the creases should not be the same. It is found that case (A) is only obtained with consideration of the second-order term; that is, it has no deformable mode. As regards case (B), the M-V combination can satisfy the condition that 'connected vector $\dot{\rho}_i l_i$ can make a closed loop (Figure 8).' Note that the condition $|\dot{\rho}_1| = |\dot{\rho}_3|$, $|\dot{\rho}_2| = |\dot{\rho}_4|$, which holds for a flat foldable four-degree crease pattern [Huffman 76] is satisfied for extracted $\dot{\rho}_i$ mode in Figure 8. Figure 9 shows an extracted result for case (C) as an example; in this case, the combination of the values of $\dot{\rho}_i$ cannot be determined as unique in each mode.

For a multi-vertex pattern, an expansion can be performed using the superposed matrix about $\boldsymbol{R}'$ with every vertex, as a numerical method. As a diagram method, the compatibility of every value of $\dot{\rho}_i$ can be checed. Figure 10 shows a sample judgement of the multi-vertex model. However, some counterexample models that are not yet deformablé exist, satisfying this compatibility conditions [Tachi 12]. For these crease patterns, which have 'spider-web' shapes, other additional constraint conditions must be considered.

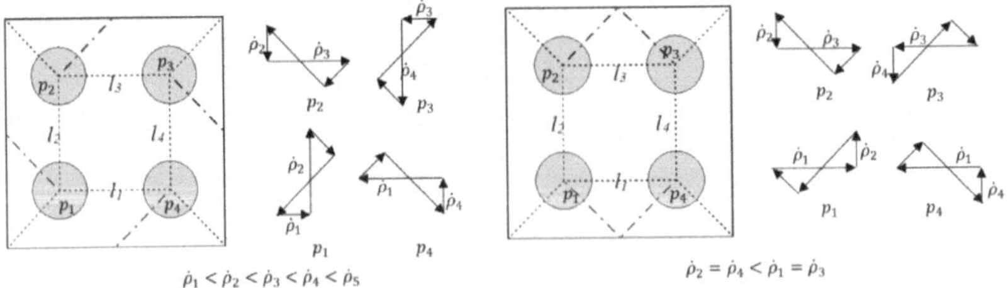

**Figure 10:** *Application to multi-vertex model.*

# 5. Characterisation of singular behaviour on folding path

## 5.1    Application to non-flat disk model

In this section, application of the rigid foldability condition in consideration to first order and second order stated in Equations 16 and 19 to a non-flat disk model for which the sum of the face angles is inequivalent to $2\pi$ around a vertex is discussed. For the deployment process, it is sufficient to apply the first condition only. However, if the non-flat disk surface is folded up flatly, both conditions must be applied, as in the case of a flat disk.

Figure 11 (model I) shows an example of folding motion mode extraction for a pyramid shape for which the face angles around vertex $(\theta_{1,2},\ \theta_{2,3},\ \theta_{3,4},\ \theta_{4,1})$ are $(\pi/3,\pi/3,\pi/3,\pi/3)$.

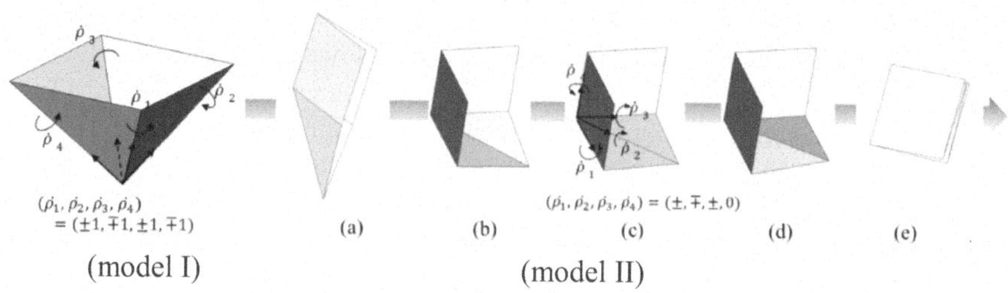

Figure 11: *Application to non-flat model.*

However, there also exists the case in which the folding motion around a particular fold line cannot be determined by applying the first condition to the deformation path. Figure 11 (model II) shows such a model, for which face angles $(\theta_{1,2},\theta_{2,3},\theta_{3,4},\theta_{4,1})$ are $(\pi/4,\pi/4,\pi/2,\pi/2)$. When the numerical procedure is applied to state (c) in this model, the solution of motion angle $\dot{\rho}_4$ converges to 0. However, state (c) is actually deformable to (b) or (d). The peculiarity of the non-flat model is shown in the asymmetry of the possible sign of folding motion angle $\dot{\rho}_i$. Here, $\dot{\rho}_4$ is deformable only in the increase direction, while $\dot{\rho}_1$, $\dot{\rho}_2$, $\dot{\rho}_3$ are deformable in both the increase and decrease directions in state (c). Focusing on the transition of the value $\pi - \rho_4$ in the (b)–(c)–(d) path, it can be noted that it increases on the (b)–(c) path and decreases on the (c)–(d) path. Then, $\dot{\rho}_4$ should be zero, because $\rho_4$ is at a minimum at state (c). This behaviour is a feature of the non-flat disk surface, which has fold lines on which the range of the dihedral angle is restricted. Thus, for the non-flat disk model on which the dihedral angle range is limited, a case exists for which the crease line contributes to rigid folding even if there exists a crease line for which the motion angle should be zero in a particular state.

## 5.2   Non-smooth point in deformation path

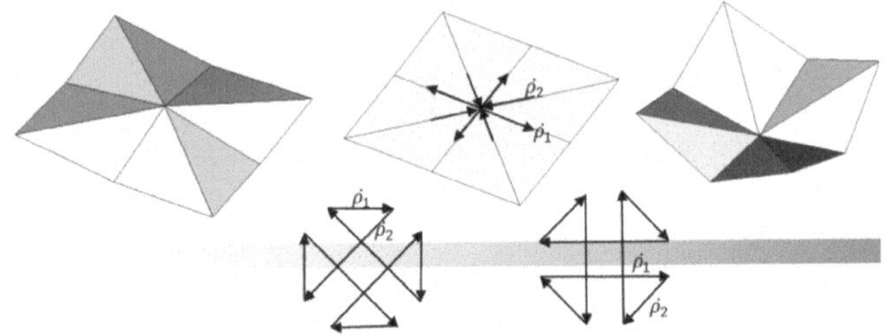

**Figure 12:** *Example of non-smooth deformation path.*

Considering the two conditions presented in section 3, the non-smooth deformation path for a flat disk surface can be charcterised. As an example, the non-smooth deformation path from pop-up to pop-down for a water-bomb crease pattern (Figure 12), even with the same M-V assignment was examined [Tachi and Hull 17]. Here, the symmetry such that eight folding angles were formed by $(\rho_1, \rho_2, \rho_1, \rho_2, \cdots, \rho_2)$ was assumed. Figure 13 shows the transition of the $(\dot{\rho}_1, \dot{\rho}_2)$ values on the folding path, and it can observed that the ratio of $\rho_1$ to $\rho_2$ changes discontinuously in the state $(\rho_1, \rho_2) = (0,0)$.

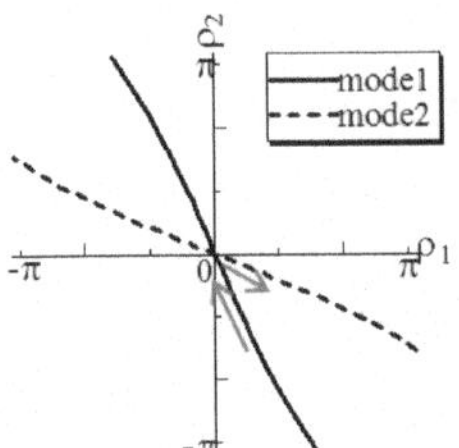


**Figure 13:** *Transition of $(\rho_1, \rho_2)$ values on folding path.*

In the flat state, two types of vector diagram that satisfy the second condition can be drawn. From the drawn diagrams, it can be determined that the possible modes of the folding motion are $(|\dot{\rho}_1|, |\dot{\rho}_2|) = (1, 1 + \sqrt{2})$ (mode 1) and $(1 + \sqrt{2}, 1)$ (mode 2). Here, it should also be noted that convexity at the centre vertex differs before and after the flat state. The concave-convex characteristics at a single vertex are determined from the folding angles around the vertex, by comparing the sum of the folding angles that contribute mountain and valley creases, respectively. For this model, the centre vertex in mode 1 was judged to be concave, because $\sum_{Mcreases} |\dot{\rho}_i| - \sum_{Vcreases} |\dot{\rho}_i| > 0$, whereas the centre vertex in mode 2 was judged to be convex, because $\sum_{Mcreases} |\dot{\rho}_i| - \sum_{Vcreases} |\dot{\rho}_i| < 0$.

As the value sign is inverted for a mountain crease compared to a valley crease, the expression $\sum_{Vcreases}|\dot{\rho}_i| - \sum_{Mcreases}|\dot{\rho}_i|$ can be given as $\sum \dot{\rho}_i$.

As stated above, it is possible for two modes to exist that force the appearance of different convexities at the centre vertex, although they have the same M-V combination. The singular behaviour of this model in the flat state can be explained by considering the fact that it is impossible to move continuously between these two diagrams by adjusting the vector lengths.

## 5  Conclusions

This paper demonstrated the correspondence of the linkage model formulation to that of the rigid origami model as regards extraction of the rigid deformable modes. Through investigation of the terms of each order obtained through Maclaurin expansion of the constraint condition, the flat state was found to be the singular state for which the second-order derivative term must be considered. A numerical method and a graphical method derived from the formulation were proposed to extract rigid foldable modes; the validity of these methods was demonstrated through examples. Certain specific models exhibiting nonconformity to the conditions were presented, and their singular behaviours on the folding path were successfully characterised.

## References

[Abel et al. 16] Abel, Z., Cantarella, J., Demaine, E.D., Eppstein, D., Hull, T.C., Ku, J. S., Lang, R. J. and Tachi, T., Rigid Origami Vertices: Conditions and Forcing sets, J. of Computational Geometry, 7(1), (2016) 171–184.

[Belcastro et al. 09] Belcastro, S.-M. and Hull, T.C., A mathematical model for non-flat origami. in Origami3: Proceedings of the Third International Meeting of Origami in Science, Mathematics, and Education, edit by Hull, T.C., A K Peters (2009). 39–51.

[Chen 11] Chen, C., The order of local mobility of mechanisms. Mechanism and Machine Theory, 46, (2011), 1251–1264.

[Connelly and Whiteley 96] Connelly, R. and Whiteley, W., Second order rigidity and pre-stress stability for tensegrity frameworks. SIAM J. Discrete Math. 9(3), (1996), 153–491.

[Demaine et al. 16] Demaine, E. D., Demaine, M. L., Huffman, D. A., Hull, T. C., Koschitz, D. and Tachi, T., Zero-Area Reciprocal Diagram of Origami, Proc. of the IASS Annual Symposium, (2016).

[Demaine and O'Rourke 07] Demaine, E. D. and O'Rourke, J., Geometric folding algorithms, Cambridge Univ. Press, (2007), 43–58.

[Filipov et al. 15] Filipov, E.T. et al, Toward Optimization of Stiffness and Flexibility of Rigid, Flat-Foldable Origami Structures, in Origami6: Proceedings of the 6th

International Meeting of Origami in Science, Mathematics, and Education, edit by Koryo, M. et al., T.C., A K Peters (2009). 409–419.

[Hangai and Kawaguchi 91] Hangai H. and Kawaguchi K., Shape Analysis – generalized inverse and its application, Baifukan, (1991) (in Japanese).

[Huffman 76] Huffman, D. A., Curvature and creases: a primer on paper. IEEE Transactions on Computers, C-25(10), (1976), 1010–1019.

[Kawasaki 97] Kawasaki, T., R($\gamma$)=I. in K. Miura (ed.): Orgami Science and Art: Proceedings of the Second International Meeting of Origami in Science and Scientific Origami, (1997), 31–40.

[Miura 91] Miura, K., A note on intrinsic geometry of origami, in H.Huzita (ed.): Proceedings of the First International Meeting of Origami in Science and Technology, (1991), 239–249.

[Tachi 09] Tachi, T., Simulation of rigid origami, in Origami4: Fourth International Conference on Origami in Science, Mathematics, and Education, edited by Robert J. Lang, A K Peters (2009), 175–187.

[Tachi 12] Tachi, T., Design of Infinitesimally and Finitely Flexible Origami Based on Reciprocal Figures. in Journal for Geometry and Graphics, 16(2), (2012) 223–234.

[Tachi and Hull 17] Tachi, T. and Hull, T.C., Self-Foldability of Rigid Origami, J. of Mechanisms and Robotics, vol.9, (2017) , DOI 10.1115/1.4035558.

[Tanaka and Hangai 85] Tanaka, H. and Hangai Y., Rigid body displacement and stabilization condition of unstable truss structures. in J. Struct. Constr. Eng., AIJ, 356 (1985) 35–43 (in Japanese).

[Vassart et al. 00] Vassart, N., Laporte, R. and Motro, R., Determination of mechanism's order for kinematically and statically indetermined systems. International Journal of Solids and Structures, 37, (2000), 3807–3839.

[Watanabe 15] Watanabe, N., Extraction of foldable mode in a singular state of panel-hinge model, in proceedings of 31th symposium on aerospace and materials, JAXA/ISAS, A6000047014, (2015).

[Watanabe and Kawaguchi 09] Watanabe, N. and Kawaguchi, K., The method for judging rigid foldability, in Origami4: Fourth International Conference on Origami in Science, Mathematics, and Education, edited by Robert J. Lang, A K Peters (2009) 165–174.

---

Naohiko Watanabe
National Institute of Technology, Gifu College, 2236-2 Kamimakuwa, Motosu-city, Gifu 501-0495, Japan, e-mail: watanabe@gifu-nct.ac.jp

# Automated Numerical Process Chain for the Design of Folded Sandwich Cores

*F. Muhs, Y. Klett, P. Middendorf*

**Abstract**: *Tessellation-based cellular structures have shown potential for application as core material in sandwich applications. Compared to other state-of-the-art materials like foams and honeycombs, the geometry of such foldcores can be designed very flexibly. The additional degrees of freedom offer a large optimization potential, but the large parameter space on the other hand makes structured evaluation of the available multitude of geometries much more demanding.*

*The focus of this study is on the development of a new, automated process chain for numerical analysis of the mechanical performance of foldcores, which can provide reliable results quickly, opening up the way to virtually screen and optimize geometries. We demonstrate the process and compare results to hardware tests to judge the prediction quality.*

## 1 Introduction

Foldcores represent a high-performance alternative to commonly used core materials for sandwich constructions. Benefits offered by foldcored application include amongst others weight saving, broad base material choice, or the integration of passive or active functionalities [Klett 13, Grzeschik et al. 11, Herrmann et al. 05]. The mechanical performance of foldcores varies in a wide range and for any given material depends heavily on the shape of unit cell geometry [Klett et al. 15, Zang et al. 16]. Given the large (basically infinite) possible parameter space, structured evaluation or optimization is a time-consuming venture. This remains true even if the tessellation type is restricted to "only" globally flat, Miura-type tessellations (as shown in Figure 1).

The most straightforward way to produce reliable results is to perform physical testing. These tests are always required to eventually evaluate the real-world performance of any construction material. Physical testing however is a time- and resource-consuming task, including the manufacture of the foldcore material, the assembly of samples, testing and analysis. For large parameter studies, or even optimization, physical testing is mostly not efficient or even feasible. [Grzeschik and Drechsler 10].

Analytical approaches to evaluate mechanical performance of foldcore material have so far not been very successful [Thorsteinsson 02], or not properly validated [Miura 72]. While analytical models would provide a very convenient way to optimize geometries, the available models do not include relevant material properties or imperfections that have a large effect on strength or modulus of resulting structures, and they are limited to very specific parameter ranges for a given geometry.

New approaches to use simplified models to numerically analyze folded structures efficiently are under development, and show good results especially for the prediction of kinematic behavior [Filipov et al. 17, Schenk and Guest 11].

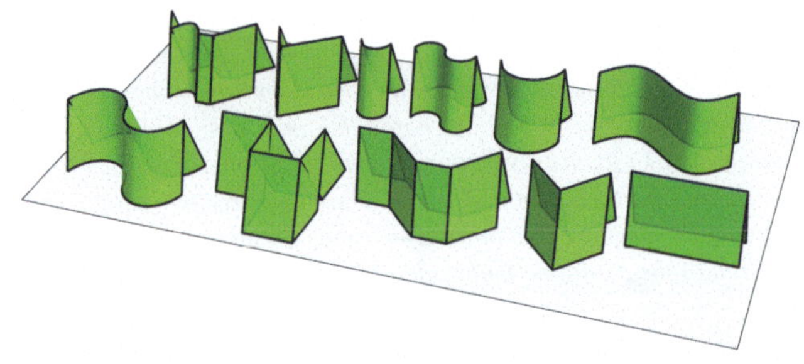

**Figure 1:** *Examples for different foldcore unit cell designs.*

However, for the accurate prediction of material properties, a fully fledged finite element analysis (FEA) is still the best available tool. FEA can incorporate arbitrary physical aspects into the simulation, and make use of realistic material models in static and dynamic scenarios [Fathers et al. 15, Fischer 15a, Heimbs et al. 10, Baranger et al. 11a, Zhou et al. 15]. FEA is generally the most expensive method in terms of computation, but also an area that constantly benefits from progress in software and hardware architectures.

FEA is not FEA however, and lavish use of computing power is no guarantee for useful results. The setup of a simulation for a folded structure is very straightforward if idealized, readily available geometries are modeled. Unfortunately, the comparison of this simple modeling methodology with physical tests shows poor accordance, because the simulation does not incorporate imperfections, which lead to wildly optimistic predictions of e.g. stiffness and strength [Fischer 12, Heimbs et al. 08].

For a better prediction of the mechanical foldcore performance, detailed models including imperfections and information of the manufacturing process have to be setup. Imperfections can be implemented by randomly manipulating the coordinates of the nodes in a user-given range (so-called node-shaking). Another approach is to superimpose a selected number of buckling modes out of a linear buckling analysis to the ideal core geometry [Heimbs 09]. For an even more physically correct consideration of the imperfections, real foldcore samples have been scanned with an optical 3D measurement system, and assembled into a FE model using these measurements [Fischer 12]. A fully numerical approach with physical background is the simulation of the folding process [Baranger et al. 11b] and to consider the imperfections with these process simulation.

All of these refined methods still require a lot of user-experience and manual tuning in the model generation process. Additionally, to capture all physical effects, especially the buckling and post-buckling of the core walls, explicit solvers are used which usually require the use of dedicated simulation hardware. [Fischer 15b, Kilchert 13, Sturm and Fischer 15].

The objective of this study is the development of a numerical process chain to design and compare foldcores in a quick and efficient way with the use of the FEA. A fully automatable workflow is implemented in conjunction with a data-base to store and compare the simulation data. To reduce the simulation effort, a representative volume element (RVE) is generated using one unit cell, which also incorporates imperfections.

## 2 Unit Cell Design

The first step to get a grip on automation of any simulation using foldcores is to implement an efficient and universally applicable geometry description. Fortunately, within the current scope of isometrically foldable, globally flat Miura-type tessellations, which can be generated by the reflection method [Mitani and Igarashi 11, Klett 13], the geometry of a unit cell can be described using just a few parameters, namely the baseline, the position of the reflection planes, and the initial extrusion direction.

Figure 2a shows these components. The initial major crease line (or baseline) is colored in red. In contrast to the usual discretization of the baseline by polylines, we use a set of piecewise defined, C0-continous quadratic Bèzier spline segments [Salomon 06]. Every spline segment is defined by three control points, shown in black in Figure 2b, and starting and ending points are shared with adjacent spline segments. In case of straight crease segments, the midpoint of each spline is on the creaselines, while for curved segments the midpoint is lies outside. The parallel reflection planes are shown in gray, and defines the Height $H$ of the core.

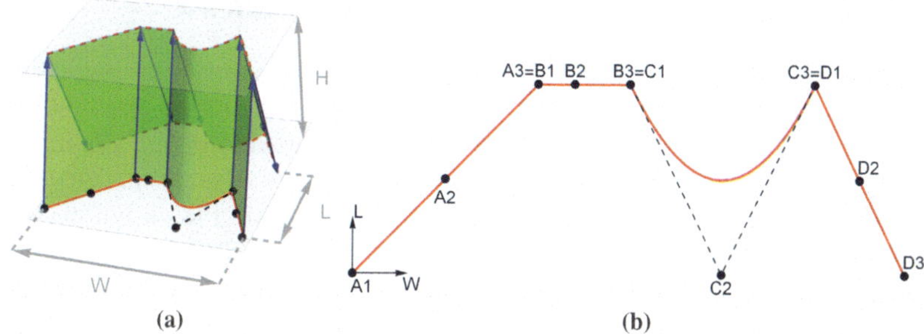


**Figure 2:** *(a) Foldcore unit cell with initial major crease lines in red, reflected major crease lines in dashed red, minor crease lines in blue and reflection planes in gray. Grid points of the spline segments colored in black and the characteristic parameters core height, width and length are indicated by H, W and L. (b) Baseline of (a) with control points for first (A1-A3) to fourth (D1-D3) spline segment.*

The final three-dimensional shape of the unit cell is created by using the reflection method [Klett 13]. The initial major crease line, formulated by the splines, is extruded along a defined direction parametrized by the height $H$ and the longitudinal extension $L$, creating the blue minor crease lines in the process. Successive reflection at the upper and lower bounding planes result in the final unit cell geometry. The reflection paradigm ensures developability, is easily implemented and can generate geometries in realtime. A detailed description of the used design method can be found in [Muhs and Middendorf 18].

Another benefit of this design methodology is the compact representation of the unit cells. All 2D coordinates of the spline control points are defined in the lower base plane in relation to the first spline point $A1$. These coordinates and the characteristic parameters $H$ and $L$ (Figure 2a) can be stored in one design vector. In addition, a unit cell can be modified on the fly to meet specific, physical or geometric criteria like overall dimensions, cell size or density.

From a numerical point of view, the Bèzier spline formulation has another advantage: Compared to already discretized linear approximations of curved segments, the Bèzier spline based surface is optimally suited for automated generation of efficient and high-quality meshes for the final FEA. Like the geometry generation, the mesh generation can be done on the fly, and is not restricted by arbitrary prior discretizations, which can lead to non-representative or inefficient meshes.

In combination, the geometry representation is well suited for automation and optimization. Whole geometry families can be described by a small number of geometric parameters. Especially the inclusion of manipulation of curved crease segments is straightforward, and opens up a large optimization space while keeping the numbers of necessary optimization parameters small. Optimization goals can include any number of properties, which e.g. result in a globally or locally optimal, lightweight structure for a given loading, while still allowing for boundary conditions like ventability or any number of secondary functionality requirements.


## 3  Numerical Simulation of Compression Loadcase

### 3.1  Setup of the Numerical Model

The numerical model setup and the calculation of the foldcores is processed with the commercial FEA software Abaqus,[1] which can be automated via Python.[2] Abaqus comes with its own API to ensure full availability of all Abaqus-specific functions within the Python environment. Data exchange within the design cycle has been implemented using a MariaDB database.[3]

After generation of the ideal geometry (section 2), the next step is to assign a specific material to the structure. In this study for the FEA and the physical tests is polyethyleneterephthalate (PET), a common thermoplastic material. The PET foil comprises a mean thickness of $0.123\,\text{mm}$ and a mean grammage of $174\,\text{g}\,\text{m}^{-2}$. The

---

[1] Abaqus 6.14-5 by Dassault Systèms
[2] Python 2.7.13 64bit with Anaconda distribution, https://anaconda.org/
[3] MariaDB Version 10.1 64bit, https://mariadb.org/

material properties of the specific foil used have been evaluated by prior tensile testing in accordance to ASTM D 822 [ASTM D882-12 12].

Under the assumption of isotropic and symmetric material behavior, the elastic-plastic material model in tabular form in Abaqus is used. The final material curve for the numerical model is shown in Figure 3 and is in good accordance with the physical test data. The parameters for the elastic region as well as the data of the plastic stress $\sigma_Y$ and strain $\varepsilon_Y$ are shown in Table 1.

For the material model only the range up to $\varepsilon_Y = 40\%$ was considered because during compressive testing no larger strains occur. Material failure is not taken into account because the foldcores out of PET fail due to local buckling of the cell walls well below the actual breaking strain of the material.

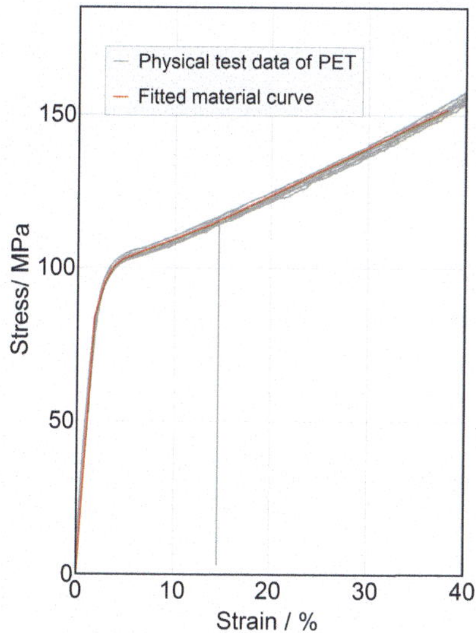

**Figure 3:** *Physical test data and fitted material model data for the PET under tension.*

**Elastic**

| $E \ / \ MPa$ | $v \ / \ -$ | $\rho \ / \ (g/cm^3)$ |
|---|---|---|
| 4897.82 | 0.25 | 1.42 |

**Plastic**

| Yield Pt | $\sigma_Y \ / \ MPa$ | $\varepsilon_Y \ / \ \%$ |
|---|---|---|
| 1 | 83.95 | 0.00 |
| 2 | 95.11 | 0.99 |
| 3 | 100.62 | 1.97 |
| 4 | 103.07 | 2.95 |
| 5 | 114.49 | 12.27 |
| 6 | 151.87 | 36.27 |

**Table 1:** *Mechanical parameters for the PET material model in the elastic and plastic region.*

The mesh element size is calculated in relation to the minimum edge length $l_{min}$ all over the core including major and minor crease lines. This minimal edge length is divided by a scale factor to receive the final element size. Within a mesh convergence study a scale factor of 50 was found to be appropriate and the element size results in $\frac{l_{min}}{50}$.

In case of curved cell walls an additional deviation factor for the mesh generation is introduced which is defined by the deviation between geometry and mesh divided by the element size. To ensure a sufficient and smooth representation of the curved surfaces with the mesh a deviation factor of 0.006 is used.

The mesh is build up with S4R shell elements with reduced integration. For the analysis a static, large-displacement approach with full Newton's method is used. This approach can be used because this study focused on the strength and stiffness of the foldcores under compression, which are evaluated in the range of the occurrence of first, local buckling. For an evaluation of the post-buckling behavior this method will not be feasible and explicit time integration approaches will have to be used.

The boundary conditions representing the compression loading plates are applied via rigid bodies on the upper and lower edges of the unit cell and represent the face sheets. The lower core edges are fixed in all degrees of freedom (DOF). On the upper edges a prescribed displacement is applied in negative Z-direction (Figure 4a) and the remaining DOFs are fixed. Five percent of the core height $H$ are prescribed as a standard displacement to ensure that maximum stress is achieved, independent of the core geometry. Displacement and reaction force are measured at the upper rigid body element.

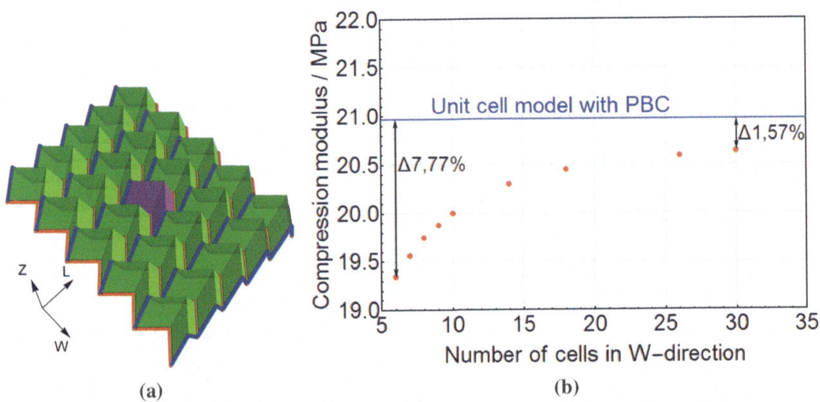


**Figure 4:** *(a) 5x5 Miura-ori foldcore with lower major crease lines in red and cell dividing minor crease lines in blue. Coordinate system indicates the W-, L- and Z-direction. (b) Dependency of compression modulus on the number of cells in W-direction without periodic boundary conditions (PBC). Blue line represents the result of a unit cell with PBC.*

To reduce the model size, which directly influences the calculation time of the analysis, it is important to take boundary conditions between adjacent cells into account. Fortunately, the smallest representative volume element (RVE) can be chosen to be one unit cell: As at the bottom facesheet a rigid connection is assumed (in red), cells do not influence adjacent cells in L-direction.

For coupling in W-direction, a periodic boundary condition (PBC) can be imposed on the blue connections, which allows reduction to one unit cell. Figure 4b shows the comparison between simulation results using a RVE with suitable PBC versus larger grids of cell. The convergence of the RVE model towards larger grids is apparent, and also indicates the danger of using small unit cell grids without proper PBC choice, which emphasize undesirable boundary effects caused by free, unsupported edges.

The calculation time is thusly reduced significantly from several days for the large model up to a few minutes for the RVE unit cell model with the same accuracy of the results.[4] This offers the opportunity to use the modeling approach in an engineering design process to check the potential of chosen foldcore geometries in a quick and easy manner.

The whole model generation and evaluation process is managed by a Python script without any user interaction. All required parameters are taken from the implemented database. This speeds up the modeling and the evaluation significantly and enables the generation and analysis of a magnitude of models in a small period of time.

The mechanical values of strength and stiffness are evaluated using the core height $H$ and the projected core area $W \times L$ (Figure 2a). Evaluation of the compressive modulus is done by taking the gradient of the secant between 30 % and 70 % of ultimate stress.

## 3.2   Imperfections

The consideration of the imperfections introduced by the manufacturing process is essential to realistically predict the mechanical properties of the foldcore. Figure 5 shows a comparison between the stress-strain curve for an ideal and an imperfect geometry, and shows the difference especially in initial compression stiffness, which is significantly overestimated for the ideal geometry.

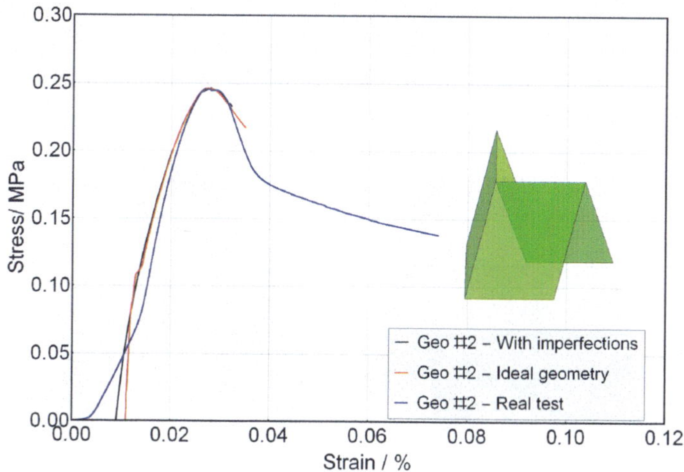


**Figure 5:** *Comparison of a flatwise compression simulation with periodic boundary conditions with and without imperfections and real test of geometry #2. For the ideal geometry, the stiffness is significantly overestimated.*

---

[4]Calculation on one Intel(R) Core(TM) i7-4700MQ CPU @ 2.40GHz with 32GB RAM

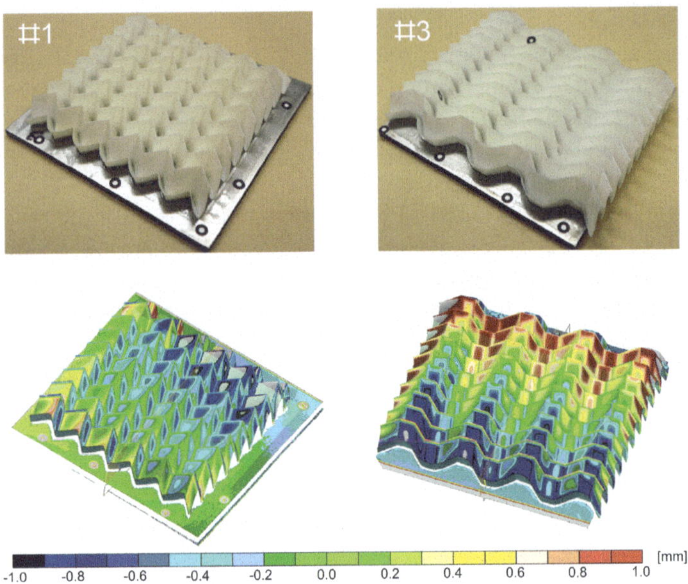

Figure 6: *Real foldcore samples for geometries #1 and #3 out of PET and results of a 3D optical measurement of these geometries compared to ideal CAD data.*

The imperfections of the core were evaluated by using a 3D optical scanner.[5] Samples were manufactured for a set of selected tessellations and adhesively bonded to a steel facesheet at the bottom 6. These samples were then scanned with the ATOS system and the results compared to computer aided design (CAD) models of the ideal geometries.

Figure 7: *FEA model of a Miura-ori unit cell with superimposed imperfections (elevated for better visualization).*

---

[5]ATOS by GOM GmbH, http://www.gom.com/de/3d-software/gom-systemsoftware/atos-professional.html

The deviation between the measured and ideal geometry can be visualized and calculated as shown in Figure 6 for two selected tessellations. In contrast to prior work, the scanning process is only done once for exemplary samples to generate templates for the imperfection modeling, which are then applied to arbitrary ideal geometries.

These deviations can be approximated with continuous functions, which are derived from the scanned data. For each cell wall type of the tessellation like squares, parallelograms or curved segments, a characteristic function can be found. The result is a set of functions, which approximate the imperfections of the foldcores introduced by the manufacturing. These functions are mapped and superimposed on the surfaces of the numerical model. The crease lines themselves are not manipulated because it is assumed that these lines match the ideal shape. The resulting FEA model for a Miura-ori unit cell with upscaled imperfections is shown in Figure 7.

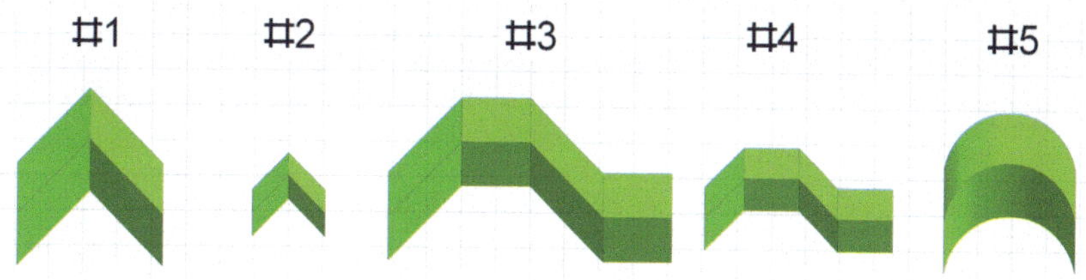

Figure 8: *Top view of selected unit cells (grid subdivision is 5mm).*

# 4   Validation of the Numerical Process Chain

## 4.1   Unit Cells

To demonstrate and validate the numerical process chain for the compression loadcase a set of five tessellations were selected. This set comprises two Miura-ori unit cells with different geometrical parameters (#1 and #2), two derivatives of the Miura-ori pattern with linear pleats (#3 and #4) and a curved fold tessellation (#5). The control points for the quadratic Bèzier spline segments of each geometry, the characteristic parameters $L$ and $H$ as well as the resulting core density $\rho_U$ are shown in Table 2. The geometrical shape of each tessellation is shown in Figure 8.

## 4.2   Sample Preparation and Test Setup

For each tessellation a minimum of six samples has been manufactured out of the PET foil described in section 2 to ensure a statistical validity. Crease patterns were scored with a flatbed laser [6], and folded manually without any tooling. To ensure the dimensional stability of the tessellations an annealing temperature cycle of $100\,°C$ for $30\,min$ was applied to the folded samples.

---

[6]Flatbed laser Speedy 400flexx with $CO_2$ laser source by Trotec GmbH.

**Table 2:** *Control point coordinates for the quadratic Bèzier spline segments with shared starting and ending points (A3 = B1, B3 = C1 etc.) and characteristic parameters L and H and resulting core density $\rho_C$ for each tessellation.*

| Geo ID | #1 | #2 | #3 | #4 | #5 |
|---|---|---|---|---|---|
| $A1_x, A1_y$ / mm | 0.0, 0.0 | 0.0, 0.0 | 0.0, 0.0 | 0.0, 0.0 | 0.0, 0.0 |
| $A2_x, A2_y$ / mm | 2.0, 2.0 | 2.7, 2.7 | 2.0, 2.0 | 2.8, 2.8 | 0.5, 3.0 |
| $A3_x, A3_y$ / mm | 10.6, 10.6 | 5.3, 5.3 | 10.6, 10.6 | 5.7, 5.7 | 2.7, 5.2 |
| $B2_x, B2_y$ / mm | 17.7, 3.5 | 8.0, 2.7 | 15.0, 10.6 | 9.7, 5.7 | 5.6, 8.0 |
| $B3_x, B3_y$ / mm | 21.2, 0.0 | 10.6, 0.0 | 20.6, 10.6 | 13.7, 5.7 | 9.7, 8.0 |
| $C2_x, C2_y$ / mm | | | 26.3, 5.0 | 16.5, 2.8 | 13.7, 8.0 |
| $C3_x, C3_y$ / mm | | | 31.2, 0.0 | 19.3, 0.0 | 16.6, 5.2 |
| $D2_x, D2_y$ / mm | | | 35.0, 0.0 | 23.3, 0.0 | 18.8, 3.0 |
| $D3_x, D3_y$ / mm | | | 41.2, 0.0 | 27.3, 0.0 | 19.3, 0.0 |
| L / mm | 14.6 | 6.7 | 12.5 | 8.9 | 14.7 |
| H / mm | 20.0 | 12.7 | 20.0 | 12.7 | 20.0 |
| $\rho_C$ / (kg/m$^3$) | 35.22 | 75.71 | 35.22 | 47.54 | 35.22 |

After cooling, the cores were bonded to 5 mm thick steel facesheets using a two component epoxy system [7]. The stiffness of the face sheets is several magnitudes higher than the stiffness of the foldcores and can be neglected. The length $L$, width $W$, core height $H$ and density $\rho_C$ have been measured for each sample to be able to calculate strength and stiffness and are shown in Table 3 as averaged values for each type of tessellation. An example of these samples for geometry #5 without the upper facesheet is shown in Figure 9.

**Table 3:** *Averaged geometrical parameters cell number in W- and L-direction $n_W$ x $n_L$, core height $\bar{H}$, width $\bar{S}$, length $\bar{L}$ and core density $\bar{\rho}_C$ for the tested samples.*

| Geo ID | #1 | #2 | #3 | #4 | #5 |
|---|---|---|---|---|---|
| $n_W$ x $n_L$ | 6x8 | 8x12 | 3x12 | 5x14 | 7x8 |
| $\bar{H}$ / mm | 19.85 | 12.51 | 19.48 | 12.58 | 20.36 |
| $\bar{W}$ / mm | 126.74 | 85.32 | 123.40 | 136.06 | 135.04 |
| $\bar{L}$ / mm | 115.31 | 79.04 | 127.51 | 122.99 | 115.29 |
| $\bar{\rho}_C$ / (kg/m$^3$) | 37.02 | 75.57 | 35.41 | 49.34 | 35.09 |

---

[7]Two component system EPIKOTE BPR A20 and EPIKURE BPH B20 by Lange+Ritter GmbH.

The flatwise compression tests have been done in accordance to the DIN 53291 [German institute for standardization 82] on a Hegewald & Peschke universal testing machine with a Class 1 2 kN load cell. The displacement was measured with a HBM WA20 inductive displacement transducer. Strength and stiffness have been calculated as described in section 3.1 for the FEA analysis.

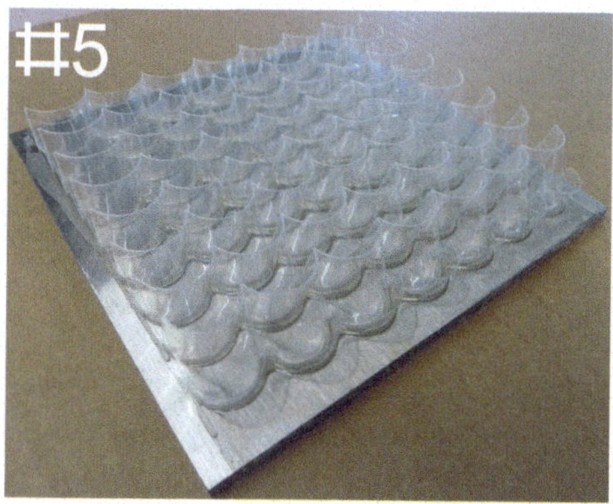


**Figure 9:** *Sample of geometry #5 for compression testing without upper facesheet.*

## 4.3 Results and Discussion

### 4.3.1 Compressive Strength

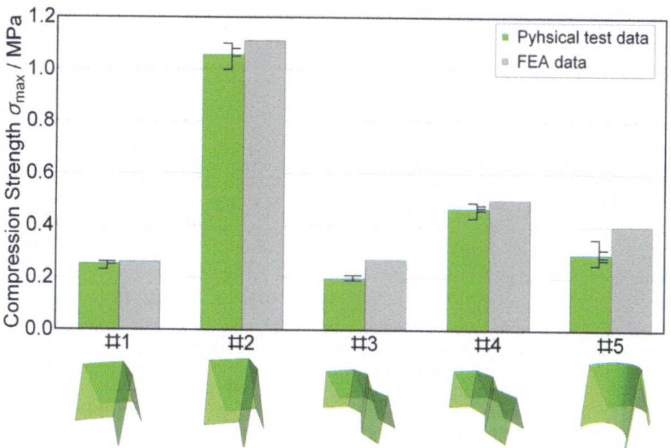


**Figure 10:** *Compression strength $\sigma_{max}$ for the FEA and physical tests. Error bars for the measured values indicate minimum/maximum values and 25%, 50% and 75% quantile.*

The results for the compression strength $\sigma_{max}$ for the physical tests and the FEA is shown in Figure 10. For all tessellations the strength is overestimated with a maximum of 27.5% for the curved fold geometry #5. The higher values for the

FEA are expected and are in accordance with literature, where simulations usually overpredict physical test data due to the reduced number of degrees of freedom in the numerical model [Fischer 15b, Johnson 08]. In addition, the implementation of the PBC suggests a model without boundary cells and will therefore be more optimistic than small cores samples.

The deviation of geometry #5 is a result of the superimposed imperfection function, which is challenging to formulate for curved foldcores, and has to be improved. For the remaining tessellations the results for strength are in an acceptable range as shown in Table 4. The overall comparison between simulation and test are shown in Figure 10.

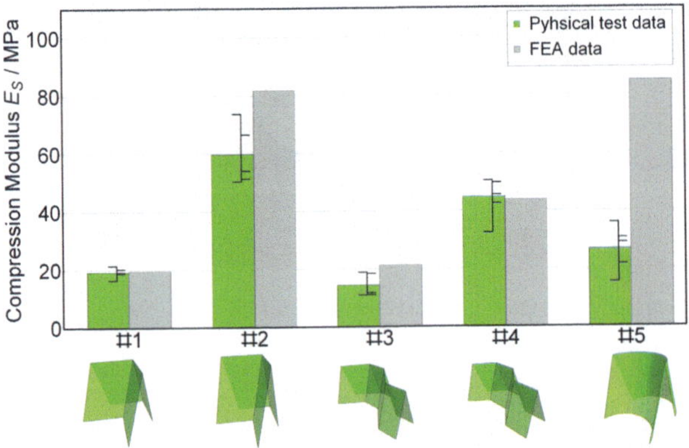


**Figure 11:** *Compression modulus $E_S$ for the FEA and physical tests. Error bars for the measured values indicate minimum/maximum values and 25%, 50% and 75% quantile.*

|  | Geo ID | #1 | #2 | #3 | #4 | #5 |
|---|---|---|---|---|---|---|
| FEA | $E_S$ / MPa | 19.68 | 81.77 | 21.30 | 43.7 | 84.81 |
|  | $\sigma_{max}$ / MPa | 0.26 | 1.11 | 0.27 | 0.50 | 0.40 |
| Physical test | $E_S$ / MPa | 19.01 | 59.74 | 14.32 | 44.84 | 26.73 |
|  | $\sigma_{E_S}$ / MPa | 1.63 | 9.36 | 3.70 | 6.76 | 6.00 |
|  | $\sigma_{max}$ / MPa | 0.25 | 1.06 | 0.20 | 0.47 | 0.29 |
|  | $\sigma_{\sigma_{max}}$ / MPa | 0.01 | 0.03 | 0.01 | 0.02 | 0.03 |
| $\left(1 - \frac{\text{Physical test}}{\text{FEA}}\right)$ | $E_S$ / % | 3.40 | 26.94 | 32.77 | −2.61 | 68.48 |
|  | $\sigma$ / % | 3.85 | 4.50 | 25.93 | 6.00 | 27.5 |

**Table 4:** *Compression secant modulus $E_S$ and maximum stress $\sigma_{max}$ for the FEA and the physical tests. For the physical tests the standard deviations $\sigma_{E_S}$ and $\sigma_{\sigma_{max}}$ for the modulus and the maximum stress as well as the deviation between test and FEA is shown.*

### 4.3.2 Compressive Stiffness

For the compression stiffness, shown in Figure 11, the values of the physical test data are overestimated by the FEA as well except for tessellation #4. This reversed behavior of geometry #4 can be a result of the higher scatter for the physical test data indicated by the error bars.

Compared to the strength results, the spread for the stiffness is larger. This is normal for flatwise compression foldcore tests, as the stiffness is much more sensitive to imperfections than the ultimate strength. Especially for tessellations #2 and #5 the stiffness is overestimated by the FEA with a maximum deviation of 69 %. This seems to be a result of the currently used superposition function for curved surfaces.

## 5 Conclusion

This study presents an automated numerical process chain for the evaluation and comparison of sandwich foldcores under compression load. The implementation comprises the design process, model setup, FEA analysis and the evaluation of the results. The workflow incorporates a number of freely available and proprietary software packages, with a focus on flexibility, easy accessibility and persistent storage of all results.

Five different core types have been analyzed by FEA and compared to physical test data to validate the method. The results of the FEA show good accordance to physical test data, especially for compressive strength. For classical Miura-ori unit cell and derivatives, the presented approach already works well, while the method has still to be improved for curved folds. Here, the generation of functions to automate superposition of imperfections are the key elements.

Further work will include improvement of the imperfection functions of the curved fold geometries and the extension of the numerical process chain to shear loadcases. Finally, the method will be incorporated with optimization algorithms to find the best geometrical foldcore shape for a given set of specifications. Given the large parameter space available, an efficient way to explore these geometry realms is extremely desirable.

## Acknowledgment

This work has been financially supported by the German Research Foundation (DFG) within the special research grant SFB 1244 ""Adaptive Skins of Tomorrow"". The authors gratefully acknowledge the support.

# References

[ASTM D882-12 12]  ASTM D882-12. "Test Method for Tensile Properties of Thin Plastic Sheeting.", 2012. doi:10.1520/D0882-12.

[Baranger et al. 11a]  E. Baranger, C. Cluzel, and P.-A. Guidault. "Modelling of the Behaviour of Aramid Folded Cores Up to Global Crushing." *Strain* 47 (2011), 170–178. doi:10.1111/j.1475-1305.2010.00753.x.

[Baranger et al. 11b]  E. Baranger, P.-A. Guidault, and C. Cluzel. "Numerical modeling of the geometrical defects of an origami-like sandwich core." *Composite Structures* 93:10 (2011), 2504–2510. doi:10.1016/j.compstruct.2011.04.011.

[Fathers et al. 15]  R. K. Fathers, J. M. Gattas, and Z. You. "Quasi-static crushing of eggbox, cube, and modified cube foldcore sandwich structures." *International Journal of Mechanical Sciences* 101-102 (2015), 421–428. doi:10.1016/j.ijmecsci.2015.08.013.

[Filipov et al. 17]  E. T. Filipov, K. Liu, T. Tachi, M. Schenk, and G. H. Paulino. "Bar and hinge models for scalable analysis of origami." *International Journal of Solids and Structures* 124 (2017), 26–45. doi:10.1016/j.ijsolstr.2017.05.028.

[Fischer 12]  S. Fischer. *Rechnerische Ermittlung der mechanischen Eigenschaften von Faltkernen.* Berichte aus der Luft- und Raumfahrttechnik, Aachen: Shaker, 2012.

[Fischer 15a]  S. Fischer. "Aluminium foldcores for sandwich structure application: Mechanical properties and FE-simulation." *Thin-Walled Structures* 90 (2015), 31–41. doi:10.1016/j.tws.2015.01.003.

[Fischer 15b]  S. Fischer. "Realistic Fe simulation of foldcore sandwich structures." *International Journal of Mechanical and Materials Engineering* 10:1. doi:10.1186/s40712-015-0041-z.

[German institute for standardization 82]  German institute for standardization. "Prüfung von Kernverbunden Druckversuch senkrecht zur Deckschicht.", February 1982.

[Grzeschik and Drechsler 10]  M. Grzeschik and K. Drechsler. "Experimental parameter studies on folded cores." In *9th International Conference on Sandwich Structures (ICSS-9)*, edited by ICSS9, 2010.

[Grzeschik et al. 11]  M. Grzeschik, M. Fach, S. Fischer, Y. Klett, R. Kehrle, and K. Drechsler. "Isometrically folded high performance core materials." In *Processing and fabrication of advanced materials*, edited by D. Bhattacharyya, R. Lin, and T. S. Srivatsan. Auckland, New Zealand: Centre for Advanced Composite Materials, University of Auckland, 2011.

[Heimbs et al. 08]  S. Heimbs, P. Middendorf, C. Hampf, F. Hähnel, and K. Wolf. "Aircraft Sandwich Structures with Folded Core Under Impact Load." In *8th International Conference on Sandwich Structures (ICSS-8)*, edited by ICSS8, pp. 369–380, 2008.

[Heimbs et al. 10]  S. Heimbs, J. Cichosz, M. Klaus, S. Kilchert, and A. F. Johnson. "Sandwich structures with textile-reinforced composite foldcores under impact loads." *Composite Structures* 92:6 (2010), 1485–1497. doi:10.1016/j.compstruct.2009.11.001.

[Heimbs 09]  S. Heimbs. "Virtual testing of sandwich core structures using dynamic finite element simulations." *Computational Materials Science* 45:2 (2009), 205–216. doi:10.1016/j.commatsci.2008.09.017.

[Herrmann et al. 05]  A. S. Herrmann, P. C. Zahlen, and I. Zuardy. "Sandwich Structures Technology in Commercial Aviation.", 29-31 August, 2005.

[Johnson 08]  A. F. Johnson. "Novel Hybrid Structural Core Sandwich Materials for Aircraft Applications.", 2008.

[Kilchert 13]  S. Kilchert. *Nonlinear finite element modelling of degradation and failure in folded core composite sandwich structures.* Dissertation, Köln: DLR and University of Stuttgart, 2013. doi:10.18419/opus-3932.

[Klett et al. 15]  Y. Klett, M. Grzeschik, and P. Middendorf. "Comparison of compressive properties of periodic non-flat tessellations." In *Miura, K. et al. (Hg.) 2016 – Origami$^6$ // Origami 6*, edited by K. Miura, T. Kawasaki, T. Tachi, R. Uehara, R. J. Lang, and P. Wang-Iverson, pp. 371–384. American Mathematical Society, 2015.

[Klett 13]  Y. Klett. *Auslegung multifunktionaler isometrischer Faltstrukturen für den technischen Einsatz*, First edition. Luftfahrt, München: Verl. Dr. Hut, 2013.

[Mitani and Igarashi 11]  J. Mitani and T. Igarashi. "Interactive Design of Planar Curved Folding by Reflection." In *Pacific Graphics Short Papers*, 2011. doi:10.2312/PE/PG/PG2011short/077-081.

[Miura 72]  K. Miura. *Zeta-Core Sandwich - Its Concept and Realization.* Report No. 480, Tokyo: Institute of Space and Aeronautical Science, University of Tokyo, 1972.

[Muhs and Middendorf 18]  F. Muhs and P. Middendorf. "New design method for sandwich foldcores based on quadratic Bèzier spline formulation." *Thin-Walled Structures* :Under Review.

[Salomon 06]  D. Salomon. *Curves and Surfaces for Computer Graphics.* New York, NY: Springer Science+Business Media Inc, 2006. doi:10.1007/0-387-28452-4.

[Schenk and Guest 11]  M. Schenk and S. D. Guest. "Origami Folding - A structural engineering approach." In *Origami$^5$*, edited by P. Wang-Iverson, R. J. Lang, and M. Yim, pp. 291–303. CRC Press, 2011.

[Sturm and Fischer 15]  R. Sturm and S. Fischer. "Virtual Design Method for Controlled Failure in Foldcore Sandwich Panels." *Applied Composite Materials* 22:6 (2015), 791–803. doi:10.1007/s10443-015-9436-5. Available online (http://dx.doi.org/10.1007/s10443-015-9436-5).

[Thorsteinsson 02]  E. B. Thorsteinsson. "Entwicklung einer Methode zur Bestimmung des E- und G-Moduls gefalteter, räumlicher Faserverbundstrukturen." thesis, University of Stuttgart, Stuttgart, 18.04.2002.

[Zang et al. 16]  S. Zang, X. Zhou, H. Wang, and Z. You. "Foldcores made of thermoplastic materials: Experimental study and finite element analysis." *Thin-Walled Structures* 100 (2016), 170–179. doi:10.1016/j.tws.2015.12.017.

[Zhou et al. 15]  X. Zhou, S. Zang, H. Wang, and Z. You. "Geometric design and mechanical properties of cylindrical foldcore sandwich structures." *Thin-Walled Structures* 89 (2015), 116–130. doi:10.1016/j.tws.2014.12.017.

---

Fabian Muhs
Institute of Aircraft Design, Pfaffenwaldring 31, 70569 Stuttgart,
e-mail:  muhs@ifb.uni-stuttgart.de

Yves Klett
Institute of Aircraft Design, Pfaffenwaldring 31, 70569 Stuttgart,
e-mail:  klett@ifb.uni-stuttgart.de

Peter Middendorf
Institute of Aircraft Design, Pfaffenwaldring 31, 70569 Stuttgart,
e-mail: middendorf@ifb.uni-stuttgart.de

# Spherical Image Analysis for Folding Templates

*D. T. Eatough, K. A. Seffen*

**Abstract**: *A method for calculating the shape of an irregular polyhedron folded from a flat polygonal net is presented. The method is based on the compatibility of linked vertices and a technique for calculating the dihedral angles of vertices. This method allows for the possibility of investigating and optimising the polyhedron by adjusting the net.*

## 1   Introduction

Unfolding of a polyhedral volume into a flat polygonal "net" of interconnected flat facets has been of of interest for centuries [Dürer 77]. While "exploding" a given polyhedron into its constituent facets is trivial, proving whether or not every convex polyhedron can be "unfolded" by first cutting along some of its edges before unfolding into a nonoverlapping, simple and flat polygon, is still an open question in mathematics [Shephard 75, Demaine and O'Rourke 07]. Furthermore, it has been shown that there exist non-convex polyhedra which cannot be unfolded [Tarasov 99]; conversely there are simple nets which can be folded into multiple and different polyhedra [Malkevitch 01]. However, it is guaranteed that any edge gluing (a joining together of all the free edges) of a polygon has a unique convex polyhedron folding [Aleksandrov 50] and, for many practical situations, constructing a polyhedron from a flat net is the simplest way to form a three dimensional shape.

The Aleksandrov Uniqueness Theorem [Aleksandrov 50] implies that for a given edge gluing, an infinite set of non-convex polyhedra can be folded. However, in prescribing the internal fold-lines as well as the glued edges, a unique and closed rigid polyhedron will almost always be formed: what then is its shape? Note that there are a small number of special cases where "Flexible Polyhedra" can be created. While these are enclosed polygons of constant volume, their dihedral angles are not fixed but can actually vary within a small range [Connelly et al. 97]. Relatively simple nets can be folded into irregular polyhedra, whose vertices are not all the same and dihedral angles are not all equal nor trivial to calculate. In practice the initial net may be constrained (for example, the net must be rectangular owing to manufacturing constraints) or we may desire to optimise the polyhedron (for example, maximising its volume).

Figure 1a shows a net which is a fairly simple arrangement of triangular facets, and Fig. 1b shows the corresponding polyhedron. Defining and parameterising the planar net geometry and thus calculating all of the internal facet angles is straightforward; however, finding the polyhedron geometry is much less so. This paper provides a method for calculating the dihedral angles of a net-folded polyhedron with rigid facets, and demonstrates its applicability by means of an example.

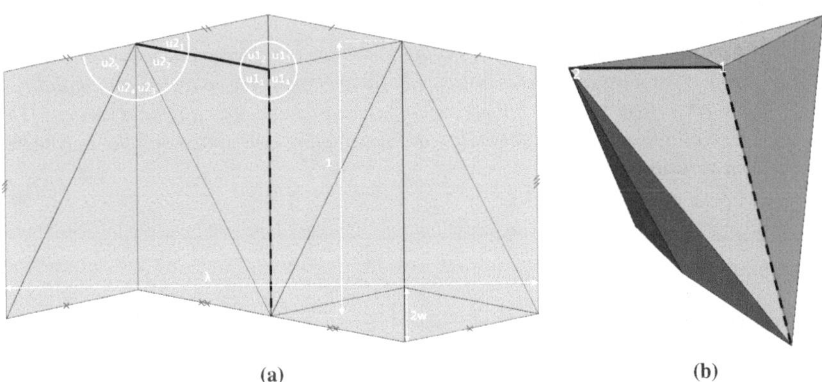

(a)       (b)

**Figure 1:** (a) *A net comprised of a dozen triangular facets. When the edges are joined according to the dashes shown (an edge gluing), the net can be folded into the polyhedron in* (b). *Additionally, for use in Section 4: the net is parameterised by the lengths w and* λ; *there are two unique vertices labeled* 1 *and* 2; *and the facet angles of each vertex are labeled on the net,* $u_1$, $u_2$ ... *etc.*

## 2  Background

It is well known that nets can be folded into closed polyhedrons. As a simple example, the "Latin-Cross" net for a cube is shown in Fig. 2a. By joining the free edges according to the specified edge gluing, the net forms a cube. From symmetry, all eight vertices of the cube are similar, and three square facets meet at each of them: the internal angles of each facet around one vertex are labeled $u_1 = u_2 = u_3 = \pi/2$. Inspection of any cube vertex tells us that the dihedral angle between facets is $\pi/2$.

Alternatively, we can use a formal vector arithmetic approach. Defining unit vectors along the edges that point away from the vertex, Fig. 2b, we can calculate the inward facing unit normals of the facets from $\bar{n}_1 = (\bar{r} \times \bar{p})/(|\bar{r} \times \bar{p}|)$ and $\bar{n}_2 = (\bar{p} \times \bar{q})/(|\bar{p} \times \bar{q}|)$. The cosine of the dihedral angle is then found from:

$$\cos\left(\text{Dihedral angle}_{12}\right) = (-\bar{n}_1 \cdot \bar{n}_2) = \left( \frac{\cos(u_3) - \cos(u_1)\cos(u_2)}{\sin(u_1)\sin(u_2)} \right)$$

Substituting for the interior angles of the facets:

$$\text{Dihedral angle}_{12} = \cos^{-1}\left(\frac{\cos(\pi/2) - \cos(\pi/2)\cos(\pi/2)}{\sin(\pi/2)\sin(\pi/2)}\right) = \pi/2$$

The vector method is not the only way to calculate the dihedral angles however; the spherical image of a vertex may also be used. Noting first that the fold angles between facets are related to the dihedral angles by the formula:

$$\text{Dihedral angle} = \pi - \text{Fold angle}$$

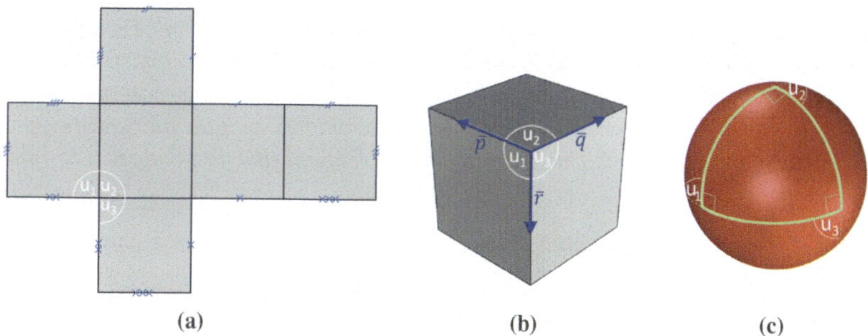

**Figure 2:** (a) *The familiar "Latin-Cross" net of a cube, after bringing together and joining edges (an edge gluing) according to the dashes this net folds into a cube.* (b) *The folded net, our cube, with identical corner vertices. The internal angles of the facets around one such vertex are labeled $u_1 = u_2 = u_3 = \pi/2$. Also shown are unit normals, $\bar{p}, \bar{q}, \bar{r}$ which are coincident to the edges pointing away from the vertex.* (c) *The spherical image of the vertex whose arc-lengths are equal to the fold angles of the vertex.*

A spherical image of a vertex can be created as follows [Calladine 83]. For an arbitrary n-facet vertex roof, we label the internal angle of each facet in a clockwise fashion $u_1$, $u_2$ ...$u_n$: Fig. 3 shows the case when $n = 3$. The outward facing unit normal for each facet is then plotted as a vector normal to the surface of a unit sphere, which defines its location on the sphere. Great arcs on the surface of the sphere connect successive points in turn; giving a closed figure is the spherical image. The external angle of a corner point on the spherical image is equal to the internal angle of the associated facet. The arc-length between two points is therefore equal to the fold angle between the corresponding two facets. Summing the internal angles of all facets for a vertex, we typically find it comes to less than $2\pi$: this difference is called the "angular defect", which can be expressed as:

$$2\pi - \sum_{i=1}^{n} u_i = \text{Angular defect}$$

A remarkable property is that the area of the spherical image is equal to the angular defect of the vertex. [Calladine 83]

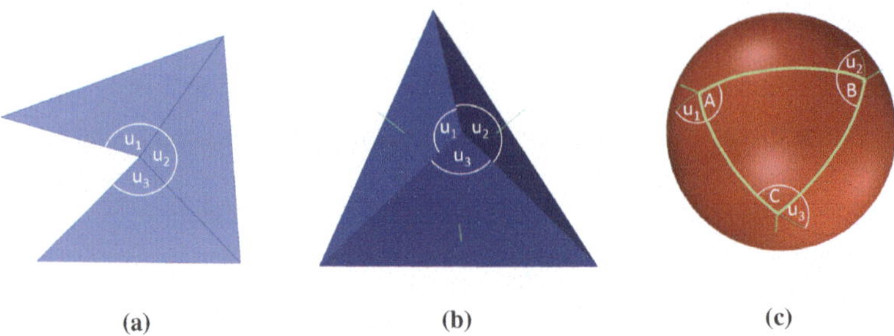

**Figure 3: (a)** *The flat net of a vertex with a positive angular defect. The internal facet angles are labeled in a clockwise fashion, $u_1, u_2 \ldots u_n$.* **(b)** *The folded shape of the vertex with the unit normals of the facets.* **(c)** *The spherical image corresponding to the vertex. The unit normals from the facets are plotted onto the unit sphere and joined up along great arcs. The following angles are defined $A = \pi - u_1$, $B = \pi - u_2$ etc.*

For the cube example, the corresponding spherical image is shown in Fig. 2c. It is a spherical triangle that covering an eighth of the sphere, giving arc-lengths as $\pi/2$ by simple inspection. Thus, the dihedral angle is $\pi/2$, as before from:

$$\text{Dihedral angle} = \pi - \text{Fold angle} = \pi - \pi/2 = \pi/2$$

For more complicated nets and polyhedrons, the vertices may have more than three facets; for four facets, the spherical image technique can be extended to calculate the ratios of the dihedral angles [Huffman 76, Lang et al. 16]; the vector method however is less amenable to such a problem and quickly becomes cumbersome.

## 3  Method

To calculate the dihedral angles of an irregular polyhedron, we must first establish the number of unique vertices in the polyhedron: symmetry will always make the unique vertices less than the total number. Having established the unique vertices, we construct their spherical images in order to calculate their dihedral angles.

However, vertices with more than three facets are not rigid but are mechanistic. In general, an $n$ facet vertex has $n - 3$ degrees of freedom. When considering the spherical image of such a vertex, it is not possible to calculate the arc-lengths uniquely given the facet angles alone. For a vertex with $n$ facets, $n - 3$ parameters must be given to fix the shape of the spherical image, to allow for calculation of the arc-lengths. Previous work seems to have focused only vertex roofs with three facets or the case of a four facet vertex with zero angular defect [Huffman 76, Lang et al. 16].

In the case of a four-facet vertex, where the fold angles are uniquely related, there is one unknown; thus, by prescribing one of the dihedral angles, we can work out the values of the other three. Equivalently, we can present this information as a ratio of dihedral angles. This approach quickly becomes unmanageable for vertices with five or more facets and a better choice of free parameter is needed. One such beneficial choice directly exploits the simplest spherical image calculations using a spherical triangle and we aim to subdivide or extend our general image as such. For each spherical triangle we can find its arc-lengths from the three internal angles using the half-side formula [Zwillinger 02], see Fig. 3c:

$$\overline{BC} = 2\tan^{-1}\left(\sqrt{\frac{-\cos(S_{ABC})\cos(S_{ABC}-A)}{\cos(S_{ABC}-B)\cos(S_{ABC}-C)}}\right) \quad \text{where } S_{ABC} = \frac{A+B+C}{2}$$

The arc-lengths (hence dihedral angles) of a vertex are now expressed in terms of the unknown parameters. As each $n$ facet vertex contributes $n-3$ unknown parameters, the total number of unknowns for a polyhedron can be found by summing over all vertices. Rigid foldability of the polyhedron ensures a constant fold angle along the entire edge, which sets the dihedral angle between the two connected vertices to be equal. Each edge therefore provides one compatibility restraint. By applying this condition around to a closed polyhedron, the total number of degrees of freedom is equal to the number of compatibility restraints.

Mathematically, we have a set of simultaneous equations, equal in number to the unknowns, and thus uniquely solvable.

## 4  Worked Example

We first calculate the geometry of an irregular net-folded polyhedron and then maximise its closed volume. We use the example in Fig. 1, where there are two unique vertices. The facet angles are trivial to find. Vertex 1 is a four faceted vertex with one concave fold-line; its spherical image is shown in Fig. 4a, which has a single free parameter, X. Vertex 2 is a five faceted, fully convex vertex and its spherical image is shown in Fig. 4b; it has two free parameters, Y and Z. As vertex 2 is symmetrical, it's spherical image is also symmetrical, and this forces the free parameters to be equal, Y=Z. The entire polyhedron can now be parameterised with just two variables, X and Y.

As before, a single subscript refers to the $n^{th}$ facet around a given vertex. A numerical suffix denotes which of the two unique vertices is being referred to, and a double numerical subscript indicates the fold-line between two facets: for example $Fold_{4,5}$ is the fold which connects facet 4 to facet 5.

There are also two distinct fold-lines which connect Vertex 1 to Vertex 2, shown as a bold and dashed bold line respectively (Fig. 1b). Rigid foldability tells us that:

$$\text{Vertex 1 Fold Angle}_{1,2} = \text{Vertex 2 Fold Angle}_{1,2} \quad \text{(Bold line)}$$

$$\text{Vertex 1 Fold Angle}_{4,1} = \text{Vertex 2 Fold Angle}_{3,4} \quad \text{(Dashed Bold line)}$$

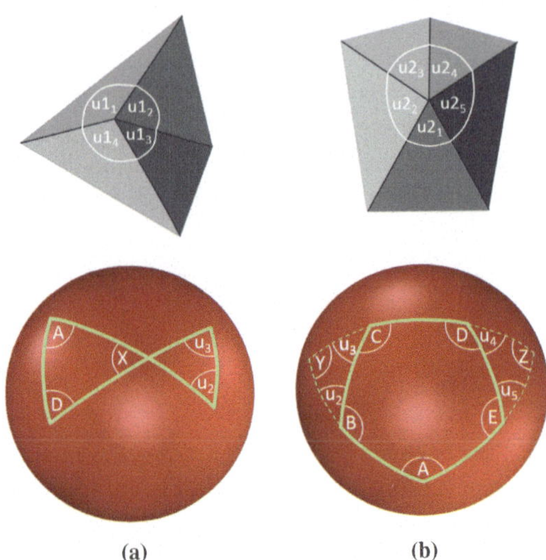


**Figure 4:** (a) *Spherical image of Vertex 1. As the vertex has four facets and a single concave fold-line, the spherical image is in the shape of a bowtie. The single free parameter is X.* (b) *Spherical image of Vertex 2. All the fold-lines are convex, and by symmetry of the facet angles, Y = Z. Close ups of Vertex 1 and Vertex 2 are shown above each corresponding spherical image.*

The fold angles can be expressed using the arc-lengths equations from the relevant spherical images, which then provides a pair of simultaneous equations in X and Y:

$$\left\| \begin{matrix} S1_{ADX} & D1 \\ A1 & X \end{matrix} \right\| + \left\| \begin{matrix} S1_{23X} & u1_3 \\ u1_2 & X \end{matrix} \right\| = \left\| \begin{matrix} S2_{AYZ} & Y \\ A2 & Y \end{matrix} \right\| - \left\| \begin{matrix} S2_{23Y} & u2_3 \\ u2_2 & Y \end{matrix} \right\|$$

$$\left\| \begin{matrix} S1_{ADX} & X \\ A1 & D1 \end{matrix} \right\| = \left\| \begin{matrix} S2_{AYZ} & A2 \\ Y & Y \end{matrix} \right\| - \left\| \begin{matrix} S2_{23Y} & u2_2 \\ u2_3 & Y \end{matrix} \right\| - \left\| \begin{matrix} S2_{45Y} & u2_5 \\ u2_4 & Y \end{matrix} \right\|$$

The triple alphanumeric subscript indicates which angles are being used in the summation terms, for example $S_{ABX} = (A + B + X)/2$. For compactness, the following notation is used for the half-side formula:

$$\left\| \begin{matrix} S & A \\ B & C \end{matrix} \right\| = 2\tan^{-1}\left( \sqrt{ \frac{-\cos(S)\cos(S-A)}{\cos(S-B)\cos(S-C)} } \right)$$

These simultaneous equations are non-linear but can be solved computationally using the MATLAB function "vpasolve" [MAT 16]. The fold angles can then be calculated by substituting these parameters into the arc-length expressions which are found from consideration of the spherical image. In this example, setting the template parameters $\lambda = 2$ and $w = 0.1$, the numerical solutions for the vertex parameters are:

$$X = 0.466450069 , \quad Y = 2.138854455$$

and the fold angles are (in radians):

| Vertex 1 | | Vertex 2 | |
|---|---|---|---|
| $Fold_{1,2}$ | 1.139871 | $Fold_{1,2}$ | 1.139871 |
| $Fold_{2,3}$ | -0.250062 | $Fold_{2,3}$ | 1.539264 |
| $Fold_{3,4}$ | 1.139871 | $Fold_{3,4}$ | 0.250062 |
| $Fold_{4,1}$ | 0.250062 | $Fold_{4,5}$ | 1.539264 |
| | | $Fold_{5,1}$ | 1.139871 |

**Table 1:** *The fold angles (in radians) around Vertex 1 and Vertex 2. Positive fold angles denote a convex fold-line and negative fold angles a concave fold-line.*

We then compute the volume of the polyhedron [Rhi 18, Gra 14] as a function of the variable parameter w, and its variation is given in Fig. 5. In varying w, we do not change the area of the net, only the volume (which has units of length cubed). A clear maximum exists where w = 0.2012.

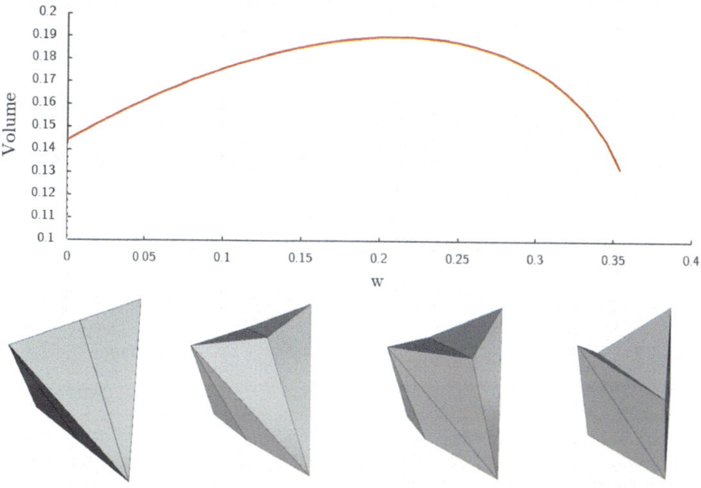

**Figure 5:** *The volume of the polyhedron from Fig. 1 as a function of the variable w, with $\lambda = 2$. There exists an optimum value of w for maximum volume; w = 0.2012. Interim shapes are indicated.*

# 5  Conclusions

This way of analysing folded vertices from their spherical images proves to be an extremely useful way of calculating the folded shape of a closed polyhedron from a flat net. The unknown parameters which control the degrees of freedom of the folded vertices with more than 3 facets were conveniently related to one another via a set of simultaneous equations arising from the compatibility of rigid fold-lines between corresponding vertices. Numerical solutions to this set of simultaneous equations were found and thus by back substituting into the expressions found and ultimately the dihedral angles of the polyhedron were calculated. From this, we may compute other shape properties in order to optimise, for example, its geometrical performance.

## Acknowledgement

This paper has been awarded the 7OSME Gabriella & Paul Rosenbaum Foundation Travel Award.

## References

[Aleksandrov 50]  A. D. Aleksandrov. *Convex Polyhedra*. Moscow: State Press of Technical and Theoretical Literature, 1950.

[Calladine 83]  C R Calladine. *Theory of Shell structures*. Cambridge University Press, 1983.

[Connelly et al. 97]  R. Connelly, I. Sabitov, and A. Walz. "The bellows conjecture." *Contributions to Algebra and Geometry* 38:1 (1997), 1–10.

[Demaine and O'Rourke 07]  E. D. Demaine and J. O'Rourke. *Geometric Folding Algorithms: Linkages, Origami, Polyhedra*. Cambridge University Press, 2007.

[Dürer 77]  A. Dürer. *The painters manual : a manual of measurement of lines, areas and solids by means of compass and ruler assembled by Albrecht Durer for the use of all lovers of art with appropriate illustrations arranged to be printed in the year MDXXV.* Translation by W. L. Strauss, New York : Abaris Books, 1977.

[Gra 14]  *Grasshopper, Build 0.9.0076*. Robert McNeel & Associates, 2014.

[Huffman 76]  D. A. Huffman. "Curvature and creases: A primer on paper." *IEEE Transactions on computers* :10 (1976), 1010–1019.

[Lang et al. 16]  R. J. Lang, S. Magleby, and L. Howell. "Single degree-of-freedom rigidly foldable cut origami flashers." *Journal of Mechanisms and Robotics* 8:3 (2016), 031005.

[Malkevitch 01]  J. Malkevitch. "Le gomtre et la paire de ciseaux." *La Recherche* 346 (2001), 62–63.

[MAT 16]  *MATLAB 2016a*. The MathWorks, Inc, 2016.

[Rhi 18] *Rhinoceros, Version 5.* Robert McNeel & Associates, 2018.

[Shephard 75] G. C. Shephard. "Convex polytopes with convex nets." *Mathematical Proceedings of the Cambridge Philosophical Society* 78:3 (1975), 389.

[Tarasov 99] A. S. Tarasov. "Polyhedra with no natural unfoldings." *Russian Mathematical Surveys* 54:3 (1999), 656–657.

[Zwillinger 02] D. Zwillinger. *CRC standard mathematical tables and formulae.* CRC press, 2002.

---

Daniel T Eatough

PhD Candidate, University of Cambridge, Advanced Structures Group, Department of Engineering, University of Cambridge, UK, CB2 1PZ, e-mail: dte22@cam.ac.uk

Keith A Seffen

Reader in Structural Mechanics, University of Cambridge, Advanced Structures Group, Department of Engineering, University of Cambridge, UK, CB2 1PZ, e-mail: kas14@cam.ac.uk

# Kinematic and Kinetostatic Classification for Motion-Task-Oriented Synthesis of Folding Mechanisms

*J. Paris, J. Merz, H. Buffart, S. Hoffmann, J. Siebrecht, C. Weigel, M. Hüsing, M. Trautz, B. Corves*

**Abstract**:

*Engineered origami-inspired folding mechanisms are advantageous over conventional motion devices when moving facets are essential for the overall function. Nevertheless, the wide use is limited due to complex kinematic and dynamic dependencies impeding the design of folding mechanisms. Thus, a structured design process and tailored methods are required to overcome these issues. Against this background, this paper proposes a motion-task-oriented design process based on a classification framework, the determination of kinematic and kinetostatic key figures and a folding mechanism toolbox for multi-body simulations.*

## 1   Introduction and Motivation

For the technical realisation of origami-based folding mechanisms and transformable folded structures, respectively, ideal thin facets are replaced by thick rigid panels and creases are substituted by revolute joints. The resulting structure is referred to as *thick rigid origami*. Whenever moving facets are essential for the overall function used for example in optical or acoustical applications, the technical implementation of engineered folding is advantageous. No additional mechanisms or supporting devices are required. In addition, other advantages of folding mechanisms such as a possible weight reduction and a compact design can be used optimally.

Yet, the design of folding mechanisms is challenging because of complex kinematic dependencies and possible collisions during motion. These obstructions are not limited to folded structures, but occur similarly in spatial mechanisms and parallel robots. In contrast to these disciplines motion-task-oriented synthesis methods, guidelines, and performance indicators are missing. Collections or catalogues for origami-based structures do not cover kinematic properties nor include classifications, which limits the broad usage of folding principles.

Design guidelines as used in various engineering fields assist scientists and engineers with the identification of innovative but feasible solutions. Therefore, this paper presents a first step towards the definition of a structured design process for folding mechanisms.

## 2 Designing Motion Devices

The theory of development and design engineering provides structured design processes for general machines. One process, that is commonly used in Germany, is the VDI Guideline 2221, see Figure 1a. Especially for complex tasks, such guidelines offer a general development methodology, which helps engineers to efficiently develop products [**VDI 93**]. This development framework can also be applied to folding mechanisms. Furthermore, specialised processes are available for certain classes of machines and devices such as mechatronics, automotive, and others. Comparably, [Hoffmann et al. 15] discusses the demand for design processes in the field of engineered origami and [Zhakypov and Paik 18] approaches design methods for particular cases of engineered origami.

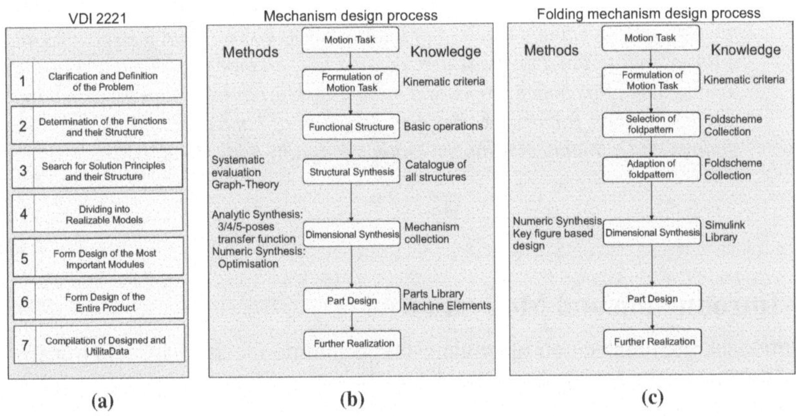


**Figure 1:** *(a) Design processes: VDI guideline 2221 (extracted) [VDI 93], (b) design of mechanisms (based on) [Corves et al. 07], (c) proposed folding mechanism design process*

Fields of research dealing with similar tasks – namely the design of coupled motion devices – are the theories of mechanisms and parallel robots. Design processes, catalogues, collections, and knowledge bases have proven to be helpful in mechanisms design, compare [Corves et al. 07]. Figure 1b shows a mechanism design process with additional information regarding available knowledge bases and synthesis methods.

Based on the mechanism design process, a corresponding folding mechanism process is proposed in Figure 1c. The aim of this paper is to introduce exemplary methods and knowledge bases for each important design process step. Herewith, this paper addresses a next step in bringing folding mechanisms closer to application. Folding mechanism related definitions are given in Section 3. The first step in designing a mechanism is the formulation of the motion task. Elementary motion tasks for folding mechanisms can be grouped in (I) exact positioning of facets (e.g. start and end pose of a specific facet of interest), (II) toleranced positioning of facets (e.g. several poses with allowed tolerance), and (III) toleranced guidance of facets (e.g. defining a target shape of the folding mechanism). Based on the defined motion task, literature, guidelines, and collections should be analysed in order to find already designed mechanisms. If no suitable design solution can be found using this heuristic method, a systematic analysis of possible kinematic structures is necessary. At this point, the structural synthesis – that is the determination of the number of joints, types of vertices, connections, etc. – is conducted in order to find the basic structure of suitable mechanisms. Using a classification and collection framework can facilitate the synthesis process as described in Section 4.

After the structural synthesis, the dimensional synthesis is carried out – that is the determination of facet angles and distances between vertices. This can be done by optimisation procedures or manually by experience. In both cases it is required to simulate the folding mechanism, which is discussed in Section 5. Therefore, a Simulink library for folding mechanisms is presented, which is an implementation of the classification framework. Before the CAD-based design of the elements, a load-oriented dimensioning of facets and foldlines is required. Therefore, Section 7 deals with the determination of material properties, the FEA study, and the experimental verification. Based on this, stresses and actuation forces for folding vertices depending on geometric dimensions are evaluated.

The design process is highly iterative, e.g. the selection of a suitable structure and the dimensional synthesis depends on the loads. Therefore, to minimise the iterations, kinematic and static key figures are introduced in Section 6 and Section 8, respectively. Key figures quantify properties approximately based only on geometric dimensions and further assumptions. They can also approximate properties, which are hard to compare otherwise. Furthermore, extending the collection with this data enables a more targeted selection of foldschemes and reduces the number of necessary iterations. An example for key figures in engineering is the transmission angle, which is used in mechanism theory. The transmission angle is calculated based only on geometric properties and gives an approximation of the expected drive torques.

The presented procedure is applied to a laserbeam guidance task. In Figure 2a, the motion task is defined by a space requirement, a laser source and two laser targets. A suitable foldscheme is then selected from the foldscheme collection. The requirements are: Only one degree of freedom (DOF) and a cylindrical or spherical motion to realise sufficient surface rotation. With this information, the Chicken-Wire foldscheme can be selected. To obtain sufficient stability, this is extended by mirroring along $\overline{CD}$. The resulting folding mechanism is adapted to the exact geometric requirements of the task by means of kinematic analysis and optimisation. The resulting structure is non-symmetric and thus two DOF are lost. An additional fold line is induced at A and F to mantain the total DOF of one, see Figure 2b. The complete folding mechanism including elastic joints, Figure 2c,

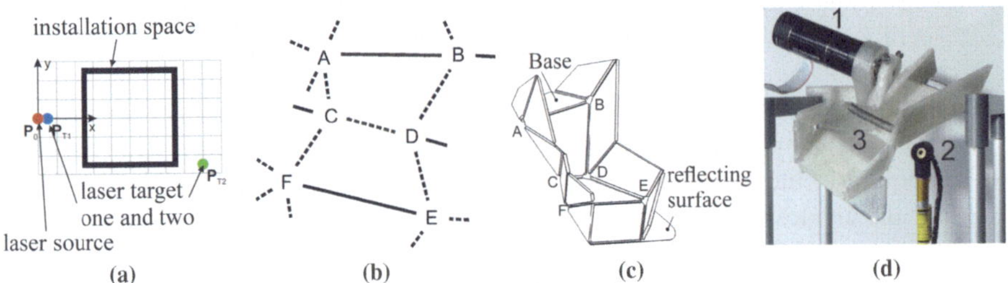



**Figure 2:** *Application of the presented process: (a) motion task; (b) kinematic manipulation of the selected foldscheme Chicken-Wire; (c) virtual prototype for the kinetostatic analysis; (d) prototype with motor (1), laser source (2), and return spring (3) based on kinetostatics*

is then analysed kinetostatically in order to dimension the actuators – a rotational drive and a return spring. The resulting prototype is shown in Figure 2d.

## 3   Folding Mechanisms

Folding mechanisms are a special class of origami-inspired foldings with a limited, small, number of facets and one or more DOF. Although, in many cases origami-inspired structures originate from symmetrical foldschemes, folding mechanisms may have asymmetries due to application-based adjustments. Amongst other examples, two successfully implemented architectural realisations are the *Soundspheres* by the University of Michigan [Thün et al. 12] and the *Al Bahar Tower* facade shading system by Aedas Architects [Aedas 12], see Figure 3.

**Figure 3:** *Examples of Folding Mechanisms: (a) Al Bahar Tower shading system (cropped) [Aedas 12] and (b) Soundspheres (cropped) [Thün et al. 12]*

In both applications, facets are essential for the overall function. They reflect sound or sunlight to achieve the desired acoustics or shading. Therefore, the best way is to position these facets with a folding mechanism, as no additional structures

are required. Aedas Architects seperated the individual vertices, to simplify the
kinematics. Researchers at University of Michigan created coupled kinematics,
however, the motion of their whole system is not constrained by any additional
boundary conditions.

The definition of all folding mechanism elements is built upon common terms
used in literature. However, new and essential is the distinction between foldpattern
and foldscheme, which allows a clear assignment of characteristic properties. Figure 4 illustrates the classification scheme of folding mechanisms.

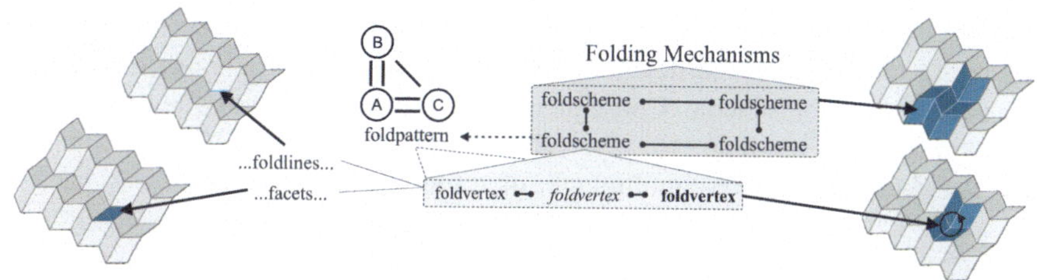


**Figure 4:** *Elements of folding mechanisms, bottom-up: Foldlines and foldfacets
build a foldvertex. A foldscheme consists of one or more vertices connected to each
other as shown by the foldpattern (these can be different vertices, denoted by differnt
fonts). One foldscheme can be combined to a folding mechanism by using the four
foldscheme operations.*

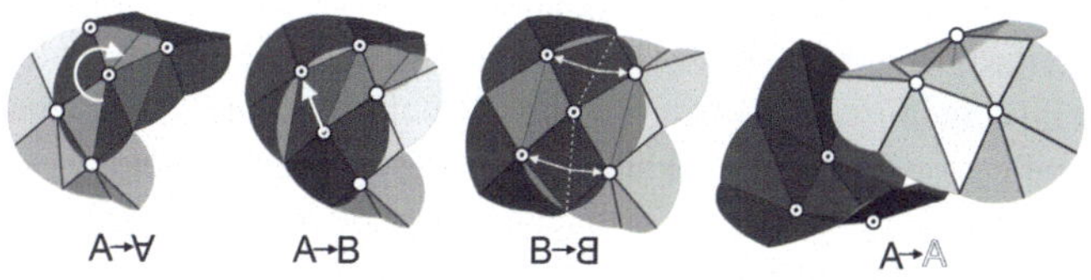


**Figure 5:** *Foldscheme operations.*

All origami-inspired structures are composed of the following basic elements:
foldline, foldfacet, and vertex. Originally, the vertex describes the intersection
point of the foldlines of a spherical vertex. However, in thick origami non-spherical
vertices occur as well. Therefore, here vertices are defined as the smallest possible
closed loop chain of foldlines and foldfacets. Vertices relevant for folding mechanisms have at least four facets and foldlines and thereby one DOF.

Commonly, a foldpattern is associated with the origami structure and defines
the smallest, repeated part of an origami model. It is not always clear, whether dihedral angles and other dimensions belong to the foldpattern. Therefore, a new determination of foldpatterns and foldschemes following the distinction between structural and dimensional synthesis is introduced. A foldpattern defines the structure
of a folding mechanism, which includes the number of foldlines, number of facets,
number of vertices, and their connection. In contrast to this, the foldscheme con-

tains all dimensional information, which include the angles between foldlines and the distances, i.e. lengths, between vertices. Consequently, different foldschemes can be based on the same foldpatterns. In mechanism theory and robotics similar distinctions are made between structural and dimensional synthesis.

Foldschemes are the minimal combination of vertices that can be assembled to the complete folding mechanism just by repetition in terms of shift, rotating, mirroring, or inverting. These actions are referred to as the four basic operations shown in Figure 5. In this context, mirroring occurs along a foldline and inverting switches all mountain to valley folds. It is beneficial, that the kinematics of the whole folding mechanism are determined by the foldscheme. At least for one DOF folding mechanisms it is sufficient to solve the kinematics of only one representative vertex of the foldscheme. Consequently, the selection of a foldscheme based on the motion task is possible. However, the transformation of a folding mechanism into a foldscheme may have multiple feasible solutions.

# 4  Classification and Collection Framework

The classification and collection of origami-inspired foldschemes has the objective to support present engineering tasks with a systematic selection and development of folding mechanisms. By structuring and categorising available foldschemes an overview of possible solutions is obtained. Nevertheless, the collection itself does not directly lead to suitable folding mechanisms for a given task. To provide selection criteria, relevant properties of the foldschemes are identified and classified below. So far, folding mechanism classifications focus on pattern affiliation [Bowen et al. 13] and design characteristics [Barej et al. 13]. Most of existing collections focus on general origami patterns from the artistic field. Besides that, [Evans et al. 15] presents a collection of rigidly foldable foldpatterns and the usage of gadgets for pattern synthesis. The foldscheme collection proposed in this paper gives a more structured framework and further classification criteria.

## 4.1  Classification

In order to classify folding mechanisms it is necessary to find suitable describing characteristics. At first, it seems appropriate to represent and analyse folding mechanisms on different levels of abstraction. In this context, the definitions given in Section 3 are helpful. Furthermore, Figure 6 shows a graphical representation of a Chicken-Wire folding mechanism. The most abstract depiction, Figure 6a, only contains information about the structure of the underlying foldpattern, that is how many vertices, here four, with how many joints, here four revolute joints denoted by R, are used. For foldschemes with vertices of different foldlines, for example with four 4-foldlines, the formula is composed of 4(4R). By adding connections and arranging vertices, the Chicken-Wire pattern is defined as shown in Figure 6b. For a spherical vertex the symbol is a letter in a circle, the lines represent the connection of vertices by means of a facet. Furthermore, the representation of the composed foldscheme embodies all kinematic dimensions and configurations, see Figure 6c.

The configuration of a foldline is either $0°$ to $180°$ or $0°$ to $-180°$, i.e. mountain or valley fold. Both representations are interchangeable. By applying the four basic operations to one foldscheme a folding mechanism as displayed in Figure 6d can be formed. Moreover, in order to get a better impression of the morphology, a three-dimensional illustration of one pose of the motion process is shown in Figure 6e. These information allow grouping and systematic identification of possible foldschemes.

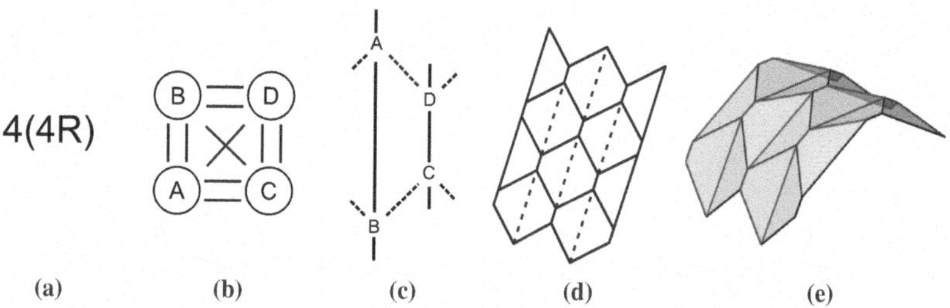

**Figure 6:** *Different levels of abstraction of the Chicken Wire*

To close the gap towards motion-task-oriented selection of foldschemes, kinematic classification criteria are required. In the following, some possible criteria are presented, however, this list is not yet complete. The DOF of a single foldscheme and the number of DOF that result from adding foldscheme elements are important kinematic properties. Applicable basic operations for the specified foldscheme help with the synthesis. For the motion process different types of spatial motion directions can be identified. These are translational, spherical, cylindrical or screwing. Based on this classification, the selection of foldschemes fulfilling the required motion-task is simplified.

## 4.2 Collection

The classification of foldschemes, which is presented in this paper, can be used to categorise identified foldschemes within the collection. Later on, it will be shown how these information and the dimensions of a foldscheme can be displayed. Further, two foldscheme collection entries are illustrated in Figure 7.

The foldscheme collection captures data on different features of foldschemes and is organised in categorising properties like main motion direction, geometric space formula, DOF, and further information on the structure. The former are explained in Section 4.1. Figure 7 shows example entries of the foldscheme collection for the Miura-Ori and the Chicken-Wire.

The *Common name* is used as reference and honours the originator. For each foldscheme a graphical representation is given showing the arrangement of vertices and indicating the sequence of pre-defined angles under *Vertices information*. Here, angles and the configuration are given. By definition, a foldscheme is only defined, if all necessary dimensions are specified. Therefore, the entries comprise a group

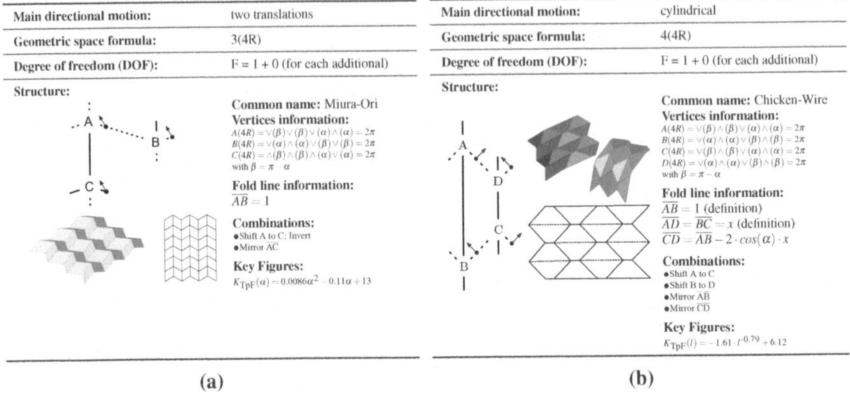

**(a)**

| Main directional motion: | two translations |
| --- | --- |
| Geometric space formula: | 3(4R) |
| Degree of freedom (DOF): | F = 1 + 0 (for each additional) |

Structure:

**Common name:** Miura-Ori
**Vertices information:**
$A(4R) = \vee(\beta) \vee (\beta) \vee (\alpha) \wedge (\alpha) = 2\pi$
$B(4R) = \vee(\alpha) \wedge (\alpha) \vee (\beta) \vee (\beta) = 2\pi$
$C(4R) = \wedge(\beta) \wedge (\beta) \wedge (\alpha) \vee (\alpha) = 2\pi$
with $\beta = \pi - \alpha$

**Fold line information:**
$\overline{AB} = 1$

**Combinations:**
- Shift A to C; Invert
- Mirror $\overline{AC}$

**Key Figures:**
$K_{\mathrm{TpF}}(\alpha) = 0.0086\alpha^2 - 0.11\alpha + 13$

**(b)**

| Main directional motion: | cylindrical |
| --- | --- |
| Geometric space formula: | 4(4R) |
| Degree of freedom (DOF): | F = 1 + 0 (for each additional) |

Structure:

**Common name:** Chicken-Wire
**Vertices information:**
$A(4R) = \vee(\beta) \wedge (\beta) \vee (\alpha) \wedge (\alpha) = 2\pi$
$B(4R) = \vee(\alpha) \wedge (\alpha) \vee (\beta) \wedge (\beta) = 2\pi$
$C(4R) = \vee(\beta) \wedge (\beta) \vee (\alpha) \wedge (\alpha) = 2\pi$
$D(4R) = \vee(\alpha) \wedge (\alpha) \vee (\beta) \wedge (\beta) = 2\pi$
with $\beta = \pi - \alpha$

**Fold line information:**
$\overline{AB} = 1$ (definition)
$\overline{AD} = \overline{BC} = x$ (definition)
$\overline{CD} = \overline{AB} - 2 \cdot cos(\alpha) \cdot x$

**Combinations:**
- Shift A to C
- Shift B to D
- Mirror $\overline{AB}$
- Mirror $\overline{CD}$

**Key Figures:**
$K_{\mathrm{TpF}}(l) = -1.61 \cdot l^{-0.79} + 6.12$

**Figure 7:** *Foldscheme collection entries: (a) Miura-Ori (3 vertices each 4 revolute joints foldpattern) and (b) Chicken-Wire. (4 vertices each 4 revolute joints foldpattern)*

of foldschemes. Length dimensions for the distance between vertices are given under *Fold line information. Combinations* list possible ways to assemble the foldscheme to complete folding mechanisms. An example of one composed folding mechanism is illustrated for easier comparison and identification of similarities or differences, respectively.

The foldscheme collection is constantly extended and additional properties of foldschemes are investigated. One objective is to define key figures, which summarise kinetostatic information to enable an easier selection of foldschemes. These key figures are application-specific. As an example the kinematic key figures $K_{TpF}$ have been derived, a parameter that determines the drive torque as a function of a geometric parameter. A more detailed explanation of the derivation of kinetostatic key figures is given in Section 6 and Section 8. Nevertheless, with the so far compiled data the foldscheme collection can be used as a base for decision making. Furthermore, it allows grouping and indicates alternative structures with similar properties.

## 5 Kinematic Analysis using a Tailored Simulink Library

In order to allow an easier synthesis of folding mechanisms, it is necessary to provide methods for the determination of kinematic and dynamic properties. Thus, the qualified evaluation of selected foldschemes and the dimensional synthesis is only possible, if a feasible method for the implementation of folding mechanisms exists. Established software tools in engineering are multi-body-simulation (MBS) and computer aided design (CAD) systems. However, standard software is difficult to apply to foldings due to spatial chain dependencies, which are time-consuming and error-prone in the implementation. Therefore, this leads to the development of spe-

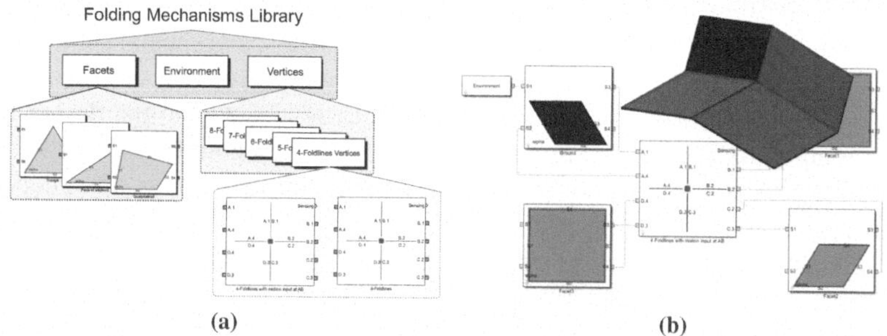


**Figure 8:** *(a) Library overview and (b) simple Simulink model with visualisation.*

cialised software tools for origami, such as [Tachi 10, Lang 04, Gattas 13]. These tools, however, mainly focus on foldpatterns and kinematics. Consequently, the software environment Mathworks MATLAB & Simulink is adopted for the MBS of folding mechanisms . This multipurpose simulation software provides libraries for different fields, e.g. mechanics, electronics and closed-loop-control. Due to the modular approach, the users are given the possibility to assemble existing functions into their own libraries. Based on the toolbox for MBS, a library for folding mechanisms is implemented. In addition, the MATLAB suite offers other valuable features and functions, for example, the fast implementation of optimisations and parameter studies for the design of folding mechanisms. In addition, a visualisation for MBS is provided directly.

The structure of the library for folding mechanisms is based on the classification described above. This structure allows an easy implementation of any foldscheme of the collection and provides a seamless integration into the design process. There are three subgroups as shown in Figure 8a. The *Environment* category, for instance contains configuration blocks for the simulation such as gravity and solver configuration. After placing those, vertices and facets from the categories *Facets* and *Vertices* can be added. Via a graphical user interface further properties are adjustable, e.g. lengths of the foldlines, facet thickness, dihedral angles, and colours. The facet and vertex blocks show 2D-visualisations of the elements, which simplifies configuration and connection. An example of a Miura-Ori vertex is shown in Figure 8b.

Using the created library, it is possible to simulate various folding mechanisms, which are rigidly foldable. Simulink allows to calculate forward and inverse dynamics even for kinetostatic overdetermined systems. However, elastic material properties cannot be taken into account. The overdetermined systems are solved kinetostatically in a pseudo-inverse way. Thus, calculated forces and torques for folding mechanisms are only permissible, if the material values do not undergo any major changes. If this requirement is met, the presented approach can be used for many applications and use cases.

## 6  Kinematic Key Figures

The determination of key figures is a round-up of larger experimental set-ups, analytical calculations, or optimisations. Key figures are used in various engineering fields and form the base of guidelines and standards to facilitate the design and dimensioning process. For folding mechanisms with their complex kinematic dependencies and greatly varying foldpatterns meaningful key figures often depend on the application. However, in this section a concept of how to determine key figures for folding mechanisms using the Simulink library is proposed.

Design properties as dihedral angles or lengths are often pre-chosen factors, which have a huge impact on motion, assembly space, or transmission outcomes. The challenge is to find a compromise between different conflicting requirements or to comply with specifications. By comparing different configurations in terms of one or more factors the optimal solution can be obtained. Consequently, as an example a key figure for the transmission of a driving torque, which is applied at the base, and an external force on another facet is calculated and its use demonstrated.

In this example a Chicken-Wire pattern of four by four facets is used. Thereby, the lengths $\overline{AD} = \overline{BC}$ as shown in Figure 7b are varied. The angle $\alpha$ has been chosen related to the length $\overline{AD}$ to $\alpha = \arccos(0.2/\overline{AD})$ and $\beta = \pi - \alpha$. The applied force is implemented using blocks from the MBS toolbox. Furthermore, the motion input is defined as a bump-less and jerk-free folding-deploying providing actuation of $0°$ to $180°$. One of the factes is fixed and acts as a base.

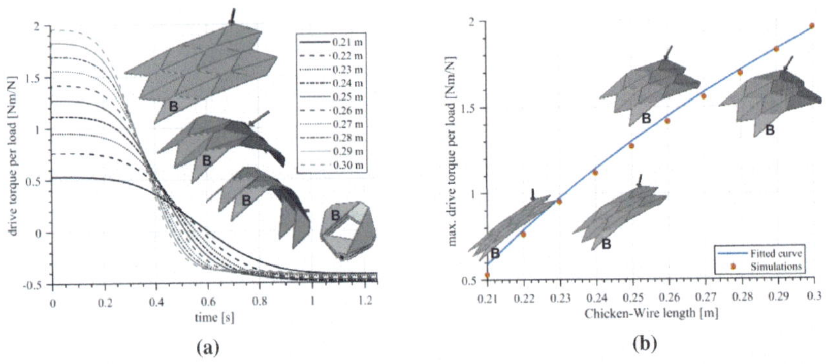


**Figure 9:** *(a) Drive torque and (b) derived key figure (max. torque per load) for varying the Chicken-Wire folding mechanisms; B indicates the base fixed facet.*

The calculation is performed by MATLAB in form of a loop, varying the lengths from $0.21\,\mathrm{m}$ to $0.30\,\mathrm{m}$. Subsequently, the drive torque is calculated and normalised to the applied force, see Figure 9a for the results. If the maxima of the drive torques are plotted over the varied lengths, the graph in Figure 9b is created. The maximum expected drive torque can thus be determined for a given system with a chosen length. Applying the curve fitting function to the determined maximum values, see Figure 9b, an equation representing the key figure Torque-

Per-Force $K_{TpF}$ can be derived, which is exemplary given in (1). Thereby, the drive torque is a function of the length and can be added as one characteristic in the fold-scheme collection, compare Figure 7. Based on such key figures, the structural and dimensional synthesis is simplified.

$$K_{\mathrm{TpF}}(l) = -1.61\frac{Nm}{N} \cdot \left(\frac{l}{1m}\right)^{-0.79} + 6.12\frac{Nm}{N} \tag{1}$$

As mentioned before there are many possible key figures for foldschemes. Most of the key figures are application specific and the other geometric parameter of the folding mechanism can be chosen arbitrary. However, the presented procedure can be applied very fast, even in multiple dimension and can lead to more insight and fast optimisation of folding mechanisms. Other meaningful ones could be the drive torques depending on the number of facets or gravity forces, and the spread of an unfolded pattern depending on the facet angles or distances between vertices.

## 7  Kinetostatic and Stress Analysis using FEA

For dimensioning of folding drive systems and structural design, an analysis of mechanical loads and motion ranges is necessary. Here, as a step towards this goal the stress distribution during folding is investigated. The results are not only taken as an indicator for areas to be strengthened, due to great loads, but are also summed up in key figures to provide a manageable factor for design applications. Finite element analysis (FEA) is an established tool for strength calculation in structural mechanics and has already been used to analyse foldings as shown in [Jiang et al. 16, Saito et al. 16, Gilewski et al. 14]. Thereby, many complex folding structure can be examined with respect to the mechanical loads. For the evaluation of foldschemes around a structured design process, an analysis of the van Mises stress and the reaction force distribution during the motion is carried out with the Abaqus CAE software, see Section 7.3. In order to obtain valid results material data and simulation results are validated by means of a physical test set-up. The tests carried out include material testing of the Young's modulus, see Section 7.1, and stress measurements during different states of motion, see Section 7.2.

### 7.1  Material Test

For the prototype assembly, a 3D printing process is used. Polylactide (PLA) is taken as material. The dependence of mechanical-physical properties of PLA on the printing parameters used, e.g. print speed, is not negligible [Tymrak et al. 14]. Therefore, a 3-point bending test is carried out to determine the Young's modulus of PLA under given 3D pressure settings. For the experiments four samples of $3\ mm$ thickness were fabricated using the filament fused deposition modelling with a nozzle diameter of $0.4\ mm$.

In the test, material samples are stored on two supports and loaded centrally by a force, see Figure 10b. Each of the four samples is loaded with three differ-

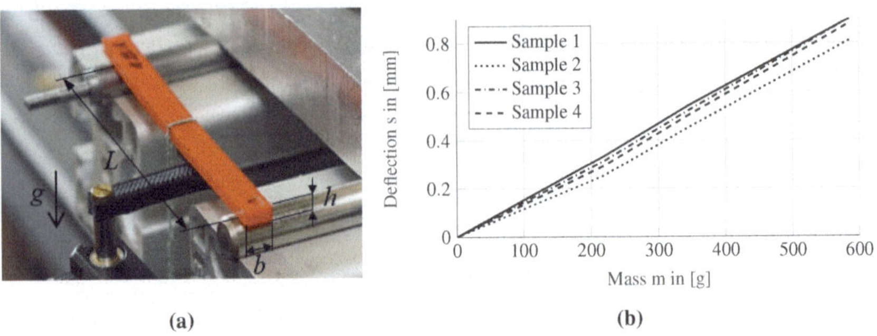

(a)           (b)

**Figure 10:** *Material test set-up (a) and (b) mass-deflection-curve.*

ent weights of 200 g, 350 g, and 550 g. Furthermore, the Young's modulus $E$ is calculated to $\frac{1}{4}\frac{L^3 mg}{bh^3 s}$ with the measured data [DIN 09]:

The change in deflection, see Figure 10b, shows an almost linear increase with mass. An average slope from deflection to mass of 1.4 mm/kg can be determined. If the mass deflection diagram is compared to a stress-strain diagram from PLA, the linear gradient can be confirmed for the linear-elastic range. According to the deflection the mean Young's modulus of the individual masses is determined as mentioned above to $E = 3540 \pm 218$ MPa. Consequently, a Young's modulus value for the simulation was determined sucessfully. The relatively high value of the standard deviation is due to measurement inaccuracies and pressure settings, but also due to material properties.

## 7.2 Static Stress Test

For validation, a prototype of a Yoshimura vertex is used. The Yoshimura vertex acts as a representative for other examined foldschemes from the collection. The measurement is performed using a strain gauge on the facets surface and compared to the Abaqus simulation results. The relationship between deformation and loading in a component is given by valid material laws, e.g. the elastic deformation range of PLA is subject to Hooke's law. The resulting stresses are determined from measured strains. Furthermore, it is assumed, that the loads are low due to inertial forces and joint friction alone. In order to generate measurable strains, the system is subjected to a static tensile force.

Revolute joints are realised by conical runners of one surface engaging in cone-shaped recesses on the other surface, see Figure 11c. This arragement approximately corresponds to the revolute joints defined for the simulation. The prototype can be printed in its final ready assempled form. Further, the experimental set-up in Figure 11 presents the Yoshimura vertex (1) as pivoted on two axles in the frame (2). The fixation on both sides of the vertex prevents axial displacement (9). Additionally, the vertex is statically loaded with an external force. The force is made up of a mass (4) acting on the folding in horizontal direction via a rope (6) and a

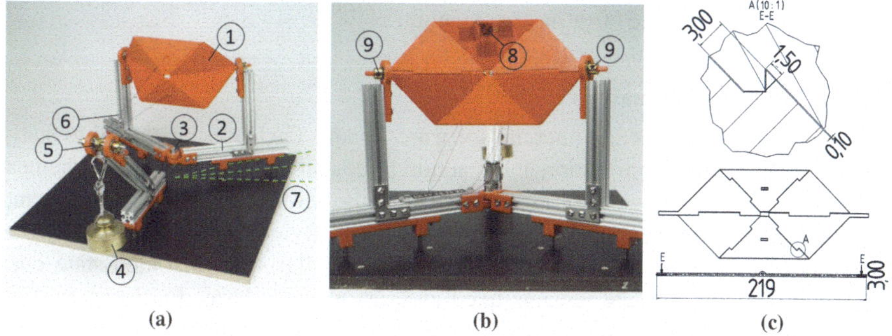

(a)       (b)       (c)

**Figure 11:** *Experimental setup of the Yoshimura-Vertex-Prototype.*

pulley (5). On the prototype's back the strain gauge is attached. It is glued on the upper surface of the vertex (8).

A static test procedure is performed for the folding poses $90°$, $120°$ and $150°$. It is possible to change from one folding position to another using the frame joint (3) and the intended insertion positions (7). The testing procedure is as follows: Before each step, the zero position is determined. Subsequently, six measurements are carried out for each folding pose. Each measured value is recorded by the measuring instrument over a period of 5 s. The mean value of the strains is calculated over time and measurement.

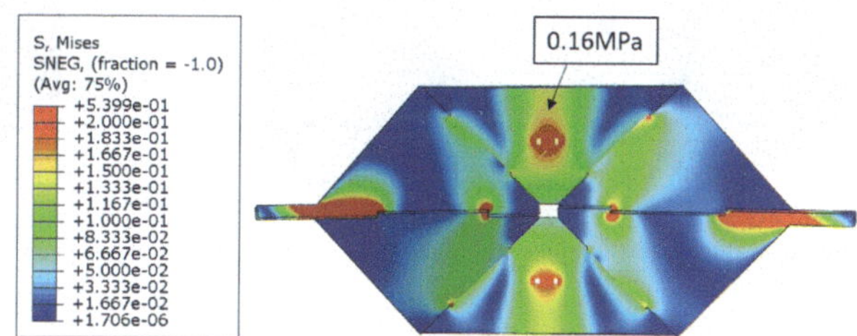

**Figure 12:** *Stress distribution in the 120° configuration.*

**Table 1:** *Results from the static stress test*

| Angles | $90°$ | $120°$ | $150°$ |
|---|---|---|---|
| $\sigma_{v,measurement}$ [MPa] | $0.18\pm0.02$ | $0.13\pm0.01$ | $0.18\pm0.02$ |
| $\sigma_{v,simulation}$ [MPa] | 0.14 | 0.16 | 0.18 |

The strains are shown with their standard deviation. Stresses are calculated from measured strains to determine the van Mises equivalent stresses, see Table 1. Overall, the static stress test was carried out successfully. Nevertheless, simplifications were made for the simulation and the model, e.g. no friction in the model, possible nonhomogenities among the sample, elasticities for the glue of the strain gauge etc. The relative error has a maximum of 28 %.

## 7.3 Kinetostatic Analysis

For the simulation, the commercial software Abaqus CAE was employed to study properties as stress distribution and reaction forces on three different foldschemes. As common examples of folding mechanisms, the Miura-Ori, the Yoshimura, and the Chicken-Wire were selected for investigation. By controlling the fold angles during the simulation the foldschemes move from flat state to folded state over 180° in 1 s. One surface remains fixed and serves as the base. Furthermore, two different load states were configured: (I) revolute joints are subject to friction with a coefficient of friction of 0.1; (II) the foldschemes are loaded with an exemplary selected external process force of $-1$ N in z-direction.

For the FEA model, homogeneous shell and standard S4R elements are used. A shell thickness of 1 mm is applied. The material model is set up for PLA according to the test before, i.e. Young's modulus of 3540 MPa, Poisson's ratio of 0.36, and density of 1240 kg/m$^3$. The revolute joints in the model are realised using the hinge tool.

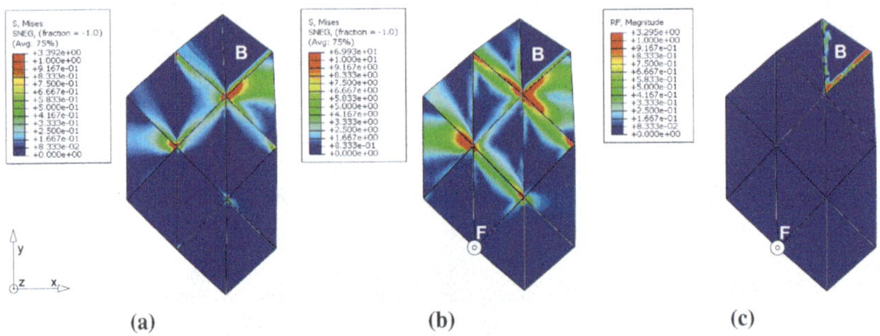

(a)       (b)       (c)

**Figure 13:** *Results from the FEA analysis: B indicates the base fixed facet, (a) stress distribution under friction, (b) stress distribution under external load and (c) reaction forces under external load: F indicates the force application point.*

Exemplarily, Figure 13a shows the stress distribution of the Yoshimura pattern for an average time of 0.5 s under friction in the joints. The base facet is here marked with a *B*. The maximum occurring stress is up to 3.4 MPa. In the vicinity of high stresses, which can be found along the joints, low stresses are mainly observed in the middle of surfaces . In accordance with the principle of flow of forces, resulting weight and inertia forces are transmitted along the shortest path via the connecting elements to the fixation. At the vertex centres, additional high loads occur. The six foldlines, that meet in the centre, lead to a local stiffening of the structure. Due to the increased stiffness, the tension increases locally along the foldline with constant deformation.

Figure 13b shows the stress distribution under an external load. Node and facet, where the force is applied, are marked with an F. It can be seen that the maximum stresses in the system are increased. In addition, the foldschemes are further deflected along the force application, resulting in partly irregular configurations. Furthermore, Figure 13c shows the reaction forces, which are caused by the external force. The applied force generates reaction forces at the fixation, especially in the case of outer edges of the fixed facet.

# 8   Kinetostatic Key Figures

Comparable to the approach chosen to define kinematic key figures, the determination of kinetostatic key figures is performed. Based on the results of tests and simulations, parameters are defined below to improve the comparability and ease of designing foldschemes. Input variables (such as geometric dimensions, e.g. lengths or dihedral angles), as well as test parameters (such as the simulation time, the fold angle, friction, or process forces), and output variables (such as the maximum reaction forces or the maximum stresses), can be used for this purpose.

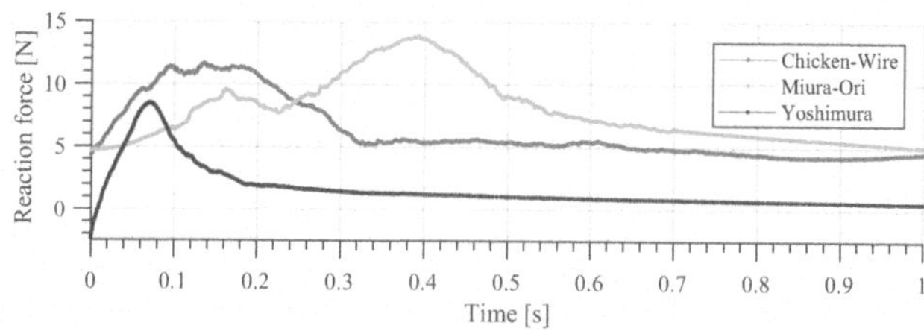


**Figure 14:** *Results from the FEA simulation for the Miura-Ori, the Chicken-Wire, and the Yoshimura foldscheme.*

Similar to mechanism theory, where the ability of a mechanism to transfer forces from the drive to the output rocker or slider is used as an indicator, the ratio of external process to internal reaction forces can be specified for folding mechanisms. Both provide a reference for force transmission, particularly in complex, kinematically closed chains. In the present case, the reaction force for the most stressed node relates to an applied force of 1 N. Figure 14 shows the time course of the ratio for the three foldschemes. At present configurations the graph shows that Yoshimura-foldscheme has the smallest ratio and is therefore subject to the lowest reaction forces. Here, too, optimisation with regard to dihedral angles or lengths can be carried out to derive kinematic key figures.

# 9   Conclusion and Outlook

This paper discusses design processes for motion devices and proposes a specific development procedure for folding mechanisms by presenting required methods and knowledge bases. These include a folding collection framework, kinematic and static analysis. Furthermore, a differentiation between foldpattern and foldscheme allows a distinctive classification of existing foldschemes and therefore simplifies the systematic synthesis of new solutions. A here proposed Simulink fold library enables engineers to directly implement foldschemes and optimise related applications. The presented procedure for static analysis with FEA is the last step, before folding mechanisms can be realised in CAD systems and as prototypes.

Future work is concerned with the extension of the foldscheme collection and the derivation of additional – and more important application-independent – key figures. For the final steps of the design process, i.e. the dimensioning and design of folding mechanisms, catalogues of specific machine components are considered advantageous. These catalogues may contain different types of line-shaped revolute joints, possibilities for the limitation of the motion range, or facets in sandwich construction.

## Additional Material and Acknowledgement

An extract from the presented Foldscheme-Collection and the introduced Simulink library for Folding Mechanisms are available on GitHub under the Eclipse Public License 1.0 (EPL). Please feel invited to download and use these materials. Pull requests are welcome: `github.com/JaschaParis/FoldingMechanisms`.

The authors would like to thank the Deutsche Forschungsgemeinschaft (German Research Foundation, DFG) for supporting the project.

## References

[Aedas 12]  Architect Team Aedas. "Al Bahar Towers Responsive Facade." In *ArchDaily by Karen Cilento*, 2012.

[Barej et al. 13]  M. Barej, Y. Safi, B. Sköck-Hartmann, T. Gries, U. Steinseifer, B. Corves, and M. Trautz. "Systematisierung gefalteter und faltbarer Strukturen in technischen Anwendungen." *Konstruktion* 1/2 (2013), pp. 69–74.

[Bowen et al. 13]  L. Bowen, C. Grames, S. Magleby, R. Lang, and L. Howell. "An Approach for Understanding Action Origami as Kinematic Mechanisms." In *ASME 2013 International Design Engineering Technical Conferences and Computers and Information in Engineering Conference*, edited by ASME, 2013.

[Corves et al. 07]  B. Corves, J. Niemeyer, and J. Kloppenburg. "IGM-Mechanism Encyclopaedia and the Digital Mechanism Library as a Knowledge Base in Mechanism Theory." In *Proceedings of the ASME International Design Engineering Technical Conferences and Computers and Information in Engineering Conference, 2006*, pp. 267–273. New York: American Society of Mechanical Engineers, 2007. doi:10.1115/DETC2006-99059.

[DIN 09]  "DIN EN ISO 178 Kunststoffe - Bestimmung der Biegeeigenschaften.", 2013-09.

[Evans et al. 15]  T. A. Evans, R. J. Lang, S. P. Magleby, and L. L. Howell. "Rigidly foldable origami gadgets and tessellations." *Royal Society Open Science* 2:9. doi:10.1098/rsos.150067.

[Gattas 13]  J. Gattas.  "Rigid Origami Toolbox.", 2013.  Available online (http://
joegattas.com/rigid-origami-toolbox/).

[Gilewski et al. 14]  W. Gilewski, J. Pelczynski, and P. Stawarz. "Origami-inspired
build-ing block and parametric design for mechanical metamaterials." *Forestry and Wood
Technology* :85.

[Hoffmann et al. 15]  S. Hoffmann, M. Barej, B. Gnther, M. Trautz, B. Corves, and J. Feld-
husen.  "Demands on an Adapted Design Process for Foldable Structures."  In
*Origami6 : Proceedings of the sixth International Meeting on Origami Science, Math-
ematics, and Education*, pp. 489–499. Sixth International Meeting on Origami Science, Mathematics, and Education, Tokyo (Japan), 10 Aug 2014 - 13 Aug 2014, Providence: American Mathematical Society, 2015.

[Jiang et al. 16]  W. Jiang, H. Ma, M. Feng, L. Yan, J. Wang, J. Wang, and S. Qu. "Origami-
inspired building block and parametric design for mechanical metamaterials." *Journal
of Physics D: Applied Physics*.

[Lang 04]  R. J. Lang.  "Treemaker 5.0.", 2004.  Available online (http://www.
langorigami.com/article/treemaker).

[Saito et al. 16]  K. Saito, A. Tsukahara, and Y. Okabe.  "Designing of self-deploying
origami structures using geometrically misaligned crease patterns.", 2016.

[Tachi 10]  T. Tachi. "Freeform Origami.", 2010. Available online (http://
origami.c.u-tokyo.ac.jp/~tachi/software/).

[Thün et al. 12]  G. Thün, K. Velikov, C. Ripley, L. Sauvé, and W. McGee. "Soundspheres:
Resonant Chamber." *Leonardo* 45:4 (2012), pp. 348–357.

[Tymrak et al. 14]  B. M. Tymrak, M. Kreiger, and J.M. Pearce. "Mechanical properties of
components fabricated with open-source 3-D printers under realistic environmental
conditions." 58, pp. 242–246, 2014.

[VDI 93]  "Methodik zum Entwickeln und Konstruieren technischer Systeme und Produkte." VDI-Guideline VDI 2221, Verein Deutscher Ingenieure, 1993.

[Zhakypov and Paik 18]  Z. Zhakypov and J. Paik. "Design Methodology for Constructing
Multimaterial Origami Robots and Machines." *IEEE Transactions on Robotics* 34:1
(2018), pp. 151–165. doi:10.1109/TRO.2017.2775655.

---

Jascha Paris, Judith Merz, Mathias Hüsing, Burkhard Corves
Institute of Mechanism Theory, Machine Dynamics and Robotics, RWTH Aachen
Univer-sity, Germany, e-mail: (paris;merz;huesing;corves)@igmr.rwth-aachen.de

Henri Buffart, Susanne Hoffmann, Martin Trautz
Chair of Structures and Structural Design, RWTH Aachen University, Germany, e-
mail:(buffart;hoffmann;trautz)@trako.arch.rwth-aachen.de

Justus Siebrecht, Chantal Weigel
Institute for Machine Elements and Systems Engineering, RWTH Aachen University,
Ger-many, e-mail: (justus.siebrecht;chantal.weigel)@imse.rwth-aachen.de

# Origami Sensitivity –
# On the Influence of Vertex Geometry

*L. Zimmermann, K. Shea, T. Stanković*

***Abstract:*** *The geometrical arrangement of crease pattern vertices, here called vertex geometry, strongly influences the kinematical behavior of the corresponding origami. This work introduces a robust simulator that enables the assessment of arbitrary vertex geometries according to rigid foldability, spatial feasibility, and self-intersection. The benefits of the simulator are two-fold, and exemplified here through the sensitivity analysis of a flasher crease pattern. First, it provides a new way to gain insight into the folding process by enabling the visual representation of the search space. Second, it enables the optimization of the vertex geometry so that the origami complies with a set of engineering constraints.*

## 1  Introduction

Today, the majority of origamis implemented in technical applications are based on a handful of crease patterns. These patterns are usually chosen based on their advantageous properties from a range of well-described patterns, and are then individually altered to fit engineering requirements [Morgan et al. 15]. The reason to revert to known patterns instead of generating new ones is the complexity of the underlying design task. While the conceptual requirements entail a deployed functional state, an enclosed space for stowage, and an actuation of the origami, a realizable solution has to also fold rigidly, stay free of self-intersections, and consist of finitely thick panels. These conflicting requirements explain why origami design for engineering applications is a tedious and time-consuming process. The application of computational methods thus holds huge potential for the facilitation of the design process by reduction of time-consumption and for the creation of previously unknown patterns that are tailored to specific engineering tasks.

The computational generation of crease patterns can be divided into two sub-problems. First, an origami graph needs to be composed whose vertices and edges form a minimal forcing set [Abel et al. 16] for a given actuation. This ensures that the pattern, once activated, is kinematically fully determined. The second sub-problem is the adjustment of the geometrical arrangement of vertices, here called vertex geometry, which determines start and end configuration, rigid foldability, and self-intersection of the origami. The aim of this paper is a step towards the computational generation of crease patterns by focusing solely on the influence of the vertex geometry, assuming that a valid origami graph is known. Ironically, this is accomplished here by reverting to known patterns to reduce the problem dimensionality.

Examining the influence of the vertex geometry equates to the assessment of the kinematical behavior by adjusting vertex locations. In the process of changing these locations, there will inevitably emerge configurations that are not rigidly foldable, i.e. that exhibit facets that either stretch, bend, or do both. For the assessment to be effective, these configurations need to be assigned with a single deterministic merit value that provides more information than a binary decision on being rigidly foldable or not. The first contribution is the introduction of a simulator that is able to reliably compute the motion of any pattern, independent of the vertex geometry and the extent of distortion.

Origami mechanisms exhibit complex three-dimensional motion when folded, making it difficult for origami design practitioners to gain detailed insight into the folding process. The simulator contributes by enabling the visual representation and interpretation of the search space and thus by providing a deeper understanding of the kinematical behavior. This is exemplified through the sensitivity analysis of an extended flasher pattern with the goal to adjust its vertex geometry so that it complies with a given set of engineering requirements. The requirements in focus are rigid foldability, spatial feasibility (i.e. if the vertices fold into a pre-defined enclosed space), and self-intersection. Finally, the flasher pattern is optimized through a generic approach as a proof of concept to show that the simulator lays the basis for the computational generation of origami.

The paper is organized as follows. The next section discusses background on computational pattern generation and rigid foldability. In the following sections we first describe the simulator, and then introduce the case study, perform the sensitivity analysis and optimize the extended flasher. The paper finishes with a discussion and conclusion.


## 2 Background

The computational generation of crease patterns has so far focused mainly on the artistic realm and corrugated surfaces. The TreeMaker [Lang 96] is the elaborate algorithm in the field of origami art. With the input of a planar graph, it provides the user with an origami base that can subsequently be folded into the desired shape, commonly an animal. Tachi's Origamizer [Tachi 10a] approximates 3D surfaces through a combination of tessellation and inclusion of molecules between surface polygons. The method is able to generate patterns for demanding shapes, but uses crimp folds that are not rigidly foldable. Tachi addresses this problem with Freeform Origami [Tachi 10b] and the generalization of Resch's pattern [Tachi 13], which both approximate 3D surfaces with less complex patterns. Lang and Howell recently presented a deterministic algorithm that generates intricate rigidly foldable single-degree of freedom origami patterns with the input of two arrays of fold and direction angles [Lang and Howell 18]. Although certainly useful within their domains, these works do not explicitly involve engineering requirements, or provide solutions that are either difficult to manufacture and control or not rigidly foldable.

Rigid foldability, the notion of a continuous path through phase space from stowed to deployed state, is an important concept in origami and has received considerable attention. Huffman derives relations between in-plane and dihedral angles using the principle of Gaussian curvature [Huffman 76], which has later been extended to general four-degree vertices [Lang et al. 15]. The same principle is used by Miura to examine properties of single vertices with varying degrees [Miura 89]. Streinu and Whiteley show that the configuration space of a single-vertex origami is connected, i.e. that one state can be transformed into the other by non-self-intersecting motion [Streinu and Whiteley 04]. Abel et al. prove a necessary and sufficient condition for the rigid foldability of a single-vertex origami that is independent of fold angles [Abel et al. 16]. Unfortunately, these analytical formulations generally consider single vertices only, and commonly only up to degree four, whereas global rigid foldability has not yet been proven. Rigid foldability for multi-vertex patterns can so far only be shown using numerical methods. Belcastro and Hull present conditions for fold angles along closed paths around vertices [Belcastro and Hull 02], and based on that work, Watanabe and Kawaguchi introduce a numerical method to judge infinitesimal rigid foldability [Watanabe and Kawaguchi 09].

# 3  Simulation

The simulator builds on earlier work of the authors [Zimmermann et al. 17] with modifications to account for crease patterns that are not rigidly foldable, as arbitrary vertex geometries do not generally lead to rigid foldability. This means that facets may experience bending and stretching, or in other words, the Euclidean distance between vertices belonging to the same facet may change. The basic idea of the simulation is to allow these changes, but to minimize the difference of actual to target Euclidean distance between vertices to obtain a measure for the total distortion of the crease pattern. In comparison to the related works, this approach does not impose any explicit conditions for rigid foldability. Instead, the distortion is presented as an error, and can be guided towards zero by optimizing the vertex geometry.

## 3.1  Input

The user inputs to the simulation are a crease pattern with vertices and edges, boundary conditions such as fixed vertices or facets, and one or multiple vertex trajectories that actuate the origami in a desired number of iterations $N$. These inputs are chosen so that an origami can be tailored to its environment, which may be limited in terms of e.g. enclosed space for stowage or deployment resources.

## 3.2 Simulation

The simulator obtains the target lengths $\mathbf{l_t}$ from the provided crease pattern, which correspond to the Euclidean distances between $n$ sides and diagonals of all facets according to Diaz [Diaz 14]. Including the diagonals ensures the flatness of facets with more than three sides. Because the actual lengths $\mathbf{l_a^{(j)}}$ may be shortened or elongated in each iteration $j$, the constraint system is composed of the absolute of the difference between $\mathbf{l_a^{(j)}}$ and $\mathbf{l_t}$ that is kept lower or equal to some small values in vector $\boldsymbol{\varepsilon}^{(j)}$. This is formulated in Eq. (1) as two sets of constraints bounded by $-\boldsymbol{\varepsilon}^{(j)}$ and $\boldsymbol{\varepsilon}^{(j)}$ from below and above, respectively. These small numbers are summed over all $n$ constraints and presented as error $E^{(j)}$, which is minimized in each iteration $j$ with respect to $\boldsymbol{\varepsilon}$ and vertex locations $\mathbf{x}$. This is stated as a constrained nonlinear optimization problem in Eq. (1) that is solved through the *FindMinimum* function in *Mathematica 10* with default settings. The starting point for the optimization in the current iteration $j$ is defined as the vertex locations in the previous iteration $j - 1$.

$$\min_{\mathbf{x},\boldsymbol{\varepsilon}}\left\{E^{(j)} = \sum_{i=1}^{n}\varepsilon_i^{(j)} \mid -\boldsymbol{\varepsilon}^{(j)} \leq \mathbf{l_a^{(j)}} - \mathbf{l_t} \leq \boldsymbol{\varepsilon}^{(j)}\right\} \tag{1}$$

## 3.3 Rigidity Error

If $E^{(j)} \leq 10^{-4}$ (we will see later how that boundary is defined) in each iteration $j$, every folding configuration exhibits numerical rigid foldability and the crease pattern is globally numerically rigidly foldable for the given actuation with reasonably high $N$. To obtain a single measure for each simulation, we define:

$$\Gamma = \frac{1}{N}\sum_{j=1}^{N}E^{(j)} \tag{2}$$

as rigidity error $\Gamma$, which is the averaged sum of all $E^{(j)}$ over $N$. Hence, the rigidity error $\Gamma$ goes to zero if a pattern is rigidly foldable, and vice versa.

## 3.4 Spatial Feasibility

The first sub-problem of composing an origami graph can be approached in one of two ways. Either the graph is generated in its deployed state and then folded into a desired enclosed space, or vice versa. Conforming a graph to a surface (which we do not constrain to be flat as in traditional origami) instead of an enclosed space is much easier in terms of possible configurations. Our approach thus starts with the deployed configuration, which is then subject of the optimization objective, e.g. maximum surface area to collect solar energy, whereas the enclosed space is enforced by a set of geometric constraints. Hence, we use the term spatial feasibility as a measure for how well the folded origami fits into an enclosed space. Assuming convex volumes and rigid foldability, it is sufficient to check only the end location of vertices to ensure that the structure lies within the predefined space.

### 3.5  Self-intersection

All facets are decomposed into triangles and subjected to a triangle-triangle intersection check [Möller 97] in each iteration. However, this check does not necessarily detect the fold angles between adjacent facets that exceed the allowed range of -180° to 180°. In addition, for adjacent pairs of facets, two vectors are calculated, each lying in its respective facet plane and pointing perpendicularly away from the shared edge. Once a folding angle approaches the boundaries of the above-mentioned range, the cross product of these vectors is compared for successive iterations, and an intersection is detected if the resulting vector reverses its direction. The measure of self-intersection provided by the simulator is defined by the number of iterations in which self-intersection is present, averaged over all iterations $N$.

### 3.6  Output

The outputs of the simulator are the rigidity error $\Gamma$, the vertex end locations to assess the spatial feasibility of the folded configuration, and the measure of self-intersection.

## 4  Sensitivity and Optimization

This section first presents a case study of an extended flasher pattern, which is subjected to a sensitivity analysis to showcase the benefits of the simulator and examine the influence of the vertex geometry on its kinematical behavior. Then, an optimization is formulated and the optimized pattern is presented. The decision process in the case study is performed manually and from a user perspective, with the intent of presenting properties of the search space to the reader. The final optimization is fully automated given the required inputs (Section 3.1).

### 4.1  Case Study

We examine the crease pattern depicted in Fig. 1a. The pattern belongs to a class first introduced into origami by Palmer and Shafer [Lang 97] called *flashers*, which denotes the principle of wrapping a piece of paper around its vertical axis. In technical applications, this principle is most often mentioned in connection with e.g. solar panels, because the pattern can be folded into an enclosed space of a launch vehicle and once deployed exhibits a large surface area, which is advantageous       for       the       collection       of       solar       energy.

   To our knowledge, there currently exist no flashers that have been shown to fold rigidly. However, if the original crease pattern is extended by additional crease lines connecting $v_2$ to $v_3$ and $v_2$ to $v_{5'}$ respectively (applied symmetrically) as shown in Fig. 1a, this new flasher pattern does fold rigidly, and moreover, still exhibits one degree of freedom (four-symmetrically). We do not provide a formal proof here for rigid foldability, leaving it open to the reader. For the remainder of the paper, the new pattern is simply referred to as the extended flasher.

With the additional crease lines, the pattern can be perceived as being composed of ring-wise *layers*. The $0^{th}$ layer consists of the central facet $f_1$, the first layer is bounded by vertices $v_2$, $v_3$ and their rotations, and the second layer by vertices $v_4$, $v_5$ and their rotations. One attribute flashers have in common is that each time an inner layer is fully folded, e.g. when $f_1$ and $f_2$ are perpendicular as shown in Fig. 1b, the folding motion must be continued by activating outer layers. This sequential folding complicates the control of a flasher because more actuators are needed, which renders it less attractive for practical applications. The goal of the case study is thus to find a vertex geometry of the extended flasher that folds into the same enclosed space the pattern in Fig. 1a would occupy once sequentially folded, but with only a single actuation until $f_1$ and $f_2$ are perpendicular, while maintaining a maximally large projected area in its deployed state.

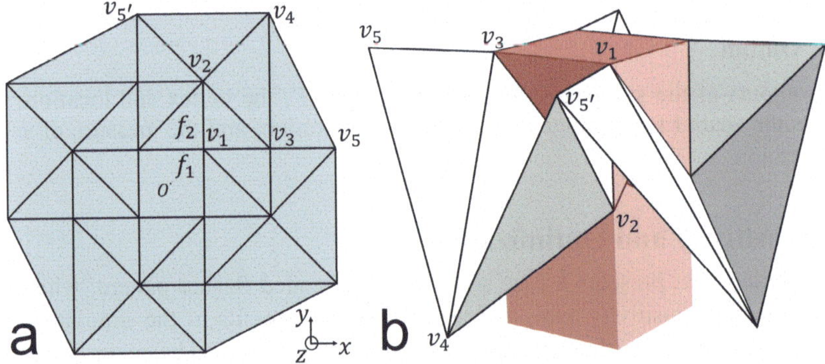


**Figure 1:** *Rigidly foldable, four-symmetrical flasher pattern (a) whose folded configuration should eventually fit into the depicted cuboid (b).*

The center of facet $f_1$ is fixed at the origin $O$ and no rotation is allowed. The vertex $v_1$ is located at $v_1^{(0)} = (1, 1, 0)$, which defines the enclosing space as a cuboid (Fig. 1b) of the following outer dimensions $-1 \leq x \leq 1, -1 \leq y \leq 1, -4 \leq z < 0$. Note that the strict inequality for the upper boundary of $z$ is to prevent self-intersection with facet $f_1$.

## 4.2 Sensitivity

The sensitivity analysis of the extended flasher is structured layer-wise from inside to outside. First, we examine two sector angles of $v_1$ in relation to rigid foldability and fix a starting location for $v_2$ that remains constant throughout the sensitivity analysis. This is a pragmatic decision considering that the actuation of $v_2$ is an input to the simulation. Second, we investigate how the starting location of $v_3$ affects the kinematical behavior of the first layer. Third, we show the influence of $v_4^{(0)}$ on the behavior of the whole crease pattern. Eventually, we draw conclusions for the optimization, which includes vertices $v_3$, $v_4$, and $v_5$.

### 4.2.1 Sector Angles of $v_1$

As origami design practitioners, our first goal is to quickly and easily assess different sector angle combinations for vertex $v_1$ to determine possible options for the starting location of $v_2$, which will be the actuated vertex for the folding process. The easiest way to demonstrate the kinematical properties of $v_1$ is by visually representing the rigidity error $\Gamma$ over two sector angles. Vertex $v_1$ shown in Fig. 2a is of degree five and developable, thus

$$360° = \alpha + \beta' + \gamma + \delta + \theta \qquad . \tag{3}$$

Facet $f_1$ is a square, hence $\delta = 90°$.

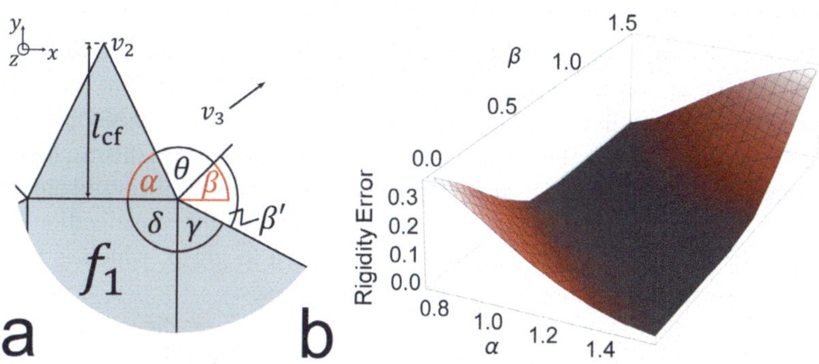

**Figure 2:** *Parametrization of sector angles around $v_1$ (a) and rigidity error $\Gamma$ for $45° \leq \alpha \leq 90°$ and $0° < \beta < 90°$ conforming to a "rigid foldability valley" (b).*

To be able to parametrize the rigidity error $\Gamma$ over two parameters $\alpha$ and $\beta$, we need to eliminate one more sector angle. To do so, we introduce a condition that requires the vertical end location of vertex $v_2$ to be identical to its end location in the original pattern, in which this location is equal to the sides of the central facet $f_1$. Hence, $l_{cf} = 2$, which for geometrical considerations of $f_2$ leads to

$$\gamma = \frac{\tan \alpha}{\tan \alpha - 1}, \tag{4}$$

and leaves the angles $\alpha$ and $\beta'$ as parameters. Instead of $\beta'$, we introduce $\beta$ that starts from the $x$ axis to make the results more comprehensible. Note that $\alpha = 90°$ and $\beta = 0°$ would result in the sector angle configuration of the extended flasher shown in Fig. 1a.

The boundaries for angles $\alpha$ and $\beta$ are set to $45° \leq \alpha \leq 90°$ and $0° < \beta < 90°$, respectively. Any smaller value for the lower limit of $\alpha$ would lead the vertex $v_2$ to lie outside the pre-defined enclosed space when folded. The boundaries for $\beta$ are chosen so that the crease lines do not intersect for any value of $\alpha$. Fig. 2b shows the outcome of the sector angle sensitivity analysis of $v_1$. It results in a "rigid foldability valley", with the rigidity error rising on both sides smoothly and monotonously. The valley is approximately straight and of constant width with a flat bottom that satisfies $\Gamma \leq 10^{-4}$, which means we have ample choice of determining a value for $\alpha$. For aesthetic reasons, i.e. reflective symmetry in addition to rotational symmetry, we decide to set $\alpha = \tan(2)$, resulting in the configuration depicted in Fig. 2a (gray facets) with $v_2^{(0)} = (0, 3, 0)$ and an actuation of

$$v_2(t) = (0, \, 1 + 2\cos t, \, -2\sin t) \text{ for } 0 \leq t \leq 90° \tag{5}$$

applied symmetrically. For the remainder of this section, the starting location $v_2^{(0)}$ remains unchanged.

### 4.2.2 Influence of $v_3^{(0)}$ on the Behavior of the First Layer

Vertex $v_3$ is the only vertex in the first layer that has not yet been considered. As we will see here, constraining the starting location of $v_3$ to the flat plane either leads to self-intersection with $f_1$ or infeasible solutions, hence we need to allow the vertex to deviate from the flat plane. Origamis are spherical mechanisms, hence the vertex geometry would be most naturally expressed in spherical coordinates using crease line length, sector, and dihedral angle. However, the visualization of the folding behavior using spherical coordinates is hard to grasp, which is why we revert to Cartesian coordinates in the following analyses.

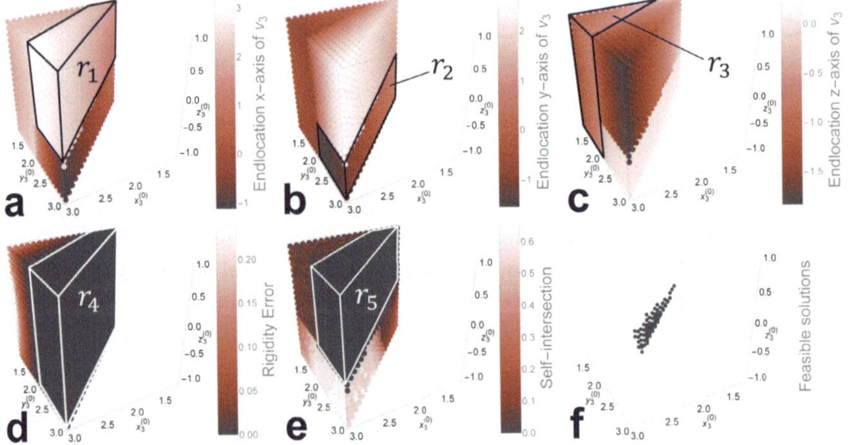

**Figure 3:** *Influence of the starting location of $v_3$ on the first layer: its absolute end location on the $x$, $y$, and $z$ axes (a-c), rigidity error $\Gamma$ (d), self-intersection (e), feasible solutions found by superposition of (a-e), and regions $r_1$ to $r_5$.*

Fig. 3 shows the sensitivity results of the influence that $v_3^{(0)}$ exerts on the behavior of the first layer. This is accomplished by iterating through different starting configurations $v_3^{(0)}$, i.e. adjusting the vertex location incrementally within the range $1.1 \leq x_3^{(0)} \leq 3, 1.1 \leq y_3^{(0)} \leq 3, -1 \leq z_3^{(0)} \leq 1$ in steps of 0.1. The starting configurations constitute the axes of all plots (a-f), while corresponding end locations of $v_3$ (a-c), rigidity error $\Gamma$ (d), self-intersection (e), and feasible solutions (f) are represented using color schemes according to Fig. 3. Due to reflective symmetry of the first layer, the plots are sliced into half-spaces to provide a better view of characteristic regions $r_1$ to $r_5$. The half-spaces shown in (c-f) can be completed by mirroring on the plane $x_3^{(0)} = y_3^{(0)}$, whereas the mirrored image of (a) would complete (b) and vice versa.

The most apparent feature of the results is the formation of three-dimensional regions $r_1$ to $r_5$ (and additional two in the case of the self-intersection which are not highlighted). Regions are bounded subspaces that signify distinct differences in the kinematical behavior of the underlying starting configuration in contrast to their neighboring subspaces. The kinematical behavior is continuous within the highlighted regions, but changes abruptly if a mutual boundary is crossed by adjusting the vertex geometry.

We start by discussing the rigidity error $\Gamma$ (Fig. 3d) since it is the linchpin of the kinematical behavior. We re-encounter the rigid foldability valley that already appeared in Fig. 2b. In contrast to Fig. 2b, we cannot visually perceive the transition from true rigid foldability to the start of distortion in Fig. 3d. However, this transition is apparent in the results of the end locations (Fig. 3a-c), where the boundary between region $r_3$ and regions $r_1$ and $r_2$ signifies the abrupt change in kinematical behavior. This determines the boundary of region $r_4$, which represents the subspace of true rigid foldability, and simultaneously determines the numerical value that connotes where the crease pattern starts to distort, which lies at approximately $10^{-4}$. The result of the rigid foldability is further independent of $z_3^{(0)}$, which means that the rigid foldability of vertex $v_1$ is only dependent on the sector angle $\beta$ once $\alpha$ is constant.

Regions $r_1$ and $r_2$ are separated horizontally by a plane with a slight negative inclination towards increasing $x_3^{(0)}$ and $y_3^{(0)}$, as can be observed best in Fig. 3c. The reason for the boundary is the existence of two rigid body modes. The factor that determines in which rigid body mode the origami "falls" into is $z_3^{(0)}$. For positive and (congruent to the inclination) slightly negative values of $z_3^{(0)}$, the crease line between $v_1$ and $v_3$ becomes a mountain (**M** in Fig. 4), and otherwise a valley crease (**V** in Fig. 4), respectively.

Fig. 4 shows starting configurations of $x_3^{(0)} = 1.5, y_3^{(0)} = 1.7$ and decreasing values of $z_3^{(0)}$ from top to bottom. If $z_3^{(0)} = -0.1$, which lies within region $r_1$, $v_3$ folds on the "outer" side of the origami and away from the origin $O$. For $z_3^{(0)} = -0.2$, which relocates the starting configuration into region $r_2$, $v_3$ folds on the "inside" and towards the origin $O$. If $z_3$ becomes even smaller, the starting configuration is still within region $r_2$ and thus exhibits the same rigid body mode, but the facets adjacent to $v_3$ intersect with facet $f_1$ (as denoted with **I** in Fig. 4). Hence, the boundaries between end locations and self-intersection are coupled, but not identical. Although regions $r_1$ and $r_5$ largely overlap, the negative inclination of the separating plane in the results of the self-intersection is bigger, which means that region $r_5$ crosses the boundary between regions $r_1$ and $r_2$. It is at this moment that the solutions start becoming feasible, i.e. are without self-intersection and inside of the cuboid that constitutes the enclosed space (Fig. 1b). Fig. 3f shows the feasible solutions, which are arranged in a wedge produced by the intersection of $r_2$ and $r_5$.

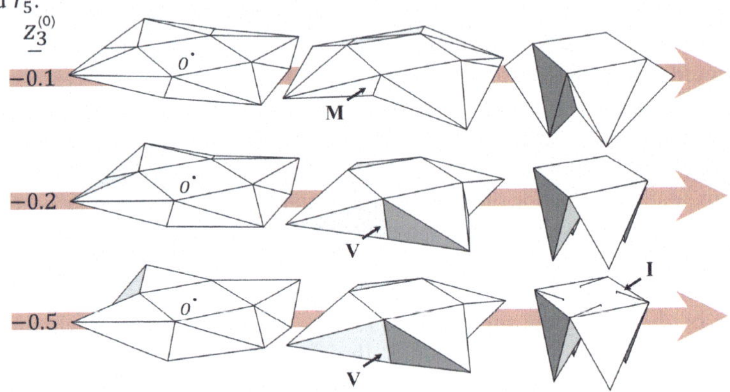


**Figure 4:** *Influence of $z_3^{(0)}$: slightly negative value leads to mountain crease* **M**, *and smaller values to valleys* **V** *and eventually to self-intersection* **I**.

### 4.2.3 Influence of $v_4^{(0)}$ on the Behavior of the Whole Pattern

In the previous section, we investigated the sensitivity of a single vertex embedded in a pattern that is otherwise fully determined. Now, we are interested in the generality of the findings, in particular of the region-specific behavior. For this purpose, we extend the sensitivity analysis to both layers of the extended flasher, and fix the starting locations of $v_3$ and $v_5$, so as to examine only the effect of $v_4^{(0)}$. Note that the choice of fixing $v_5^{(0)}$ and adjusting $v_4^{(0)}$ is arbitrary; both belong to the second layer, are of degree four, are connected to the same vertices in the first layer, and thus interchangeable for this analysis.

The starting location of $v_3$ is fixed in its feasible solution space, $v_3^{(0)} = (1.2, 1.3, 0)$. This is necessary if we are interested in both rigidly and non-rigidly foldable regions for the following reason. If both layers are rigidly foldable, trivially, the whole pattern is rigidly foldable. If the first layer is rigidly foldable but the second is not, the second layer may or may not distort the first layer. This follows from Eq. (1), which minimizes the *sum* of the errors. However, if the first layer is not rigidly foldable, no vertex geometry of the second layer can guide the first towards rigid foldability because the constraints in the first layer are already unsatisfied. Hence, to be able to detect both rigidly and non-rigidly foldable solutions, the first layer must be rigidly foldable. The starting location of $v_5$ is randomly set to $v_5^{(0)} = (3.1, 2, -0.2)$, and the range of the starting locations of $v_4$ is $-1 \leq x_4^{(0)} \leq 1, 3 \leq y_4^{(0)} \leq 4, -0.5 \leq z_4^{(0)} \leq 0.5$. Fig. 5 shows the results for the end locations of all vertices $v_3$ to $v_5$ (a-i), the rigidity error $\Gamma$ (j), the self-intersection (k), and the feasible solutions, which are inexistent.

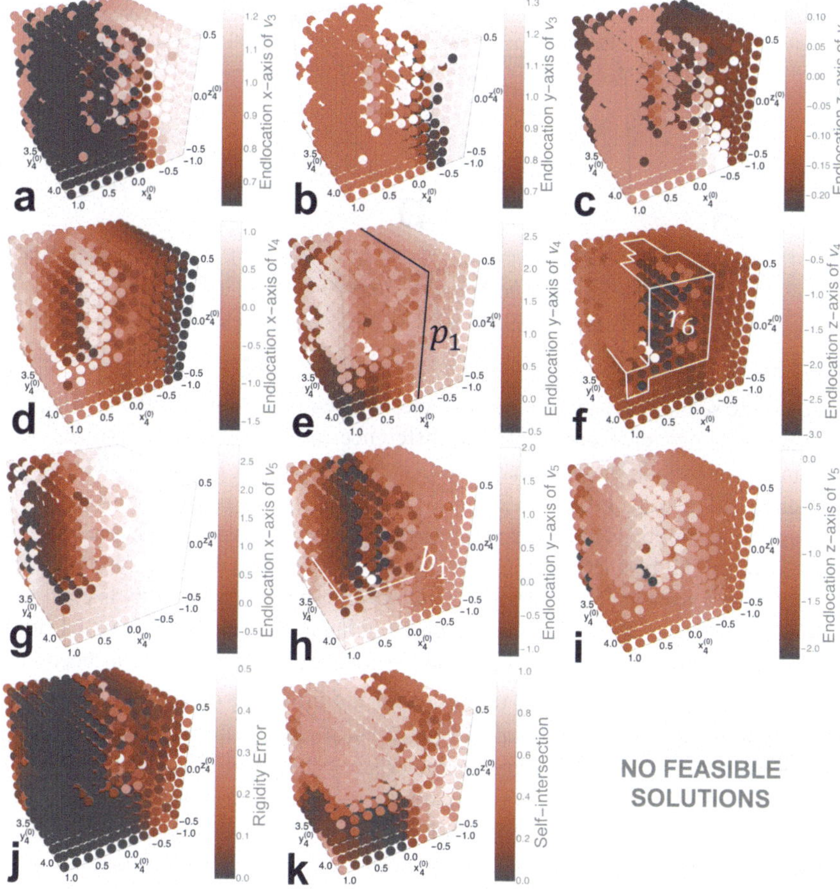

**Figure 5:** *Influence of the starting location of $v_4$ on the behavior of the whole pattern: absolute end location of vertices $v_3$, $v_4$, and $v_5$ on the x, y, and z axes (a-i), rigidity error $\Gamma$ (j), and self-intersection (k).*

All plots seem to be roughly divided into two half-spaces by the plane $p_1$ highlighted in Fig. 5e at $x_4^{(0)} = -0.2$. Within the half-space in the negative $x$ direction, the result of the end locations behaves widely continuous, while the configurations are not rigidly foldable, and the self-intersection is indistinct for positive $z_4^{(0)}$ but continuous otherwise. These are the exact same features region $r_3$ in Fig. 3 exhibits. In the other half-space, the configurations are largely rigidly foldable, and the results for the vertex end locations as well as the self-intersection are divided by the $xy$ plane, which is the same characteristic that regions $r_1$ and $r_2$

exhibited in Fig. 3. There exists a change of rigid body mode between $r_1$ and $r_2$, and indeed, the same is true for configurations divided by boundary $b_1$ (Fig. 5h), as shown in Fig. 6.

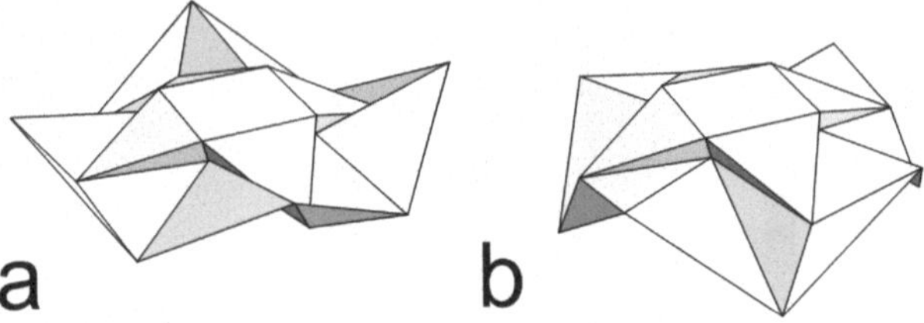

**Figure 6:** *Two rigid body modes at $x_4^{(0)} = 0.8$, $y_4^{(0)} = 3.6$, for $z_4^{(0)} = 0.2$ (a), and $z_4^{(0)} = -0.2$ (b), respectively.*

Hence, there seem to exist the same characteristic regions, though with less distinct boundaries. However, there is one big difference; in this sensitivity analysis, we encounter the formation of a noisy region $r_6$ highlighted in Fig. 5f, which shows throughout all results of the vertex end locations, as well as in the rigidity error $\Gamma$. We have so far not been able to determine the exact circumstances that lead to this phenomenon. The best explanation we can provide is related to the symmetry of the underlying pattern. The examined range of the sensitivity analysis is centered around the $yz$ plane, which coincides with the actuation of vertex $v_2$. Usually, rigid body modes are divided by the $xy$ plane because the constraint system becomes singular in the flat state. In fact, the result for the influence of $x_4^{(0)}$ on itself (Fig. 5d), which in this case corresponds to the vertical coordinate for "conventional" rigid body modes, shows the same characteristic behavior of an abrupt change in the end location. However, in comparison to actual rigid body modes, the configurations in this case are only rigidly foldable in one half-space, which might explain the noisy region $r_6$.

### 4.3  Optimization

The optimization serves as an illustrative example to provide the application context for the presented simulator. Hence, we optimize the extended flasher pattern here as a proof of concept. The method used for the optimization of the vertex geometry is the simulated annealing algorithm. Let the objective of the optimization be to maximize the projected area $A_p$ of the starting configuration onto the flat plane and let us define the variables of the optimization as the starting locations of vertices $v_3, v_4$, and $v_5$. It is of advantage here to apply spherical coordinates $r, \varphi, \rho$, so that if e.g. a specific in-plane angle leads to rigid foldability, it is not influenced by changing other variables, which would be the case for Cartesian coordinates. The starting location of $v_3$ is related to $v_1$, $v_4$ to $v_2$, and $v_5$ to $v_{2'}$, respectively, to make sure there is no dependency between variables. The actuation is given in Eq. (5). The simulated annealing parameters are tuned according to Downsland and Thompson [Downsland and Thompson 2012]. The starting temperature is defined as the maximum change of the objective function in early iterations, and the temperature schedule is logarithmic with a reduction factor of 0.89. The number of inner loops is 60, and the number of outer loops is 50. The size of the neighborhood is one tenth of the allowable range of the variables, respectively (see Eq. (6)). Rigid foldability, spatial feasibility, and self-intersection are embedded in the objective function as weighted penalty functions $\Omega_{rf}$, $\Omega_{sf}$, and $\Omega_{si}$, respectively. The weights are determined heuristically so that all penalty functions equally contribute to the objective function in the approximate magnitude of the projected area, if the constraints are violated. Within the cuboid described in Section 4.1, all penalty values are zero, so that the only influencing factor within the spatially feasible region is the size of the projected area. The spatial feasibility of each vertex is expressed by the sum of distances between the end location and cuboid in direction of $x, y$, and $z$ separately, or zero if within the cuboid. The optimization problem is formally stated as:

$$\min_{\mathbf{r}^{(0)}, \boldsymbol{\varphi}^{(0)}, \boldsymbol{\rho}^{(0)}} (-A_{\mathrm{p}} + \Omega_{rf} + \Omega_{sf} + \Omega_{si})$$

$$s.t.$$

$$0.4 \leq r_3^{(0)} \leq 1.5, 20° \leq \varphi_3^{(0)} \leq 70°, \; -45° \leq \rho_3^{(0)} \leq 0° \qquad (6)$$
$$0.5 \leq r_4^{(0)} \leq 3.0, 45° \leq \varphi_4^{(0)} \leq 135°, -30° \leq \rho_4^{(0)} \leq 30°$$
$$0.5 \leq r_5^{(0)} \leq 3.0, 45° \leq \varphi_5^{(0)} \leq 135°, -30° \leq \rho_5^{(0)} \leq 30°$$

The best solution is qualitatively shown in Fig. 7. The size of the projected area $A_{\mathrm{p}}$ is 44.0, and the rigidity error $\Gamma = 1.6 * 10^{-6}$ with all $E^{(j)} \leq 10^{-4}$, hence $\Omega_{rf} = 0$. The vertex end locations are all within the cuboid defined in Section 4.1, i.e. $\Omega_{sf} = 0$, and there are no self-intersections, $\Omega_{si} = 0$.

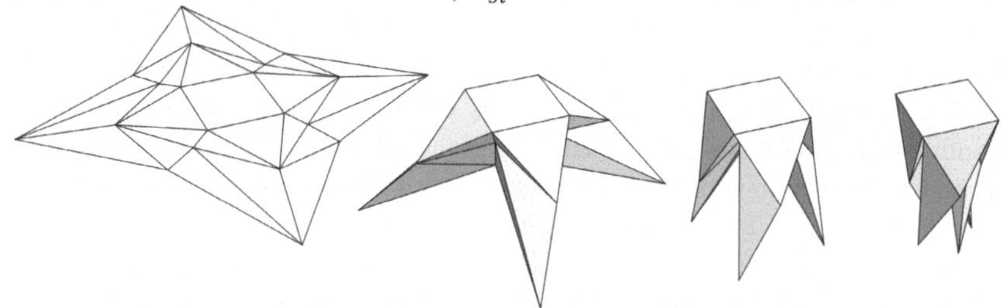

**Figure 7:** *Discrete folding states of the optimized extended flasher at $j = 0, 12, 24, 30$ from left to right.*

## 5  Discussion

The formulation of Eq. (1) as an optimization problem enables the presented simulator to compute the folding process of any vertex geometry independent of the extent of distortion. It also makes the simulator easily extensible to consider different design constraints. Further, the results are repeatable due to the deterministic solving procedure. The total running time for a simulation with $N = 30$ iterations in e.g. the optimization in Section 4.3 is about 10 seconds on an Intel i7 processor with 16GB RAM.

The sensitivity analyses show the existence of three-dimensional bounded regions within the configuration space of starting locations of vertices. In the case of a single vertex influence on its own behavior (Fig. 3), these regions show well-defined boundaries and continuous behavior within. The rigidity error $\Gamma$ is continuous and resembles the shape of a valley with a smooth transition from rigid foldability to the distortion of the origami. This transition further coincides with the boundary between regions $r_1$ and $r_2$ with $r_3$, which determines the upper numerical limit for rigid foldability to be approximately $10^{-4}$. This value must strictly only be used in connection with the error $E^{(j)}$ in Eq. (1), as the rigidity error $\Gamma$ is an averaged number that may contain bigger errors in certain iterations. The boundary between regions $r_1$ and $r_2$, which shows the abrupt change in the vertex end location, is not identical but related to the boundary between configurations with self-intersections and without ($r_5$). A boundary in the vertex end location further suggests the existence of rigid body modes if adjacent regions are rigidly foldable. In the case of the influence of $v_4^{(0)}$ on multiple vertices, the regions exhibit similar characteristics, but with weaker manifestation of region boundaries. There also exists a noisy region, whose interpretation requires further investigation. The results shown provide a new way for users to perceive the kinematical behavior of an origami and facilitate the interpretation of the folding process.

Although the simulator lays a basis for the optimization of the vertex geometry through a generic approach, the stochastic approach in Section 4.3 with 3'000 evaluations and a total runtime of approximately 8.5 hours is not very efficient. A more efficient procedure is plausible and definitely worth considering in the future. The missing link for such procedure is the detection of the location of region boundaries and noisy regions. If noisy regions were avoided, gradient based local optimization methods could be used to capitalize on the smoothness of the search space within regions with well-defined boundaries. If successful, this would signify a huge step forward in the computational approach. We would also like to mention that the optimization can be guided efficiently with user input. Seeing the origami fold, and in particular seeing how non-rigidly foldable patterns distort, experienced users are able to adjust the vertex geometry towards feasible solutions. This could be utilized to find suitable starting points for the optimization.

While the optimized solution of the extended flasher is able to comply with the given set of engineering requirements, it has lost some of the advantages of the flasher principle. In conventional flashers, each consecutive folded layer only adds its thickness to the outer dimension of the cuboid, which increases the folded diameter much less rapidly than the surface area of the deployed state. This is not achievable with the adapted pattern, where more layers could only be added by increasing the height of the enclosed space. Also, whereas there are no acute vertices in the original pattern, we find them in the optimized version of the extended flasher, which would complicate the realization with finitely thick materials. Nonetheless, the purpose of the case study, next to the sensitivity analysis, was to show that it is possible to optimize the vertex geometry based on the presented simulator, which was successful.

# 6  Conclusion

A robust simulator was presented that enables the kinematical simulation of arbitrary vertex geometries. We introduced the rigidity error $\Gamma$ as a measure for rigid foldability, and gave an upper numerical limit for the error $E^{(j)}$ of approximately $10^{-4}$ in each iteration $j$. The simulator was then used to illustrate the influence of the vertex geometry of an extended flasher pattern on its rigid foldability, spatial feasibility, and self-intersection. The visualization of the search space shows the existence of regions with different kinematical behavior, which provides new possibilities to perceive and interpret the folding process of an origami. As a proof of concept, we showed that the simulator can be used to optimize the vertex geometry through stochastic optimization, which is a step towards the computational generation of origami crease patterns.

# References


[Abel et al. 16] Zachary Abel, Jason Cantarella, Erik D. Demaine, David Eppstein, Thomas C. Hull, Jason S. Ku, Robert J. Lang, and Tomohiro Tachi. "Rigid Origami Vertices: Conditions and Forcing Sets." Journal of Computational Geometry, 7(1), 171-184, 2016.

[Belcastro and Hull 02] Sarah-Marie Belcastro, and Thomas C. Hull. "A mathematical model for non-flat origami." Origami 3: Proceedings of the 3rd International Meeting of Origami Mathematics, Science, and Education, 2002.

[Diaz 14] Alejandra R. Diaz. "Origami Folding and Bar Frameworks." ASME 2014 International Design Engineering Technical Conferences and Computers and Information in Engineering Conference, 2014.

[Downsland and Thompson 2012] Kathryn A. Downsland, and Jonathan M. Thompson. "Simulated annealing." Handbook of natural computing. Springer Berlin Heidelberg, 2012.

[Huffman 76] D. Huffman. "Curvature and Creases: A Primer on Paper." IEEE Transactions on Computers C-25:10 (1976), 1010–1019.

[Lang 96] Robert J. Lang. "A computational algorithm for origami design." Proceedings of the twelfth annual symposium on Computational geometry, 1996.

[Lang 97] Robert J. Lang., Origami in Action: Paper Toys That Fly, Flag, Gobble and Inflate! St. Martin's Griffin, 1997.

[Lang and Howell 18] Robert J. Lang, and Larry Howell. "Rigidly Foldable Quadrilateral Meshes From Angle Arrays.", J. Mechanisms Robotics, 10.2 (2018): 021004.

[Lang et al. 15] Robert J. Lang, Spencer Magleby, and Larry Howell. "Single-degree-of-freedom rigidly foldable origami flashers." ASME 2015 International Design Engineering Technical Conferences and Computers and Information in Engineering Conference. American Society of Mechanical Engineers, 2015.

[Miura 89] Miura, Koryo. "A note on intrinsic geometry of origami." Research of Pattern Formation, KTK Scientific Publishers, Tokyo, Japan (1989), 91-102.

[Möller 97] Möller, Tomas. "A fast triangle-triangle intersection test." Journal of graphics tools 2.2 (1997): 25-30.

[Morgan et al. 15] Jessica Morgan, Spencer Magleby, Robert J. Lang, and Larry L. Howell. "A Preliminary Process for Origami-Adapted Design." ASME 2015 International Design Engineering Technical Conferences and Computers and Information in Engineering Conference. American Society of Mechanical Engineers, 2015.

[Streinu and Whiteley 2004] Ileana Streinu, and Walter Whiteley. "Single-vertex origami and spherical expansive motions." Japanese Conference on Discrete and Computational Geometry. Springer, Berlin, Heidelberg, 2004.

[Tachi 09] Tomohiro Tachi, "Simulation of rigid origami." Origami 4 (2009), 175-187.

[Tachi 10a] Tomohiro Tachi, "Origamizing polyhedral surfaces." IEEE transactions on visualization and computer graphics 16:2 (2010), 298-311.

[Tachi 10b] Tomohiro Tachi, "Freeform variations of origami." J. Geom. Graph 14:2 (2010), 203-215.

[Tachi 13] Tomohiro Tachi, "Designing freeform origami tessellations by generalizing Resch's patterns." Journal of mechanical design 135:11 (2013), 111006.

[Watanabe and Kawaguchi 09] Naohiko Watanabe, and Ken-ichi Kawaguchi. "The method for judging rigid foldability". Origami 4 (2009), 165-174.

[Zimmermann et al. 17] Luca Zimmermann, Tino Stanković, and Kristina Shea. "Finding Rigid Body Modes of Rigid-Foldable Origami Through the Simulation of Vertex Motion." ASME 2017 International Design Engineering Technical Conferences and Computers and Information in Engineering Conference. American Society of Mechanical Engineers, 2017.

---

Luca Zimmermann,      lucaz@ethz.ch
Kristina Shea,         kshea@ethz.ch
Tino Stanković,        tinos@ethz.ch

Engineering Design and Computing Laboratory
Dept. of Mechanical and Process Engineering
ETH Zürich
Tannenstrasse 3
8092 Zürich
Switzerland

# On Using Tessellation Properties for the Development of Classifying Criteria for Foldable Mechanisms

*J. Siebrecht, G. Jacobs, C. Weigel, S. Dehn,*
*H. Buffart, S. Hoffmann, J. Paris, M. Trautz, B. Corves*

**Abstract**: *Using catalogues in engineering design offers a compact way to summarize and present possible solutions regarding a given task. To find the needed classifying criteria for rigid foldable foldings, one possible way is presented here through the use of tilings as an analogy. This foundation of an enclosed quantity for general tiles ensures that the specialisation to foldings also remains enclosed and therefore contains all possible folding variants. The reduction of the general solutions to the specialised foldings is done by applying boundary conditions for rigid foldable foldings and reducing the number of possible facets by applying engineering objectives.*

## 1 Introduction

Using the design principle of foldings allows the function integration of both transformability and rigidity into the same application. As a solution for engineering development tasks, such as moving and positioning objects in an automated production line or realising a self-supporting arrangement of facets, either for optical and acoustic tasks or covers, they are more relevant than ever. However, the previous use of folds are largely singularly developed solutions.

One way to provide known solutions to the developer, is to introduce a design catalogue. Here, foldschemes are structured on the basis of different objectively as-certainable criteria and prepared for systematic access. From the viewpoint of the application, the number of folding vertices or the number of degrees of freedom for a certain foldscheme can be used as criteria. However, these criteria do not allow any statement about the completeness of the catalogue, since the characteristics are unlimited.

The aim of this contribution is to establish further criteria, where the maximum number of criteria manifestations is known and therefore allows the evaluation of the seclusiveness for given solutions. Known geometrical symmetry groups are possible ways to categorise foldschemes and are therefore analysed. This contribution proposes a way to categorise foldschemes via criteria with countable numbers of variants and presents the way in which the criteria are developed.

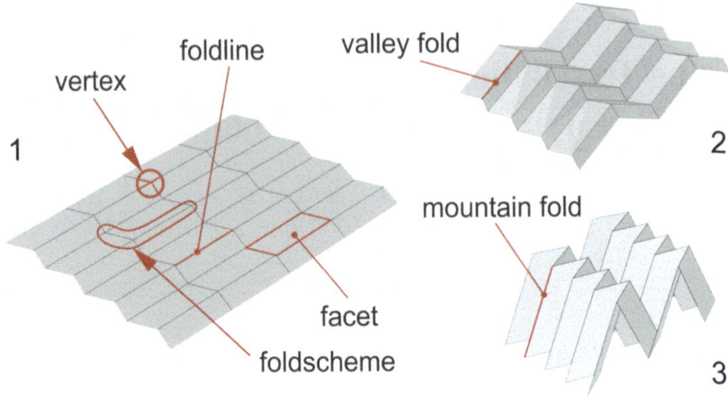


**Figure 1:** *Definitions for foldings.*

## 2    Definitions

A folding describes the complete connected mechanism which is analysed. It is assembled by one or more foldschemes which are equal in structure and dimensions. The foldscheme shows the type and position of the foldlines and defines the entire fold through lengths and angles. These foldschemes also contain vertices and facets. A vertex describes the point at which several foldlines intersect, whereas facets are the two surfaces besides the foldline. Depending on the movement of the adjacent facets, it is either a mountain or a valley fold. If the surface normals of the considered areas point away from each other, it is a mountain fold, whereas a valley fold exists when the surface normals point to a common point. The representations with the numbers 1 to 3 from figure 1 show the movement form of the folding and the described parts.

## 3    Catalogues

When designing a product, it is advisable to follow a structured process. For technical products the German standard VDI 2221 [VDI 93], the methodology for the development of technical systems, is a recommended starting point. In a methodological design process various steps are needed to develop a product, however two are particularly interesting for this contribution. Firstly the clarification of the task and secondly the search for solution principles, i. e. foldschemes and their structures.

The first step defines how the foldings have to behave. It defines the properties. In [Paris et al. 17] a classification scheme is introduced, which allows a uniform presentation of foldings with a growing dataset of properties. The agreement of requirements and folding properties, for example the form of the assembly space, required movement pattern or the degrees of freedom, defines the success of the

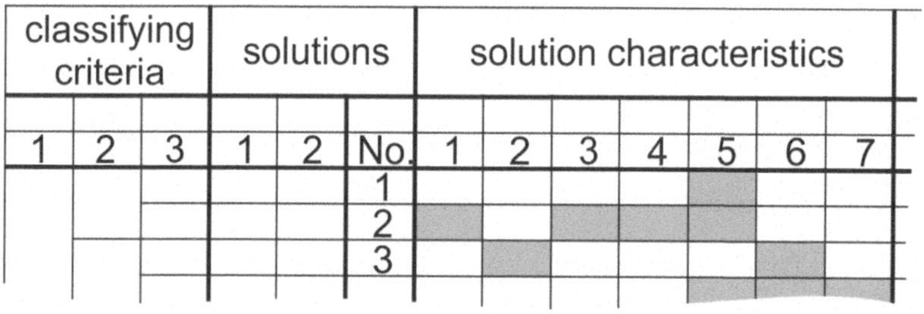

| classifying criteria | | | solutions | | | solution characteristics | | | | | | |
|---|---|---|---|---|---|---|---|---|---|---|---|---|
| 1 | 2 | 3 | 1 | 2 | No. | 1 | 2 | 3 | 4 | 5 | 6 | 7 |
|  |  |  |  |  | 1 |  |  |  |  |  |  |  |
|  |  |  |  |  | 2 |  |  |  |  |  |  |  |
|  |  |  |  |  | 3 |  |  |  |  |  |  |  |

**Figure 2:** *Basic structure of a catalogue.*

development process. To find the optimal solutions, choosing the right foldscheme in the second development step and confirming that all possible solutions were regarded, is of great importance. Negligence of solutions can reduce the success of the development process. Both tasks, holding the solution properties and the completeness of the solution space, are addressed by design catalogues.

Design catalogues are a useful tool to support systematic work. Figure 2 shows the basic structure as defined by [Roth 82]. They allow the user to access an information set for various solutions regarding a defined topic. The structure is always organised into one part sorting the solutions, one naming the solutions and one presenting the properties of each solution. Since [Paris et al. 17] defined relevant solutions characteristics, the classifying criteria for structuring and evaluation of the completeness need to be defined.

## 3.1 Properties of Catalogues

In [Roth 82] and [VDI 82], design catalogues are divided into four groups. The different groups of design catalogues differ in their intended use. Variations are presented below on the basis of the work of [Roth 82].

- Object catalogues contain basic facts and represent properties of the presented content independently of tasks. They can be viewed as a collection of expressions within a subject area.

- Operation catalogues represent procedural steps and sequences of operations that are used in a methodical design process.

- Solution catalogues are task-dependent. They assign physical effects to a function, suitable effect carriers to physical effects as well as contour suggestions and manufacturing processes to effect carriers.

- Relationship catalogues reflect the relationship between two objects.

## 3.2   Object Catalogues

For the examination of foldings without further knowing the application, the object catalogue is chosen as the framework. In order to assess the suitability of each classification criteria, their function in the structure of an object catalogue must be taken into account.

[Roth 82] recommends the use of a one-dimensional design catalogue for an object catalogue. It consists of the classification criteria, solution and solution characteristics part as defined in figure 2. The classification part serves to classify the elements listed in the solution part. For this purpose, the classification part according to [Roth 82] must be constructed so that the completeness of the criteria can be validated by the user. In the main part, in the case of an object catalogue, the treated objects are displayed in an abstracted form so that they can be captured quickly by the user. The access part allows an overview of the decisive features of the object shown in the main part. This type of structure makes it possible to prepare objects within a framework (the classification part) and make them accessible through the access part, i. e. solution characteristics. According to [Roth 82] two requirements present in the classifying criteria are:

- The subdivisions of an object catalogue must be complete within their considered boundary conditions.

- The catalogue or subdivision must be systematically expandable.

# 4   Requirements on the Structure of Foldings

For the usefulness of a design catalogue in the field of architecture or engineering, further reductions have to be made. In this work, foldings are to be considered which are rigid foldable. The term rigid refers to the property of the facets, which are assumed to be rigid. The term foldable describes the changeability of the entire folding.

With regard to the analogy of tilings, it is necessary to set structural requirements for foldschemes. In the publication by [Abel et al. 16], necessary conditions are formulated that must be met for a vertex to be rigidly foldable:

1. At least four foldlines must converge on a vertex.

2. All surface angles $\alpha_i$ must be smaller than $\pi$.

3. The foldlines must not form an unspecified cross.

Figure 3 shows four different vertices. The first vertex violates the first condition: fewer than four foldlines meet. The second vertex has a surface angle $\alpha_i \geq 180°$. This violates condition 2. The third vertex is not rigidly foldable because the converging foldlines form an unspecified cross. An unspecified cross occurs when four foldlines converge on a vertex and two of the adjacent face angles add up to an angle $\alpha_i = 180°$. Vertex 4, on the other hand, fulfils all conditions

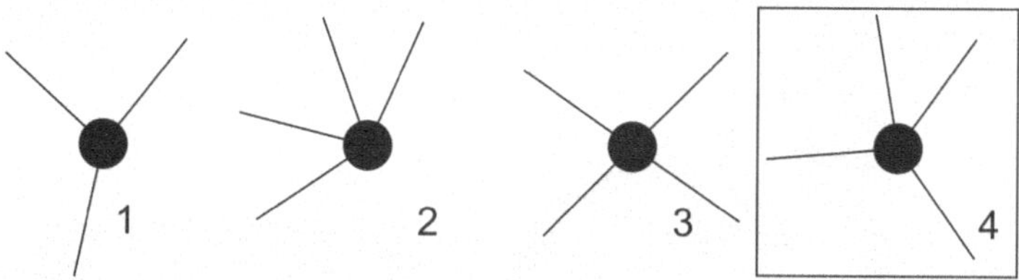

**Figure 3:** *Vertex configurations.*

and can therefore be considered as a rigid foldable vertex.

## 5  Isohedral Tessellations with Convex Polygons

Since the ability to evaluate the seclusion is important, existing analogies are used. The approach followed in this work establishes an analogy between tessellations and foldschemes. Tessellations in a two dimensional plane consist of tiles and are therefore also called tilings.

A general tiling $P(K_i)$ consist of an infinite number of tiles with $i$ different tile forms $K_i$. These tiles are closed topological disks, i.e. they have no holes. The tiles can be put together in such a way that they fill the plane seamlessly without covering themselves. Figure 4 shows three different tilings, comprising one, two or three different tile forms, designated by $P(K_1)$, $P(K_1,K_2)$ and $P(K_1,K_2,K_3)$.

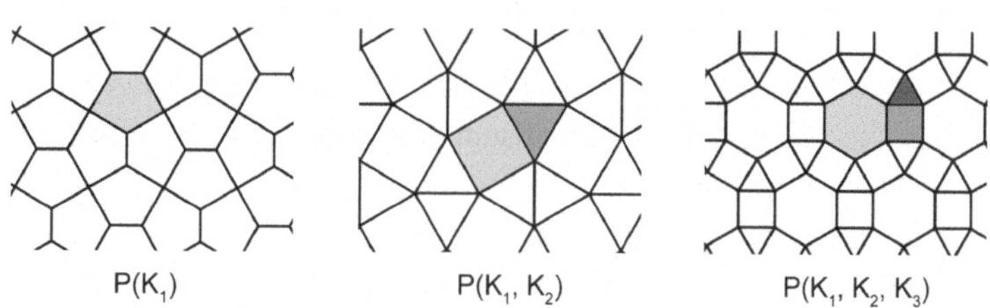

**Figure 4:** *Tesselations with different number of tile forms.*

A more complex tiling is shown in 5. When the colour in neglected, only one tile form $K_1$ exists, since every tile is a congruent mapping of tile $K_1$. The tiling is therefore characterised by $P(K_1)$.

As an object catalogue has the property to be subdivided, only one subdivision is needed to show the applicability of tiling classification on foldschemes before expanding it. As a starting point for this subdivision, the engineering design approach for common parts is used. Common parts reduce costs, for example through

**Figure 5:** *Artistic tiling [Escher 55]*

reducing development and manufacturing time, therefore making foldings more applicable. This validates the reduction to tilings with only one tile form $K_1$ as a proof of concept.

If a tiling consists only of one tile form $K_1$ or congruent mappings of this tile, it is called monohedral. Furthermore, if each of these tiles can be projected onto each other using the same symmetry operation, it is called isohedral. Thus, every isohedral tiling is also monohedral [Kaplan and Salesin 00].

To evaluate the usefulness of isohedral tilings to classify foldings, all possible variations need to be considered. [Grünbaum and Shephard 78] researched a list of all tilings, which fulfil the following conditions:

1. The tiling is isohedral: It consists only of the congruent tiles of type $K_1$. Each tile $K_1$ can be projected onto each other using the same set of symmetry operations.

2. The tiling consists of convex polygons: The tiles are convex and their edges are straight.

The first condition is already covered with the properties of the object catalogue and the respective subdivision. The second condition is equal to the condition for rigid foldable vertices, to have no angle $\alpha \geq 180°$. Since there is only a finite number of isohedral tessellations with convex polygons, they fulfil the requirement to be secluded. Within these limits [Grünbaum and Shephard 78] merge all possible tiles and arrangements. This is achieved by a combination of classification systems, most importantly the permutation of so called Laves nets and incidence symbols. Applying the named conditions and classification systems leads to 107 different types of tilings [Grünbaum and Shephard 78].

Laves nets describe the structure of a tiling by counting the number of coinciding edges at each vertex [**Laves 31**]. Figure 6 shows two different tilings, each based on a five sided isohedral tile. For each tiling, the isohedral tile is shown along with the number of edges coinciding at the vertices. The Laves net names the occurrence of these numbers, starting counter-clockwise at the vertices with the smallest number of edges.

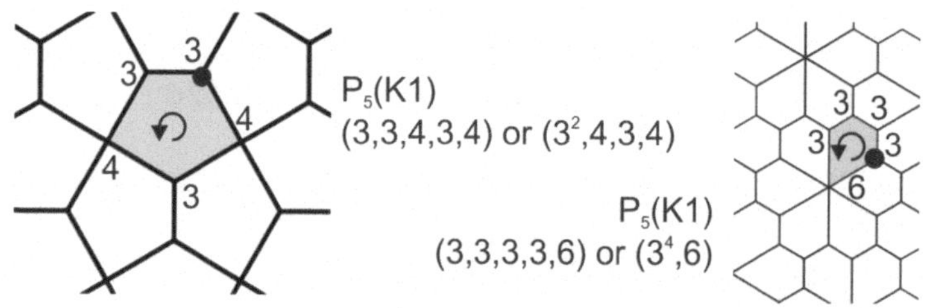


**Figure 6:** *Examples for Laves nets according to [Bongartz et al. 88].*

Contrary to the Laves nets, which sort tilings by vertices, the incidence symbol classifies isohedral tilings by describing the tiles' edges. This allows to characterise the symmetry properties of the tile as well as the symmetry operation which has to be applied at each edge to fill the plane entirely [**Grünbaum and Shephard 79**]. In figure 7, examples for incidence symbols are shown. Each tile has a starting point, indicated by a black dot, from which the edges are named counter-clockwise.

Each new edge receives an ascending letter, starting with $a$. If an edge can be derived from an already named edge by means of mirror symmetry, it receives the same letter. The original edge is marked with an upper case $+$, the mirrored one with an upper case $-$. On the left side in figure 7, the first edge from the starting point is named $a$. Since the second edge is a mirrored version of edge $a$, it is marked as $a^-$ and the original (the first edge) with $a^+$. This first part of the incidence symbol is called the tile symbol.

The second part describes the symmetry operation that is needed, to map the initial tile onto the adjacent tile and is called the adjacency symbol. For each edge it is analysed, if the direction of rotation of the adjacent tile changes in regard to the initial tile (here counter-clockwise). If the adjacent tile is a rotated version of the initial tile, the direction of rotation stays the same and an upper case $+$ is assigned to the connecting edge. If the adjacent tile is mirrored, the direction changes to a clockwise rotation, which is indicated by an upper case $-$ at the edges' letter of the adjacency symbol. From the tile symbol in the third example of figure 7 it can be derived, that the tile consists of four different edges $a^+$, $b^+$, $c^+$ and $d^+$. The adjacency symbol $(a^- b^- c^+ d^-)$ indicates by an upper-case $+$ that the tiling is created by rotating the initial tile around the midpoint of edge $c$ and mirroring it at the edges $a$, $b$ and $d$. This is shown by the upper-case $-$.

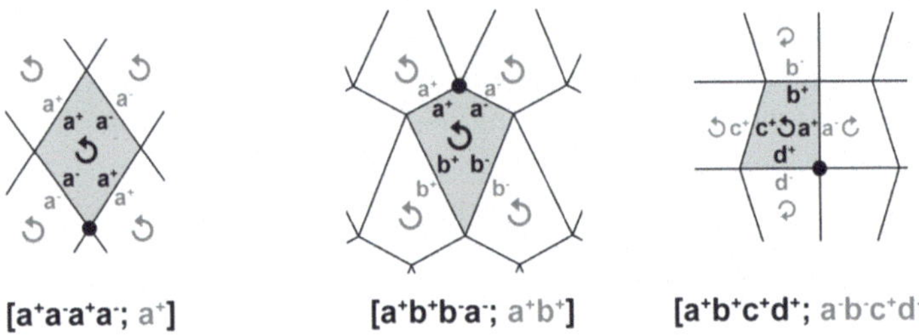


**Figure 7:** *Incidence symbols for different tiles [Grünbaum and Shephard 78].*

Figure 8 shows part of the tabular list of 107 different convex-isohedral tilings by [Grünbaum and Shephard 78] with additional information. The first column of the table (see number 1) names the tiling. The given name consists of a $P$ for a general polygon, a subscript number indicates the number of edges of the considered polygon and a number to enumerate the different tilings. Marked by number 2 in figure 8 is the net type according to Laves. Further information on the angle relations is given by the column marked with number 3.

**Table I**
Isohedral tilings by triangles.

| List number (1) | Net (2) | Incidence symbol (3) | Symmetry group (4) | Tile group (5) | Vertex transitivity (6) | Angle relations (7) | Edge transitivity (8) |
|---|---|---|---|---|---|---|---|
| *$P_3$-1 | $[3.12^2]$ | $[ab^+b^-;\ ab^-]$ | p6m | d1 | $\alpha\alpha\beta$ | $A=C,\ B=D$ | $\alpha\beta\alpha\gamma$ |
| $P_3$-2 | $[4^4]$ | $[a^+b^+c^+d^+;\ a^+b^+c^+d^+]$ | p2 | e | $(\alpha)\alpha\alpha\alpha$ | $-$ | $\alpha\beta\alpha\gamma$ |
| $P_3$-3 | | $[a^+b^+c^+d^+;\ c^-b^+a^-d^+]$ | pgg | e | $(\alpha)\alpha\alpha\alpha$ | $-$ | $\alpha\alpha\beta\gamma$ |
| $P_3$-4 | | $[a^+b^+c^+d^+;\ b^-a^-c^+d^+]$ | pgg | e | $\alpha\alpha\alpha(\alpha)$ | $B=C=\frac{\pi}{2},\ A+D=\pi$ | |
| $P_3$-5 | | | pgg | e | $(\alpha)\alpha\alpha\alpha$ | $A+C=\pi,\ B=D=\frac{\pi}{2}$ | |
| $P_3$-6 | | | pgg | e | $\alpha(\alpha)\alpha\alpha$ | $A=C,\ B=D$ | |
| $P_3$-7 | | $[a^+b^+b^-a^-;\ a^+b^+]$ | pmg | d1(l) | $(\alpha)\alpha\alpha\alpha$ | $A=C,\ B=D$ | |
| *$P_3$-8 | $[4.6.12]$ | $[a^+b^+c^+;\ a^-b^-c^-]$ | p6m | e | $\alpha\beta\gamma$ | $A=B,\ C=D$ | |
| *$P_3$-9 | $[4.8^2]$ | $[a^+b^+c^+;\ a^+b^-c^-]$ | cmm | e | $\alpha\alpha\beta$ | $B=D$ | |
| *$P_3$-10 | | $[ab^+b^-;\ ab^-]$ | p4m | d1 | $\alpha\alpha\beta$ | $A=B=C=D=\frac{\pi}{2}$ | |
| *$P_3$-11 | $[6^3]$ | $[a^+b^+c^+;\ a^+b^+c^+]$ | p2 | e | $\alpha\alpha\alpha$ | $A=C,\ B=D$ | |
| *$P_3$-12 | | $[a^+b^+c^+;\ a^-b^+c^+]$ | pmg | e | | $A=B=C=D=\frac{\pi}{2}$ | |

**Figure 8:** *Excerpt from the tabular list of convex isohedral tilings [Grünbaum and Shephard 78].*

Not all tilings meet the conditions necessary to be considered as a foldscheme that represents a rigid foldable folding. Using the conditions that need to be fulfilled by the foldscheme as defined in section 4, the amount of convex-isohedral tilings can be reduced. With the first requirement, all tilings whose Laves net contains at least one vertex with three edges (see figure 3) are out of consideration, as

the vertex would not be rigidly foldable. This step drastically reduces the amount
of tilings to 29.

All tilings, that have an edge-to-edge relation, are marked with an asterisk ($^*$) in
front of the name [Grünbaum and Shephard 78]. The difference to a corner-to-edge
relation is shown in figure 9. A corner-to-edge relation defines an angle $\alpha_i = 180°$
trough a T-shaped Vertex. This violates requirement 2 ($\alpha_i \geq 180°$). The angle can
also be found in the Laves nets. A Laves net which describes more vertices than
corners of the tile (see figure 9), directly leads to an angle $\alpha_i = 180°$. Considering
this requirement, 18 tilings remain.

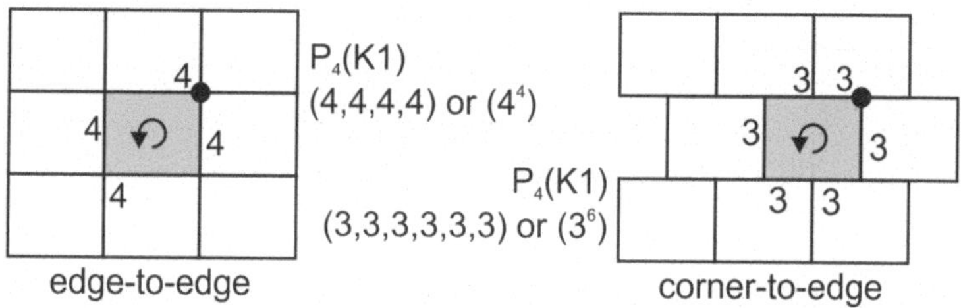


**Figure 9:** *Corner-on-corner and corner-on-edge condition.*

The third and final requirement for the foldability states, that the foldlines at
a vertex must not form an unspecified cross. This requirement can be read from
the angular relationships in column (7) (see number 3 in figure 8). The angular
relationships

$$A = C, \qquad\qquad B = D$$

indicate an equality of the respective opposite angles $A$ and $C$ or $B$ and $D$,
thus resulting in an unspecified cross. This also appears in the case of an angular
relationship of

$$A = B = C = D = \frac{\pi}{2}$$

Eleven types of tilings ultimately remain, which are generally suitable as a
configuration for a foldscheme. Of these isohedral tilings, seven consist of quadrangular polygons ($P_4 - 43, P_4 - 44, P_4 - 46, P_4 - 47, P_4 - 51, P_4 - 52$ and $P_4 - 53$)
and four of triangular polygons ($P_3 - 11$ to $P_3 - 14$).

Figure 11 shows the seven tilings with four-sided tiles, figure 10 the four tilings
with triangular tiles. These overviews are ordered according to increasing demands
on symmetry. The schematic representation of the tiling shows the designation
by [Grünbaum and Shephard 78], a common name and the incidence symbol. The
tilings $P_4 - 43$ (figure 11) and $P_3 - 11$ and $P_3 - 12$ (figure 10) are classified by the

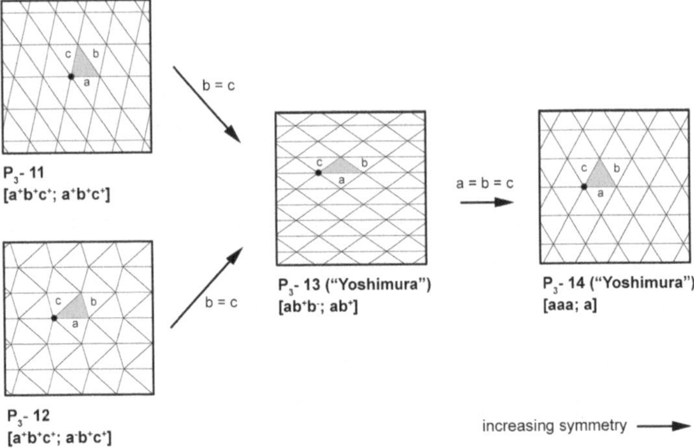


**Figure 10:** *Remaining three-sided polygons suitable for foldschemes.*

Laves nets $(4^4)$ and $(6^3)$, which stay the same, even when changing lengths and angles. Therefore the Laves nets do not sufficiently differentiate the shown tilings.

The incidence symbol however considers these changes. The inscription on the arrows indicates the respective restriction of the shape, which results in a changed incidence symbol. Tilings with different incidence symbols therefore have at least one geometric difference (angle, length or symmetry).

In their structure, $P_4 - 51$ corresponds to the foldscheme "Miura Ori" [Miura and Kenkyjo 85] and $P_4 - 52$ to the foldscheme "Chicken Wire" [Miura 69]. $P_4 - 47$ and $P_4 - 53$ represent two different variants of the "Huffman Grid" [Huffman 76]. The tilings $P_3 - 13$ and $P_3 - 14$ can be seen as variants of the "Yoshimura" folding [Yoshimura 55]. These foldschemes have already been studied in literature in various contexts. $P_4 - 44$ is treated in [Evans et al. 15] as a variation of the Miura Ori folding. In [Wang 16], for example, foldschemes based on parquets $P_4 - 43$, $P_4 - 44$, and $P_4 - 52$ are examined in the context of rigidity and degrees of freedom.

## 5.1 Specification of Mountain and Valley Folds

The last feature that distinguishes the tilings from a foldscheme at this point, is the specification of mountain and valley foldlines. For the structures of already known foldschemes (Chicken Wire, Miura Ori, Huffman Grid, Yoshimura) the usual mountain and valley foldline configuration is used in this work (see figure 12).

Of tilings $P_4 - 44$ and $P_4 - 46$ different tiling shapes are possible within the geometric constraints. Depending on the tile shape, the vertices of the tiling $P_4 - 44$ assume a structure that is similar to either the vertices of "Chicken Wire" or those

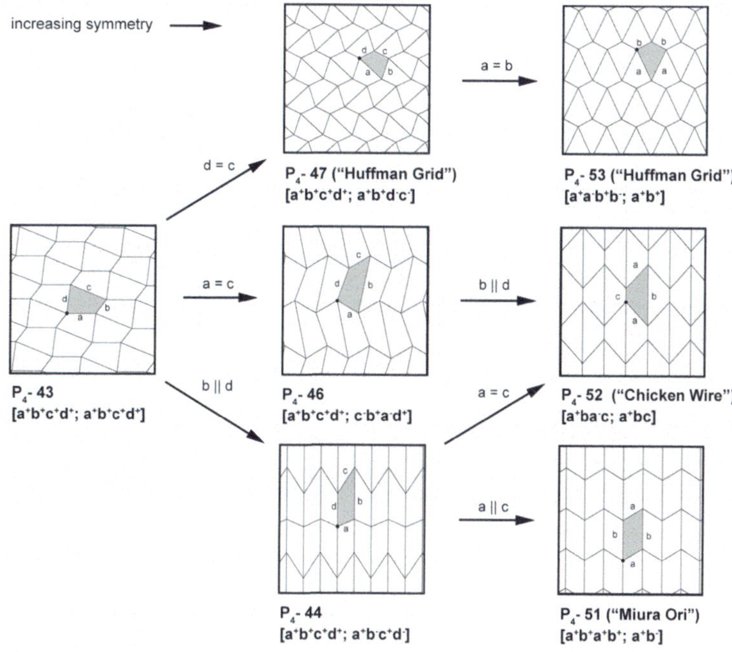

**Figure 11:** *Remaining four-sided polygons suitable for foldschemes.*

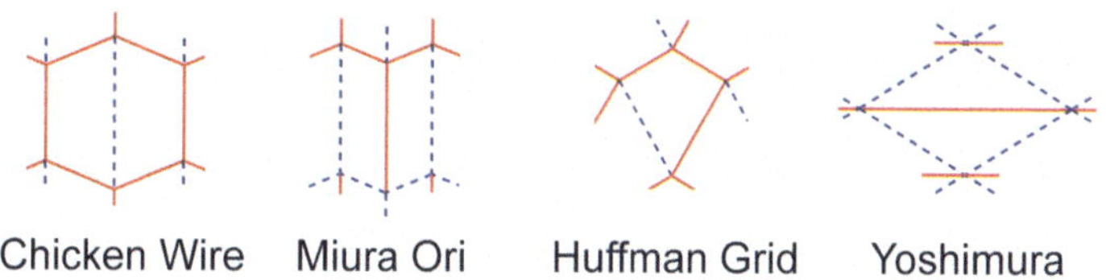


**Figure 12:** *Mountain and valley folds of known foldings.*

of the "Miura Ori" folding (see Figure 13). The distribution of valley and mountain foldlines for this tiling introduces the subtypes $P_4 - 44/1$ ("Chicken Wire" type) and $P_4 - 44/2$ ("Miura Ori" type). This similarly applies to $P_4 - 46$. The different distribution of mountain and valley folds is in turn expressed by subtypes $P_4 - 46/1$ and $P_4 - 46/2$.

Regarding triangular tilings, the mountain and valley folds of the tiling $P_3 - 12$ can be assigned in two different ways. For this purpose, types $P_3 - 12/1$ and $P_3 - 12/2$ are introduced as shown in figure 14.

The intersection of all convex-isohedral tilings and the requirements for fold-

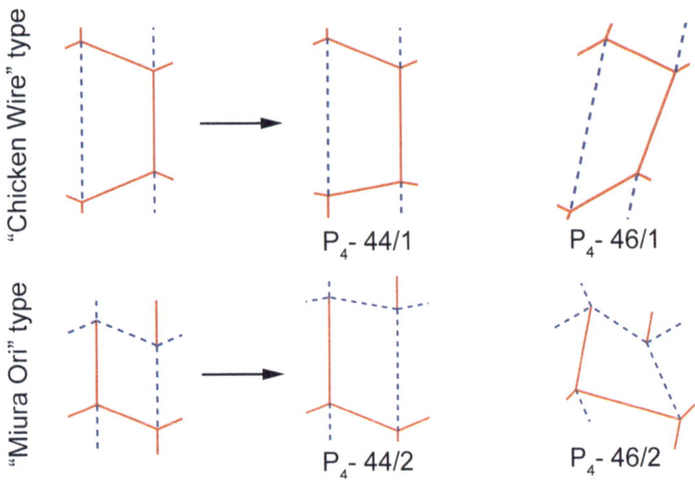



**Figure 13:** *Subtypes for $P_4 - 44$ and $P_4 - 46$.*

**Figure 14:** *Subtypes for $P_3 - 12$.*

ability yields 11 different tilings. Taking into account the mountain and valley foldline configuration, there are 14 foldable tiling patterns.

# 6  Result

Utilising the analogy to tilings, classifying criteria for foldings can be derived. The characteristic features of a tiling can be expressed in an order of increasing precision. In this case, the restriction to the initial form being a plane (two dimensional space) represents the highest classification criterion. An other starting form can be 'spatial'. Tilings in the plane can be further subdivided into the number of unique tiles forming the group. At the last level of refinement, the shape of the tiles can be differentiated.

Due to the closeness of the set of convex polygons shown in [Grünbaum and Shephard 78], this choice of classifying criteria for foldschemes is also limited. Through the requirements for the design of vertices according to section 4 the quantity can be reduced to eleven tilings suitable for foldschemes. As shown in subsection 5.1, different mountain and valley fold configurations have to be considered in the case of foldings. Despite the same structural design, the different configuration results in three additional classes.

The classification criteria presented thus meet the requirements for the subdivision part of an object catalogue (see section 3.2). They allow the mapping of all possible objects and are at the same time systematically expandable. In sum-mary, a possible structure for an object catalogue can be developed, as presented in figure 15. In the outlined part, the subdivision is based on geometric properties of the folding and the form of the tiles. The main part contains the 14 elaborated foldschemes, named after [Gr¨unbaum and Shephard 78] and represented by the tile and tiling. Only two examples are given here due to the limited space. In the access part, solution characteristics such as the type of movement and the degree of freedom of the folding can be found and expanded, for example through [Paris et al. 17].

# 7  Conclusion

In this paper the challenges of finding suitable solutions for foldings are raised and a solution was presented. Through the analogy to tilings, it is possible to identify a closed number of elements possibly relevant for foldings. Forming an intersection of these convex-isohedral tilings and requirements for rigid foldable vertices limits the number of different tilings. Considering the distribution of mountain and valley folds, 14 different types of foldschemes can be generated. These 14 classes fulfil the criteria established by the object catalogue.

# 8  Outlook

A comparison of the developed classification criteria with the requirements for the construction of a design catalogue has shown, that the derived criteria are suitable. Merging the proposed classifying criteria with the existing solution properties as shown in [Paris et al. 17] is the next step.

Further research however needs to be done to verify the consistency of the presented classifying criteria with respect to general folds. Therefore the catalogue needs to be expanded by using foldings with two or more equal tiles (2-isohedral, n-isohedral). For spatial foldings and foldings with curved foldlines however new classifying criteria have to be found since the analogy of tilings regarded here con-tains only tiles with straight edges defined in two dimensional space.

# 9  Acknowledgements

The authors would like to thank the German Research Foundation (Deutsche Forschungs-gemeinschaft DFG) for supporting the interdisciplinary research project ”E$^2$F-Designing Deployable Folded Plate Structures”.

| classifying criteria | | | solutions | | | solution characteristics | | |
|---|---|---|---|---|---|---|---|---|
| starting form | number of tiles | tile form | no. | name | tile and tiling | movement type | DOF | ... |
| 1 | 3 | 4 | 1 | 2 | 3 | 1 | 2 | 3 |
| planar | 1-isohedral | ▲ | 1 | P₃-11 | 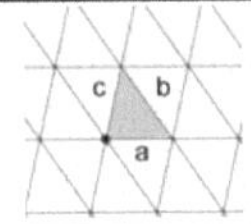  | twisting | ... | ... |
| | | | 2 | P₃-12 / 1 | ... | cylindrical | ... | ... |
| | | | 3 | P₃-12 / 2 | .... | spherical | ... | ... |
| | | | 4 | P₃-13 | ... | cylindrical | ... | ... |
| | | | 5 | P₃-14 | ... | cylindrical | ... | ... |
| | | ■ | 6 | P₄-43 | 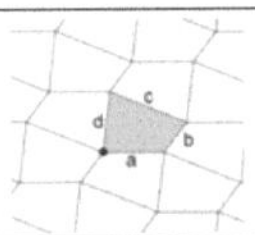  | twisting | ... | ... |
| | | | 7 | P₄-44 / 1 | ... | cylindrical | ... | ... |
| | | | 8 | P₄-44 / 2 | ... | cylindrical | ... | ... |
| | | | 9 | P₄-46 / 1 | ... | twisting | ... | ... |
| | | | 10 | P₄-46 / 2 | ... | torsion | ... | ... |
| | | | 11 | P₄-47 | ... | twisting | ... | ... |
| | | | 12 | P₄-51 | ... | planar | ... | ... |
| | | | 13 | P₄-52 | ... | cylindrical | ... | ... |
| | | | 14 | P₄-53 | ... | cylindrical | ... | ... |
| | 2-isohedral | ⋮ | ... | | ... | ... | ... | ... |
| | n-isohedral | ⋮ | ... | | | | | |
| spatial | | | | | | | | |

**Figure 15:** *Proposed catalogue with the expansion to n-isohedral tiles and spatial foldings.*

# References

[Abel et al. 16] Zachary Abel, Jason Cantarella, Erik D. Demaine, David Eppstein, Thomas C. Hull, Jason S. Ku, Robert J. Lang, and Tomohiro Tachi. "Rigid origami vertices: conditions and forcing sets." 2016. doi:10.20382/jocg.v7i1a9.

[Bongartz et al. 88] Klaus Bongartz, Walter Borho, Detlef Mertens, and Andreas Steins. *Farbige Parkette: Mathematische Theorie und Ausführung mit dem Computer. Vier Aufsätze zur ebenen Kristallographie*, Mathematische Miniaturen, 4. Basel: Birkhäuser Basel, 1988.

[Escher 55] Maurits Cornelis Escher. *Swan*, 1955.

[Evans et al. 15] Thomas A. Evans, Robert J. Lang, Spencer P. Magleby, and Larry L. Howell. "Rigidly foldable origami gadgets and tessellations." *Royal Society Open Science* 2:9. doi:10.1098/rsos.150067.

[Grünbaum and Shephard 78] Branko Grünbaum and G. C. Shephard. "Isohedral tilings of the plane by polygons." *Commentarii Mathematici Helvetici* 53:1 (1978), 542–571. doi:10.1007/BF02566098. Available online (`http://www.springerlink.com/index/10.1007/BF02566098`).

[Grünbaum and Shephard 79] Branko Grünbaum and G. C. Shephard. "Incidence symbols and their applications." In *Relations between Combinatorics and Other Parts of Mathematics*, Proceedings of Symposia in Pure Mathematics, 34, edited by D. Ray-Chaudhuri, Proceedings of Symposia in Pure Mathematics, 34, pp. 199–244. Providence, Rhode Island: American Mathematical Society, 1979. doi:10.1090/pspum/034. Available online (`http://www.ams.org/pspum/034`).

[Huffman 76] David A. Huffman. "Curvature and Creases: A Primer on Paper." *IEEE Transactions on Computers* C-25:10 (1976), 1010–1019. doi:10.1109/TC.1976.1674542. Available online (`http://ieeexplore.ieee.org/document/1674542/`).

[Kaplan and Salesin 00] Craig S. Kaplan and David H. Salesin. "Escherization." pp. 499–510. ACM Press, 2000. doi:10.1145/344779.345022. Available online (`http://portal.acm.org/citation.cfm?doid=344779.345022`).

[Laves 31] Fritz Laves. "Ebenenteilung in Wirkungsbereiche." *Zeitschrift für Kristallographie* 76:1-6 (1931), 277–283. doi:10.1524/zkri.1931.76.1.277.

[Miura and Kenkyjo 85] Koryo Miura and Uch Kagaku Kenkyjo. *Method of Packaging and Deployment of Large Membranes in Space*. Institute of Space and Astronautical Science report, Institute of Space and Astronautical Sciences, 1985.

[Miura 69] Koryo Miura. "Proposition of Pseudo-Cylindrical Concave Polyhedral Shells." *ISAS report* 34:9 (1969), 141–163. Available online (`http://ci.nii.ac.jp/naid/110001101617/en/`).

[Paris et al. 17] Jascha Norman Paris, Marcel Gottschalk, Henri Buffart, Susanne Hoffmann, Justus Siebrecht, Mathias Hüsing, Georg Jacobs, Martin Trautz, and Burkhard Corves. "Klassifikation u nd z ielgerichtete E ntwicklung v on w andelbaren Faltkonstruktionen." In *[12. Kolloquium Getriebetechnik, 2017-09-20 - 2017-09-22, Dresden, Germany / Michael Beitelschmidt (Herausgeber)]*, 2017. Available online (`http://publications.rwth-aachen.de/record/707264`).

[Roth 82] Karlheinz Roth. *Konstruieren mit Konstruktionskatalogen: Systematisierung und zweckmäßige Aufbereitung technischer Sachverhalte für das methodische Konstruieren*. Springer Berlin Heidelberg, 1982.

[VDI 82]  VDI. *VDI 2222 Bl. 2 Konstruktionsmethodik: Erstellung und Anwendung von Konstruktionskatalogen*, Februar edition. Berlin: Beuth, 1982.

[VDI 93]  VDI. *VDI 2221/Methodik zum Entwickeln und Konstruieren technischer Systeme und Produkte*. VDI-Richtlinien, Berlin: Beuth, 1993.

[Wang 16]  Kunfeng Wang. "Folding a Patterned Cylinder by Rigid Origami." In *Origami 5: Fifth International Meeting of Origami Science, Mathematics, and Education*, 2016.

[Yoshimura 55]  Y. Yoshimura. *On the Mechanism of Buckling of a Circular Cylindrical Shell Under Axial Compression*. National Advisory Committee for Aeronautics, 1955.

---

Justus Siebrecht, Chantal Weigel, Simon Dehn and Georg Jacobs
Institute for Machine Elementes and Systems Engineering (MSE), RWTH Aachen University, Germany,
e-mail: justus.siebrecht@imse.rwth-aachen.de
e-mail: chantal.weigel@imse.rwth-aachen.de

Henri Buffart, Susanne Hoffmann, Martin Trautz
Chair of Structures and Structural Design (trako), RWTH Aachen University, Germany, e-mail: buffart@trako.arch.rwth-aachen.de
e-mail: hoffmann@trako.arch.rwth-aachen.de

Jascha Paris, Burkhard Corves
Institute of Mechanism Theory, Machine Dynamics and Robotics (IGMR), RWTH Aachen University, Germany,
e-mail: paris@igmr.rwth-aachen.de

# Simulating Pleated Tension Folds

*G. Konjevod*

**Abstract**: *Origami tension folds are models that achieve a three-dimensional form due to tension induced in the folded sheet by the folds themselves. We describe an approach to simulating tension folds with flat-foldable crease patterns. We focus primarily on pureland-folded pleat tessellations. Our representation for the folded state explicitly represents the three-dimensional structure even though the folds we consider are theoretically flat. By updating the z-coordinates through the folding process, we can keep track of the arrangements of layers. To simulate the dynamics under fold tension, we use the physics model of Ghassaei's Origami Simulator. We avoid complex collision handling by making a simplifying assumption (high friction). We notice that under this assumption, we can modify the mesh that represents the folded sheet into a new mesh (which we call the free complex) that can be simulated without worrying about collisions. We also give a procedure for local layer reordering, allowing the simulation of a larger class of patterns that are not simply foldable.*

## 1   Introduction

We present an approach to simulating curved shapes created by tension in a family of origami models derived from Paul Jackson's *Bulge* [Jackson 91] (Fig. 1), and explored primarily by the author. A major feature of most of the folds in question is the alternation of horizontal and vertical pleats. The pleats interlock, but in some areas of the folded sheet, the arrangement may allow some movement. Since real paper creases don't fold flat but prefer to stay at a non-zero angle, alternating pleats create tension, which curves the models into the third dimension. We refer to these as *tension folds* to stress the fact that their three-dimensional form is induced by the tension in the folded sheet created by the interaction among multiple layers of paper. The term *dry tension folds* [Sternberg 11] has been used to emphasize the distinction from wet folded models whose three-dimensional form is achieved through shaping the paper while wet. However, the three-dimensional form of typical dry tension folds is determined by the geometric arrangement of creases and the tension is used only to prevent the model from unfolding. We are not aware of other families of three-dimensional origami whose crease patterns imply flatness.

The Bulge is the basic form and the simplest example in the family. It consists of two pleat sequences, alternating horizontal and vertical pleats, and an optional

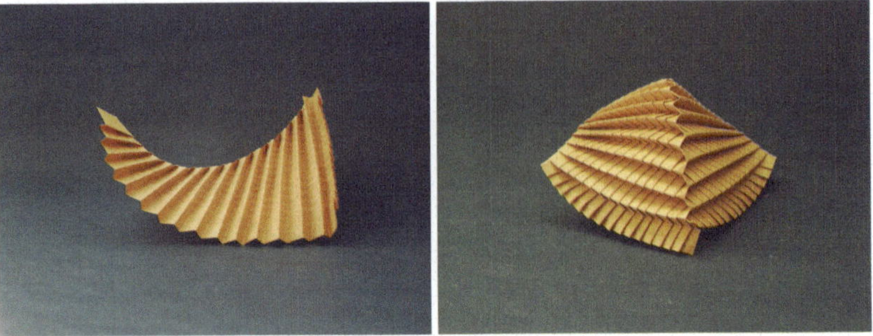

**Figure 1:** *Basic Form (Bulge) in two variants: just after pleating, and with two edges locked. In the view on the left, it is the zig-zag edge in the foreground that gets locked by folding it over and over while the pleats are kept folded.*

lock along two of the edges. Notice that without the lock, each pleat has an open end where its creases can flex freely.

Wave and Double Wave (Fig. 2, right) are examples of pleat sequences that were designed to produce a particular form. However, this is generally not easy to do. Thus, it would be be of interest to save time by deferring the time-consuming part of the work to the machine through a model that allows simulation of the equilibrium shape of a folded paper sheet. An earlier paper [**Konjevod and Kuprešanin 09**] gives the process for folding these and other related models, as well as expressions generating the corresponding folding sequences.

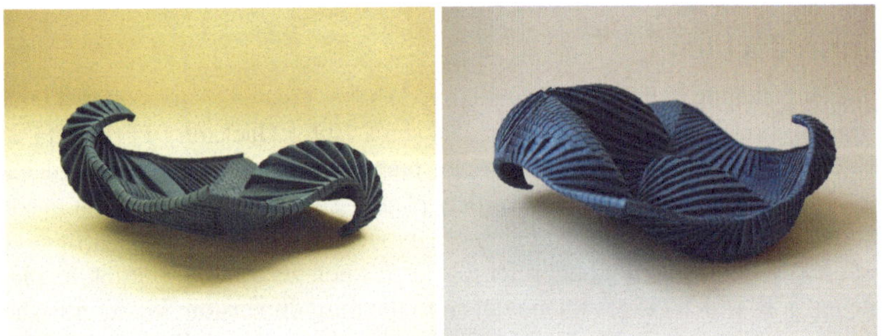

**Figure 2:** *Wave and Double wave. Wave consists of two simultaneously folded basic forms, starting from opposite corners, so that the corners curve in opposite directions. Double wave is a two-by-two tessellation of the wave, consisting of four simultaneously folded basic form sequences.*

There exist a number of computer programs to help visualize or simulate various aspects of the origami design and folding process. Treemaker [**Lang 11**] was the first broadly available general tool for origami design, and Oripa [**Mitani 12**] has been popular among origami tessellation artists for its capabilities in visualizing and editing flat-folded origami patterns. Tachi [**Tachi 09**] has developed a simulator for rigid origami and subsequently also the Origamizer [Tachi 10b], which computes a crease pattern that folds into a given polyhedron. Despite the impressive progress made through these works, most simulations of origami focus only on geometry and not dynamics. This is not surprising, because throughout the origami history, most of the models designed were so done with a purely geometric approach and with a static figure in m ind. While there h ave been b ooks focused on *Action Origami* [Lang 97, Shafer 01], not a lot of effort has been expended on its systematic study until recently. Notable exceptions to the rule can be found in the work of Schenk and Guest [**Schenk and Guest 11**] and Tachi [**Tachi 10a**]. A more recent example that further builds upon these two is Ghassaei's Origami Simulator [**Ghassaei 17**]. (Another very recent paper [**Filipov et al. 17**] came to our attention only after the original submission; it promises a more realistic model as a basis for analyzing the behavior of folded structures.)

In Origami Simulator, the sheet is represented by a triangulated polyhedral complex. To each vertex of the complex is assigned a point mass. Each edge is assigned a stiff spring to help model the inextensibility of the sheet. Further springs are associated with each face vertex, to prevent in-plane deformations (although given that the complex is triangulated, these may not be strictly necessary). Finally, each crease has a folding spring associated with it, which pulls the faces at the two sides of the crease towards a preferred fold angle (e.g., for an unfolded crease, the 0 degree angle, or for a valley fold, 180 degrees). Given a crease pattern together with each crease's preferred fold angle, the model enables the computation of forces on individual nodes and, using an iterative process, a simulation of the motion towards the equilibrium state.

Origami Simulator stands out for its speed. The simulation loop distributes the work of calculating forces and updating node displacements across multiple processes, and the code uses a javascript library to run the computation on a GPU, which makes it possible to interact with the system in real-time, even for large crease patterns. Another extremely useful feature is the availability of source code, without which the work presented here would have been more difficult.

To model tension-folded origami, it is not enough to treat each node independently; in such models, there are places where multiple layers of paper touch and exert forces on each other. Thus, it is necessary to somehow handle the behavior of paper in contact with itself, and avoid self-intersection. The collision detection (and resolution) problem has been studied in computer graphics, usually in the context of solid body or cloth simulation. Available codes typically handle a relatively small number of interacting layers in arbitrary—but usually less complex—arrangements [**Narain et al. 12**]. We do not rely on these algorithms, although incorporating one in this work may be useful. In any case, we leave that as a challenge for future work.

Instead, we precompute the "important" contact constraints among sets of nodes in our folded sheet model. This relies on making some assumptions. The first is that paper trapped within pleat intersections cannot slip out or move easily. We make this assumption more formal in the definition and discussion of the *free complex*.

The second assumption is that every occurence of contact during the relaxation is accounted for by the free complex. This may be seen as avoiding the problem of collision detection and resolution, and that view is fair, but as we'll see, our approach covers a lot of interesting origami models.

Given the two assumptions outlined here, it suffices to "glue" together sets of nodes in contact. The forces on each node are computed as usual, but for each "supernode" consisting of a collection of nodes glued together, we compute the resultant—the sum of forces on each individual node in the collection. The resultant force is then applied to the supernode and its constituent nodes are moved together as one.

In addition to simulating simply folded pleat sequences, we show how to extend our folded sheet model to pleat rearrangements. This involves computing feasible positions in three dimensional space for nodes of the grid in which have been folded a collection of axis-parallel pleats. The node positions must be compatible with local layer orderings at every location, and so the problem is not completely trivial. Some more general versions of this problem are NP-complete [Arkin et al. 04], but for the family of models we consider, a feasible embedding can be computed by solving a linear programming problem.

## 2   Flat-folded grid

Most of the origami models we are interested in are constructed by folding sequences of axis-aligned pleats, starting from a square sheet. For this reason, as well as for simplicity, we use a square grid as the starting point for our model of a folded sheet. Most of what we do would work in a more general setting, but some steps in the process would become more complicated. We note some of these issues in Section 8 and leave them as challenges for future work.

We further restrict our consideration to flat-foldable models. Even though the point of this work is to simplify the manipulation and improve our understanding of the three-dimensional forms of tension-folded origami, all the models we study here are theoretically flat-foldable. In practice, especially with thicker sheets, the word "theoretically" is important here. Models folded from heavy watercolor paper or from sheet metal are not really flat in any practical sense. In any case, we model our flat-folded origami as lying roughly in the $xy$ plane and treat the layer ordering separately.

Consider making a simple fold along a line in a square grid. We fix the sheet on one side of the crease, and let the remainder move by rotation through the third dimension around the axis formed by the crease. The new position of each grid point is easy to calculate. Iterating between valley and mountain folds, we can fold a sequence of pleats. The position of each grid point in the $xy$ plane remains easy to calculate, but what about the layer ordering? With a real sheet of paper, its thickness presents constraints on the amount of flattening possible, and each point can be assigned a third coordinate. Our model is motivated by this, and similar considerations, and illustrated in Figures 3 and 4. We maintain the conceptual separation of the position of a point with respect to the $xy$-plane, and the rank of

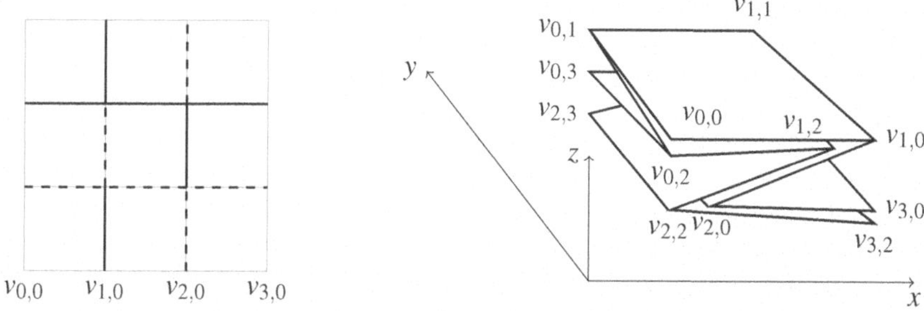

**Figure 3:** *Two intersecting pleats. As indicated in the crease pattern, the node at coordinates $(i, j)$ is labeled $v_{i,j}$. In the drawing of the folded state, nodes $v_{1,3}$, $v_{3,1}$ and $v_{3,3}$ are hidden behind the layers in the upper right and node $v_{1,2}$ behind the layers in the upper left. Note that in the folded state, $v_{2,2}$ lies to the left of $v_{2,0}$, and so its $x$ coordinate should be smaller. Similarly, $v_{0,0}$ is above $v_{0,2}$, which in turn is above $v_{2,0}$ and $v_{2,2}$. On the other hand, the latter two have equal $z$ coordinates and so their locations in $\mathbf{R}^3$ are instead disambiguated through the $x$-offset. We use offset vectors to represent such relationships.*

the point in the ordering of layers at the corresponding plane position. To represent the layer ordering, we associate with each grid point an *offset* vector in $\mathbf{R}^3$.

As each fold is made, some of the layers on the moving side of the crease rotate to land above (or below) the fixed side. We compute the new folded state in two steps. First, the coordinates in $xy$ plane are calculated easily by reference to the crease. Second, the offset of a point not on the crease is calculated by considering the rank in the layer ordering and the existing layers at the position to which the point moves. Finally, the points on the crease are a special case and they are offset not only in the $z$ direction, but also in the direction normal to the crease. See Figure 4 for illustration.

The simple offset algorithm described in Figure 4 can over the course of a pleat sequence lead to offset values that are quite large. Especially if folding a long sequence of pleats, it may be useful to reduce the $z$ components of offset vectors as much as possible, while preserving the local layer orderings. This procedure can be applied at any point during the folding process.

While the offsets, and thus layer orderings are calculated for individual grid points, perhaps of more interest is the layer ordering of grid faces (which are unit squares in the grid) at the various locations in the folded model. By examining the offset calculation steps and potential $z$-reduction steps, it can be seen that under reasonable assumptions on the offset values, the layer ordering of grid faces induced by the three-dimensional coordinates given by shifting folded points by their offset vectors is consistent with the pleat sequence. It follows that we can use simple geometric reasoning to infer the layer ordering of faces at any location in the folded model. This further shows the dual roles of offset vectors: they allow recovery of local layer ordering (both for nodes and for faces) and, when combined

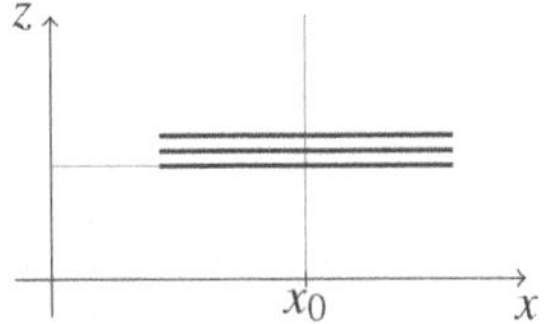 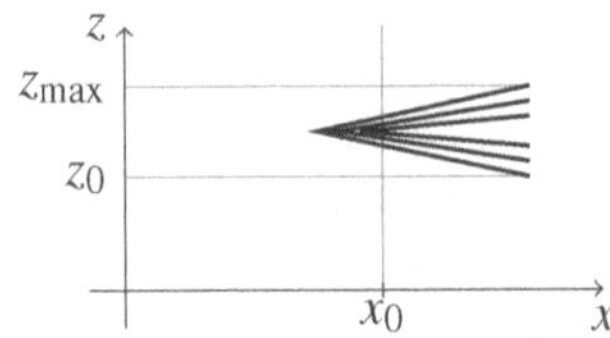

**Figure 4:** *A valley fold through several layers along the crease $x = x_0$: on the left, three layers before the fold is made; on the right, after a valley fold. The view is in the direction of the $y$ axis. In this example the part of the sheet with $x > x_0$ stays in place. To conserve the layer ordering, the $z$-offset of each moving point is compatible with the existing layers at the point's new locations. Points on the crease are further shifted along the $x$ axis. The $x$-offset applied to each point depends on its $z$-offset before the fold. The full algorithm is as follows: (1) Assign $z$-offset $z_{max} + 1$ to each point on the crease, where $z_{max}$ is the current maximum $z$-offset. (2) For each such point (with $z$-offset, say, $z_0$), subtract $z_{max} - z_0$ from its $x$-offset. (3) For each point with $x$ coordinate $x_1 < x_0$ and $z$-offset $z_1$, set its new $x$ coordinate to $x_0 + (x_0 - x_1)$ and its new $z$-offset to $z_{max} + 1 + z_{max} - z_1 + 1$. If only a top subset of layers is to be folded, then the previous applies only to points in such layers and all other points stay where they were.*

with flat coordinates, they provide a reasonable starting point for the relaxation.

## 3   Free complex

In order to avoid modeling contact forces, instead of the mesh derived directly from the flat-folded grid, we find and identify groups of nodes that can be assumed to move together during the relaxation.

Consider a fold through two or more layers. For example, consider $v_{2,2}$ and $v_{2,0}$ in Figure 3. Let $a$ (e.g. $v_{2,2}$) be a point on the crease $c$ on the outermost layer. We say $a$ *traps* the point $b$ (e.g. $v_{2,0}$) if $b$ is folded to the same $xy$ position and belongs to a layer of paper inside the two layers formed by the crease $c$. Similarly, $b$ traps $a$. Intuitively, $b$ cannot move independently of $a$ unless either $c$ is unfolded or the two layers slide against each other. This definition can be extended to an equivalence relation: we say that $a$ and $b$ are *joined* in a flat fold if either $a$ traps $b$ or vice versa. Figure 5 illustrates the point by showing a simple two-pleat example in simulation.

Now we can describe the *free complex* of a flat-folded origami. Start with the two-dimensional polyhedral complex formed by the original square grid. For each maximal set of joined vertices, merge them into a single vertex, assigning it the position in $\mathbf{R}^3$ of any of the underlying vertices (in our case, the offsets —typically around 0.01 in $x$ and $y$ dimensions, and 0.0001 in $z$—are small enough that the vertices being joined are close enough that the choice here makes no difference). In Figure 6 we show wireframe visualizations of the free complex of a simple two-pleat fold and of the basic form with pleat sequence of length eight.

To use the free complex in a physics model, we modify the resulting structure by combining the relevant features of joined vertices. More precisely, for each node

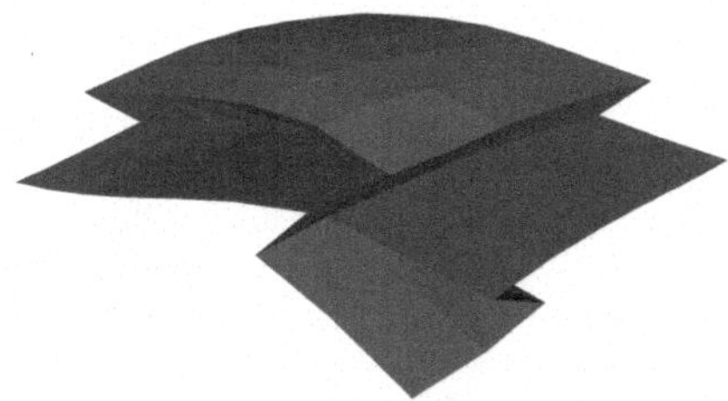

**Figure 5:** *A pair of pleats in the process of unfolding. Note that the pleat running left to right cannot even begin unfolding at the intersection until the pleat running front to back is mostly open. More generally, the trapped vertices aligned by an earlier pleat cannot separate unless either all the subsequent pleats are unfolded first, or unless they can move along the layers trapping them. This can be achieved with real paper, with some effort (see Section 5), but does not typically happen on its own. On the other hand, the layers of the first-folded pleat on the far left side of this view are already separating.*

resulting from a merge, we assign it the mass equal to the sum of its constituents. Similarly, the sum of all forces on the merged vertices gives the the resultant force on the supernode.

## 4  Physics model

The physics model is directly based on Ghassaei's Origami Simulator. The folded sheet is represented by a triangulated mesh. The nodes of the mesh are the points of the folded grid. Each square of the grid is triangulated by adding as an edge one of its diagonals. Each node represents a unit mass. We assume three internal forces act on each node of the triangulated grid.

**Edge springs.** Each edge incident on the node acts as a spring with known resting length (1 for square edges or $\sqrt{2}$ for diagonals) and fixed spring constant $k$ and damping factor $D$.

**Face springs.** Each face incident on the node acts as a spring with known resting angle (in our case either $\pi/4$ or $\pi/2$). Figure 7 shows the expression for the force applied to each node by its incident faces.

**Crease springs.** For each (triangular) face incident on the node, if an edge of the face is a crease (either a mountain or a valley), the crease acts as a spring with known resting dihedral angle. For a node incident on the crease, the crease applies the force in the direction normal to the crease and lying in the plane that bisects the dihedral angle. For a node opposite the crease, the force depends on the distance from the node to the crease as well. In each case, the force is proportional to the deviation of the current dihedral angle from the given resting angle for this crease.

One might ask if it is really necessary to give creases a resting angle away from

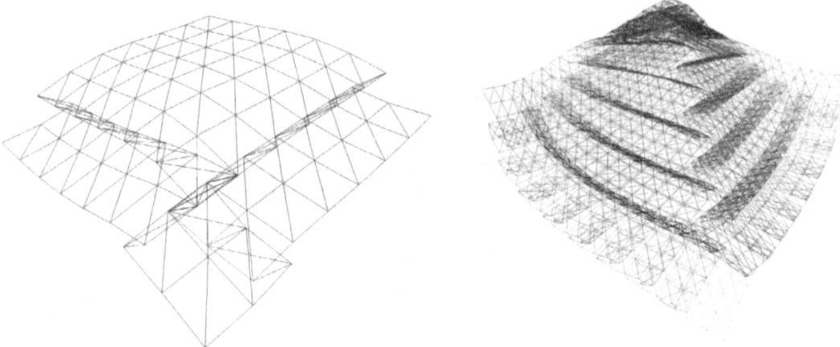

**Figure 6:** *Visualization of the free complex for a single pair of pleats, and for the basic form from an eight-pleat sequence. In each case, the merged node sets are in darker regions of the image.*

180 degrees. It turns out that it is: otherwise, no springs act as hinges to open up the folds, and the model doesn't curve into the third dimension. All that happens is minute shifts of nodes in the $xy$ plane.

**Simulation step.** After all the forces have been computed and added up at each node, the simulation proceeds by iterating the standard Verlet integration step $x_{t+1} = 2x_{t-1} - x_{t-2} + a\Delta t^2$, where $t$ is the timestep and $a$ the acceleration. For the simulations in this paper we used the timestep of 0.008s. We use $k = 0.7$ for all stiffness constants and $d = 0.85$ for edge damping. In the current version of the model, only the edge springs have damping. It would be of interest to consider the effects of adding damping to the other two spring types, and further to try calibrating the constants to more realistic models of paper [Filipov et al. 17].

## 5 Local layer rearrangements

It turns out that further interesting shapes can be obtained by locally rearranging layers in a pleat folded model. In previous work [Konjevod 15] we have considered the problem of minimizing the number of required pleat rearrangements. For an example of a model that requires pleat rearrangements, consider Figure 8. In the current setting, pleat rearrangements present a challenge because there is no way to fold such models using a sequence of simple folds (pleats). Thus, in order to simulate such models, we need a way to start with a pre-folded grid, and then rearrange the layers in a given set of locations (through redefining the three-dimensional coordinates of appropriate points), but again end up with a feasible and non-self-intersecting arrangement of layers.

To motivate the following discussion, consider again Figure 3. The figure shows the outcome of folding a horizontal pleat followed by a vertical pleat. All sixteen nodes of the pattern are arranged at four locations in the $xy$-plane, four at each. Consider for example the lower left, where we have $v_{2,2}$, $v_{0,2}$, $v_{2,0}$ and $v_{0,0}$. As discussed above, the $z$ offsets of these four nodes satisfy $z_{0,0} > z_{0,2} > z_{2,0} = z_{2,2}$. We also pointed out previously that $x_{2,2} < x_{2,0}$. Analogous constraints can be written

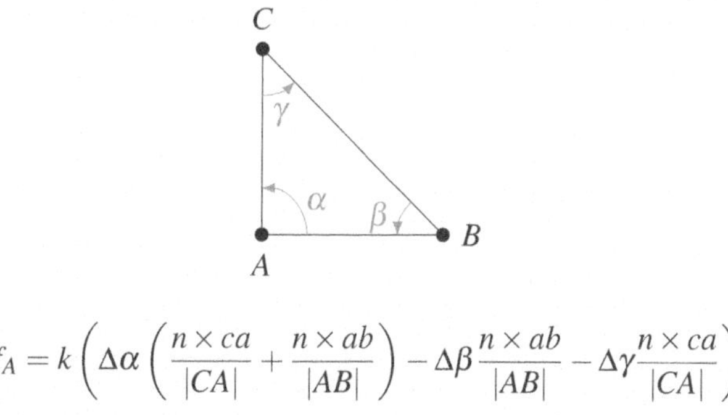

$$f_A = k \left( \Delta\alpha \left( \frac{n \times ca}{|CA|} + \frac{n \times ab}{|AB|} \right) - \Delta\beta \frac{n \times ab}{|AB|} - \Delta\gamma \frac{n \times ca}{|CA|} \right)$$

**Figure 7:** *The distortion of a face induces forces on its vertices. Here, AB and CA denote vectors between the respective points, while ab and ca denote the same vectors scaled to unit length. The vector n is this face's normal vector. The differences between the resting and actual angles are denoted by $\Delta\alpha$, $\Delta\beta$ and $\Delta\gamma$. The stiffness constant is k. The expressions for forces on B and C are analogous.*

for each of the four corners of the folded state. In order to rearrange the local layer ordering, we can enforce alternate constraints on the offsets of the sixteen vertices. Next we describe the general approach.

We consider a folded (pleated) rectangle of $n$ rows and $m$ columns. For each square $(i, j)$, we are given as input the layer ordering at that square.

Note that, assuming each pleat is of unit width and follows the lines of a regular square grid, there are either 1, 3, or 9 layers at any grid square. We use $\lambda_{i,j}$ to denote the number of layers over square $(i, j)$ of the folded state. There are four possible 3-layer orders and eight possible 9-layer orders.

The case $\lambda_{i,j} = 1$ is trivial and requires no additional constraints.

For $\lambda_{i,j} = 3$, there is a single pleat passing through the square $(i, j)$. If this is a vertical pleat, we take $v_{i,j} \in \{-1, 1\}$ and $h_{i,j} = 0$. If this is a horizontal pleat, we set $v_{i,j} = 0$ and take $h_{i,j} \in \{-1, 1\}$. If this pleat's mountain-fold coordinate is smaller than that of its valley-fold, we choose 1 for the nonzero value, otherwise we choose $-1$.

Finally, for $\lambda_{i,j} = 9$, there are two pleats crossing at $(i, j)$ and a total of eight possibilities. We specify the pleats folded by using $v_{i,j} \in \{-1, 1\}$ and $h_{i,j} \in \{-1, 1\}$ as in the 3-layer case, and in addition we use $hv_{i,j}$ to specify the order: $hv_{i,j} = 1$ if the horizontal pleat is folded first, and $hv_{i,j} = -1$ if the vertical pleat is folded first.

The values of the parameters $\lambda$, h, v and hv can be recorded while "folding" the pleats. After the folding is complete, we modify the values of these parameters for squares where the layers need to be rearranged. Then we save all of them to a data file used by the linear programming solver to generate the specific problem instance. After the LP solution is produced, it is used to modify the grid point offsets, and the relaxation simulation can be run.

Next, we outline the linear constraints used to compute the appropriate offsets for points mapped to each folded square. The variables $x_{a,b}$, $y_{a,b}$ and $z_{a,b}$ will denote the offset values at grid point $(a,b)$. For each $(i,j)$,

- if $\lambda_{i,j} = 3$, there are 4 constraints on $z$, determined by the combination of $v_{i,j}$ and $h_{i,j}$. All $x$ and $y$ variables here can be set to 0. For example, a horizontal pleat with mountain fold at $y = 1$ and valley at $y = 2$ requires (among others) the constraint $z_{1,0} \geq z_{3,0} + 1$.

- if $\lambda_{i,j} = 9$, there are 12 constraints on $z$ and 4 constraints on $x$ (if $hv_{i,j} = 1$) or on $y$ (if $hv_{i,j} = -1$). The $z$-constraints include 8 constraints of the form $z_{0,2} \geq z_{2,2} + 1$ (as in Figure 3) and 4 constraints of the form $z_{2,2} = z_{2,0}$. The $x$ (or $y$) constraints are of the form, for an example referring again to Figure 3, $x_{2,0} \geq x_{2,2} + 1$.

Finally, we upper-bound each $x$, $y$ and $z$ variable by $M$, for a variable $M$ which also serves as the objective function to minimize.

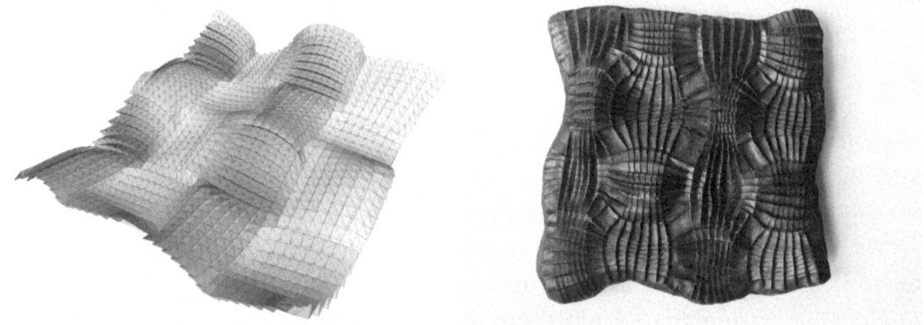

**Figure 8:** *The simulation result of a pattern formed by rearranging pleats in four of the sixteen fields. For comparison, a real model is also shown, made in cast iron from a paperfolded original.*

## 6   Results and discussion

Here we show some results obtained by running our simulation for various pleat-based folds. Global shapes output by the simulation generally match the ones folded from real paper very well. Figure 9 shows the results of simulating the basic form with crease resting angles varying between around 140 degrees and just over 160 degrees. For smaller resting angles our approach doesn't produce sensible results without additional constraints (this happens even with the smallest angle shown in this plot, just under 140 degrees). Figures 11–16 show additional renderings and comparisons to folded models.

Some of the folded models have multiple local optima, and in a few situations, the simulation outputs a different one from the one that we get by folding. In some cases (for example, Simple Bowl, Figure 10), it may be possible to simply start from a different point configuration. In others, it may be necessary to apply some forces partway during the relaxation. While folding the corresponding pieces from

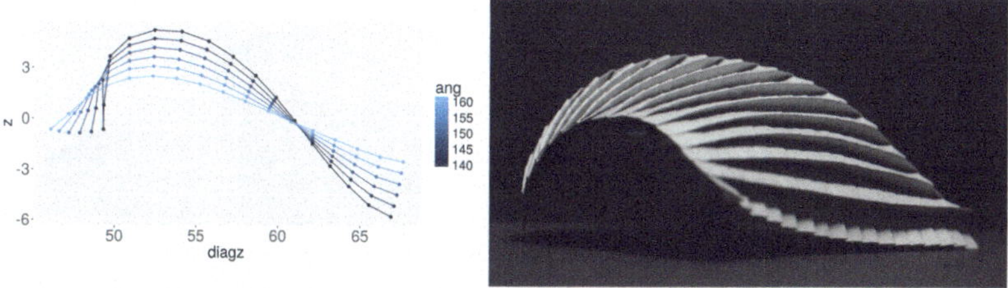

**Figure 9:** *Left, simulation results for varying resting angles. Curves shown represent the diagonal of the simulated sheet. Right, analogous view of a model folded from elephanthide paper.*

actual paper, varying the process by folding on the table or in the air, or with one or the other side of the sheet pointed up or down tends to produce minor variations in the results.

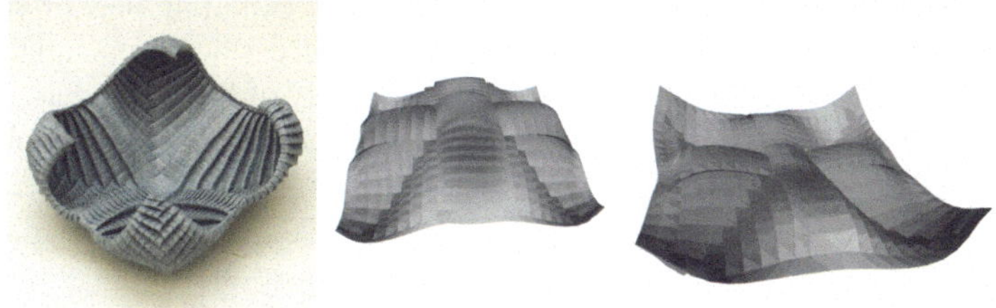

**Figure 10:** *Simple Bowl comparison. Center image shows a computed local optimum different from the folded one. Image on the right is the result of a simulation started from a different initial position. Instead of starting in the xy plane, the points were "draped" over a half-sphere (concave side up).*

However, in most of the produced meshes, minor local intersections between some pairs of faces can be observed. Some of this is easily explained: during the simulation the normal vector of a face in general changes direction. Now consider two faces that are glued together in the construction of the free complex. The pairs of corresponding nodes belonging to these faces remain in exactly the same relative position throughout because of the way their motion is currently handled by the simulation. Thus, this may create minor collisions that do not get resolved. It should be possible to eliminate this problem by allowing the axes joining merged node pairs to rotate during the simulation, together with the faces containing the nodes.

Further exploration of the influence of the spring constants and other parameters would be useful. In real examples, this angle heavily depends on multiple factors, including the force applied when making each fold, as well as the thickness of the paper and is relation to pleat width. The former typically varies quite a bit over the different parts of individual models and so a uniform setting as here is a coarse approximation.

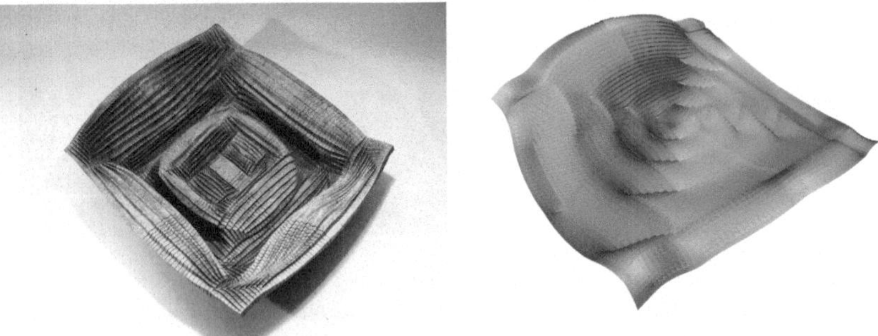

**Figure 11:** *Comparison of a real and simulated fold based on similar pleat sequences. The real origami piece was folded in 2008 from elephant hide paper. It has several locally optimal states and can be switched between them.*

However, as an approach to reducing the effort in exploring new or complex pleat sequences, we consider the current results a success and a potentially very useful tool for understanding the behavior of this family of folded models.

More examples, as well as some mesh files, a video, and in the near future the implementation itself, can be found at

`http://organicorigami.com/tensionfoldsimulation.`

## 7   Implementation

The current implementation has been written in pure Python, with the exception of the computation of local layer rearrangement, which, as described above, invokes a linear programming solver. The linear program is formulated on demand by extracting local layer orderings from the known sequence of folded pleats. Instead of directly generating the constraint matrix, the LP solver uses a generic model description written in the GMPL modeling language and a separate data file generated by the folding code.

Being sequential, our code is much slower than Ghassaei's GPU-based Origami Simulator.

The flat-folded patterns can be saved in the FOLD format [Demaine et al. 16], but doing so with the simulated meshes and the free complex requires a custom format extension.

The use of the free complex and the merging of nodes doesn't influence the running time of the simulation, because we can redefine the internal methods of the force-calculation classes on-the-fly. This means that exactly the same functions are called to compute forces on each node. The forces are nevertheless added up appropriately, according to the structure of the free complex. The same trick allows us to avoid a more complex implementation of the edge-locking step. Instead, we glue together the pairs of nodes we'd like to keep joined, and proceed without changes.

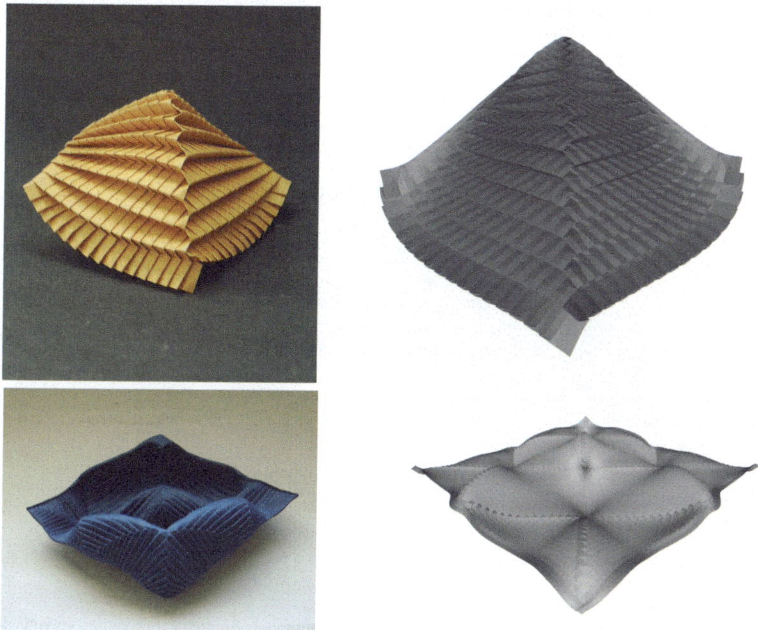

**Figure 12:** *Basic Form (Bulge) and Double Wave—comparison of photograph and rendering.*

# 8   Future work

Our physics model is simplistic. Recent work [Filipov et al. 17] offers a more realistic approach that may allow a quantitative validation. Figure 9 shows the initial results of varying the crease resting angle to generate surfaces of varying curvature, but it is only a very small step towards understanding the parameters that govern this process. As the same figure shows, with smaller resting angles, the free complex we use is not sufficient to prevent self-intersection in the simulated mesh. Furthermore, we don't consider the influence of thickness on the behavior of the sheet around the folds.

Both the initial folding and the free complex computation assume axis-aligned folds. It should be possible to avoid making these assumptions, but more general crease patterns may complicate the derivation of the free complex by requiring new nodes wherever a node gets glued to a non-node point of a face.

Another useful extension would be a mechanism to apply external forces at various moments during the relaxation process. This may take the form of fixing certain nodes in space, or moving them along a prespecified trajectory, thus simulating a sculpting-like shaping process after a sequence of pleats has been folded. Alternatively (or in addition), it may take the form of changing target angles of particular creases, even to the extreme of allowing a procedural unfolding of the folded state back to a flat unfolded sheet (this would in also require the ability to

**Figure 13:** *Hierarchical double pleated model.*

dynamically update the free complex, which would be somewhat complicated with the current code).

Finally, a dynamic general collision detector, even if applied only every few hundred iterations (most renderings shown here took between 10 and 20 thousand steps) would be useful to prevent very small self-intersections and would help produce meshes easier to manipulate afterwards.

# References


[Arkin et al. 04] Esther M. Arkin, Michael A. Bender, Erik D. Demaine, Martin L. Demaine, Joseph S. B. Mitchell, Saurabh Sethia, and Steven S. Skiena. "When can you fold a map?" *Computational Geometry* 29 (2004), 23–46.

[Demaine et al. 16] Erik D. Demaine, Jason S. Ku, and Robert J. Lang. "A New File Standard to Represent Folded Structures." In *Abstracts from the 26th Fall Workshop on Computational Geometry*, 2016.

[Filipov et al. 17] E.T. Filipov, K. Liu, T. Tachi, M. Schenk, and G.H. Paulino. "Bar and hinge models for scalable analysis of origami." *International Journal of Solids and Structures* 124 (2017), 26–45.

[Ghassaei 17] Amanda Ghassaei. "Origami Simulator." `http://apps.amandaghassaei.com/OrigamiSimulator`, 2017. Accessed: March 15, 2018.

[Jackson 91] Paul Jackson. *The Encyclopedia of origami & papercraft techniques*. Quarto Publishing, 1991.

[Konjevod and Kurešanin 09] Goran Konjevod and Ana Maria Kurešanin. "Notation for a class of paperfolded models." In *Proceedings of the 12th Annual Bridges Conference, Banff*, pp. 47–54, 2009.

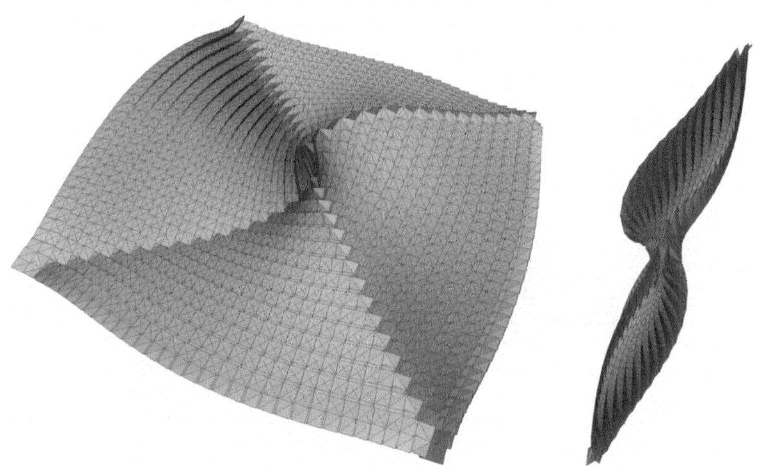

**Figure 14:** *Two views of the simulation result of another wave-like model. This is composed of two pleat sequences simultaneously crossing the sheet diagonally, each of them skipping every other potential pleat location. This model doesn't require locking the edges.*

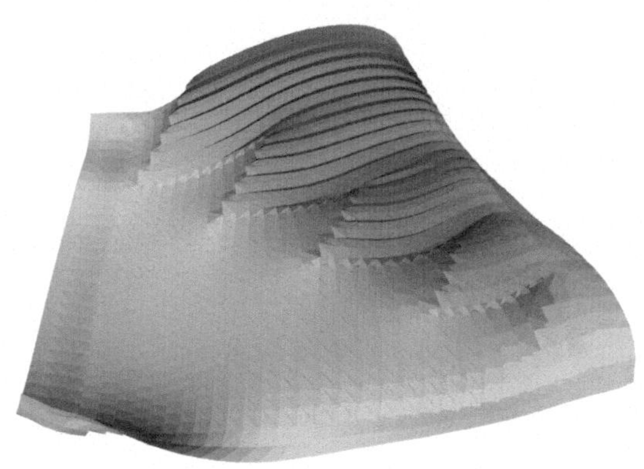

**Figure 15:** *A sequence of pleats resulting in a repeated pattern.*

[Konjevod 15]  Goran Konjevod.  "On pleat rearrangements in pureland tessellations."  In *Origami[6] I:Mathematics*. AMS, 2015.

[Lang 97]  Robert J. Lang. *Origami in Action: Paper Toys that Fly, Flap, Gobble and Inflate.* St. Martin's Griifin, 1997.

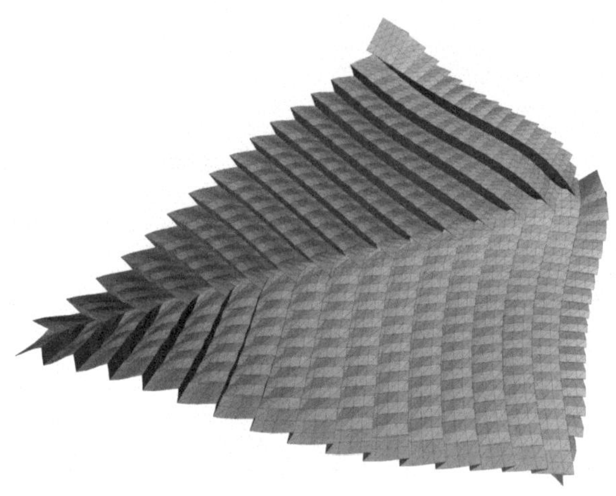

**Figure 16:** *A simple pleat sequence resulting in a folded sheet that slowly twists around the axis formed by the diagonal.*


[Lang 11] Robert J. Lang. *Origami Design Secrets, 2nd edition*. AKPeters / CRC Press, 2011.

[Mitani 12] Jun Mitani. "ORIPA: Origami Pattern Editor." `http://mitani.cs.tsukuba.ac.jp/oripa`, 2012. Accessed: March 15, 2018.

[Narain et al. 12] Rahul Narain, Armin Samii, and James F. O'Brien. "Adaptive Anisotropic Remeshing for Cloth Simulation." *ACM Trans. Graphics* 31 (2012), 147:110.

[Schenk and Guest 11] Mark Schenk and Simon D. Guest. "Origami Folding: A Structural Engineering Approach." In *Origami*$^5$, pp. 291–304. CRC Press, 2011.

[Shafer 01] Jeremy Shafer. *Origami to Astonish and Amuse*. St. Martin's Press, 2001.

[Sternberg 11] Saadya Sternberg. *Sculptural Origami*. Dover, 2011.

[Tachi 09] Tomohiro Tachi. "Simulation of Rigid Origami." In *Origami*$^4$, pp. 175–188. AK Peters, 2009.

[Tachi 10a] Tomohiro Tachi. "Freeform variations of origami." *J. Geom. Graphics* 14 (2010), 203–215.

[Tachi 10b] Tomohiro Tachi. "Origamizing Polyhedral Surfaces." *IEEE Trans. Vis. Comp. Graphics* 16 (2010), 298–311.

# Modelling the Folding Motions of a Curved Fold

*Y. Watanabe, J. Mitani*

**Abstract:** *A user interface system to design a curved folding and to simulate its folding motion is proposed. The system simulates a sheet of paper with a single curved crease, and supports ruling transition which occurs during the folding motion. To handle a curve with a zero curvature point, rectification of the torsion and the folding angle is carried out. The system was evaluated visually and by numerical verification, to confirm its usability and reliability.*

## 1 Introduction

Curved folding is one of the most popular fields of origami, which attracts people by its beautiful appearance. But the behaviour of the paper is not easy to understand because the curved surface has a flexibility to change its bending directions as well as the amount of bending. For further development of the curved fold designs, it would be helpful for a designer to understand the behaviour of the rulings while the paper is being folded. To this end, we propose a user interface to model and visualize curved folding with one curved crease on a sheet of paper, supporting the ruling transition which occurs in the motion of folding and twisting the fold curve. Figure 1 shows an example of folding motion in 3D space (top row) and in 2D space, which represents the crease and the rulings projected on flat paper (bottom row). The ruling directions change as the paper is folded or unfolded. In our system, the folded paper is represented as two adjacent quad strips with a curved crease in between. Through the graphical user interface (GUI), a user can design the shape of the folded paper by adjusting the curvatures, torsions, and folding angles of some control points placed evenly on the curve. From the input parameters on the control points, the shape of the crease curve in 2D and 3D, the folding angles on all points on the crease, and the ruling directions are derived, which defines the 3D shape of the paper with a curved fold. In simulating the motion of the curved fold, the geometry of the quads are updated at every step of the motion, as the ruling directions are recalculated according to the curve shape and folding angle at that moment.

In this paper, related work is first introduced in section 2, followed by the theoretical framework of this system in section 3. The detailed description of the system and the procedure to design the crease curve, its 3D shape, and the folding angle are written out in section 4, along with the method to simulate its folding/unfolding motion. The results of subjective evaluation and numerical verification are discussed in section 5, and the conclusions are given in section 6.

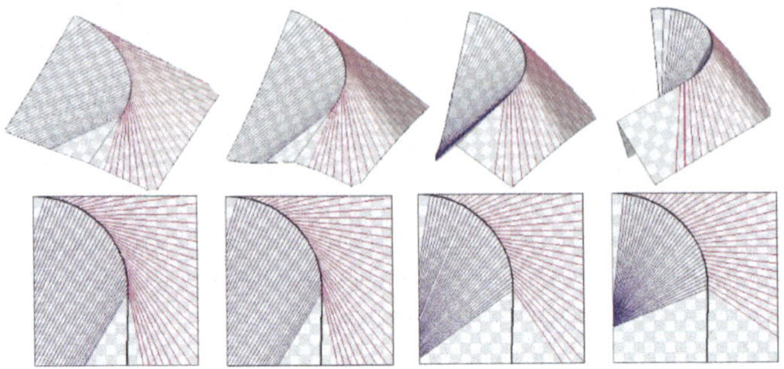

**Figure 1:** *Transition of rulings as the paper being folded, in 3D space (top row) and in 2D space (bottom row). The surface of the paper is textured with light checkerboard pattern to make the curved shape visible.*

## 2　Related Work

In recent studies on paper modelling and simulation, 3D surfaces generated by curved folding are often represented by a polygon model of fixed geometry composed of a group of quad strips and planar polygons. Kilian proposed a method to reconstruct a 3D polygon mesh of a developable surface from 3D data of a curved folding as an input, by initially fitting planar polygons on the 3D shape and then optimizing their shapes and orientations [Kilian et al. 08]. Their method is successful in designing the developable surface of a user-intended shape but does not design nor consider its folding motion. Tang et al. introduced an interactive design system of a curved-creased origami, which defines each surface by a pair of spline curves and solves the constraints for its global developability through an iterative process [Tang et al. 16]. As to the folding motion, Tachi developed a software to simulate the folding motion of the paper using a rigid folding method. His method represents paper by planar polygons with fixed geometry and mainly deals with straight fold lines [Tachi 09]. He also proposed a design method for flat-foldable vault structures composed of flat-foldable tubes assembled by welding two sheets [Tachi 13]. While dealing with the folding motion of the curved crease, this structure restricts the folding angle to be constant throughout the curve, which makes the rulings to be fixed, so that the structure is folded rigidly. It does not support deformation where the directions of rulings change while the paper is folding. Another method to simulate a sheet of thin materials by adaptive mesh refinement [Narain et al. 13] models irregular deformations of paper such as crumpled paper or ruling transitions, but it results in complicated mesh data composed of thousands of faces.

In our system, the paper is modelled as a quad strip with relatively small number of faces. In interactive design and the simulation of folding motion, the 3D shape of the paper is calculated instantly according to the shape and the folding angle of the curve, with no iterative process. Unlike in the surface subdivision method, our approach allows the deformation of paper by updating the ruling directions, which makes the number of quads and their connectivity to be consistent.

## 3 Developable Surface with Single Curved Crease

Our system is based on the relations between the curve shape in 2D and 3D, the folding angle, and the ruling directions, induced by Fuchs and Tabachnikov [Fuchs and Tabachnikov. 99], and applied by Tachi in his work [Tachi 11] [Tachi 13]. Section 3.1 introduces this relationship and the mathematical framework used for it, which is also described thoroughly in [Demine et al. 14]. In section 3.2, the discrete expression of the curve, vectors and parameters and the calculation method used in our system are explained. Finally, in section 3.3, rectification of the torsion and folding angle is discussed. This rectification process modifies the shape of the curve and the surfaces to be a developable surface.

### 3.1 Curved Folds and Rulings

Given a space curve $X(s)$ parameterized by the arc length $s$, its tangent vector $T(s)$, normal vector $N(s)$, and binormal vector $B(s)$ are given by the following equation, with $k(s)$ indicating curvature of the curve.

$$\begin{bmatrix} T(s) \\ N(s) \\ B(s) \end{bmatrix} = \begin{bmatrix} X'(s) \\ X''(s)/k(s) \\ X'(s) \times X''(s)/k(s) \end{bmatrix}. \tag{1}$$

$T(s)$, $N(s)$, $B(s)$, curvature $k(s)$ and torsion $\tau(s)$ are integrated using Frenet–Serret formula,

$$\begin{bmatrix} T'(s) \\ N'(s) \\ B'(s) \end{bmatrix} = \begin{bmatrix} 0 & k(s) & 0 \\ -k(s) & 0 & \tau(s) \\ 0 & -\tau(s) & 0 \end{bmatrix} \begin{bmatrix} T(s) \\ N(s) \\ B(s) \end{bmatrix}. \tag{2}$$

The definitions of the vectors on the curve and the angles which define the rulings are shown in Fig. 2. For the intuitive understandings of the curved crease and the rulings on the paper, we use the coordinate system in which the binormal vector $B(s)$ always orient to the bottom side of the paper. Then, looking in the direction of the tangent vector $T(s)$, the normal vector $N(s)$ always orients to the right side of the crease curve and the curvature $k(s)$ takes a positive or negative value depending on whether the curve is clock wise of counter clock wise. The ideas of such orientation fixed vectors and the signed curvature are also described in [Demine et al. 14] as the "top-side Frenet frame". As in Fig. 2, $\alpha(s)$ is the folding angle, which is always the same for the left and the right side, as proved in [Demine et al. 14]. $\beta_L(s)$ and $\beta_R(s)$ are the angles between the tangent vector $T(s)$ and the rulings in 2D space of the left and the right side. Using this coordinate system, the relationship between the folding angle, the 2D curve, and the 3D curve is

$$k_{2D}(s) = k(s) \cos \alpha(s), \tag{3}$$

where $k_{2D}(s)$ is the curvature of the 2D curved crease. The geometric relationships between the curve shape defined by the curvature and the torsion, folding angle, and the ruling direction in 2D space are

$$\cot \beta_L(s) = \frac{-\alpha(s)' + \tau(s)}{k(s) \, \sin \alpha(s)}, \tag{4}$$

$$\cot \beta_R(s) = \frac{\alpha(s)' + \tau(s)}{k(s) \sin \alpha(s)}. \tag{5}$$

Note that the signs of some parameters in equations (4) and (5) are different from those of Tachi and Fuchs due to the difference in the vector orientations. The rulings in 3D space are calculated as follows:

$$r_L = \cos \beta_L \, T - \sin \beta_L \cos \alpha N + \sin \beta_L \sin \alpha \, B, \tag{6}$$

$$r_R = \cos \beta_R \, T + \sin \beta_R \cos \alpha N + \sin \beta_R \sin \alpha \, B. \tag{7}$$

From equation (3), we could roughly say that two of the three elements, 3D curve, 2D curve, and folding angle, define the other element. In other words, a 2D curve may be derived from the 3D curve and folding angle (case A), a 3D curve may be derived from the 2D curve, folding angle, and torsion (case B), and a folding angle may be derived from the 2D and 3D curves (case C). This is because the 3D space curve is defined by curvature $k(s)$ and torsion $\tau(s)$, and the 2D curve is defined by 2D curvature $k_{2D}(s)$. In case B, the torsion $\tau(s)$ also needs to be given as it is not given by equation (3). Then, by equation (4) and (5), the ruling directions in 2D space, using $\beta_L(s)$ and $\beta_R(s)$, are derived. By equation (6) and (7), the 3D ruling directions are obtained using vectors $T(s)$, $N(s)$, $B(s)$ derived from the 3D space curve.

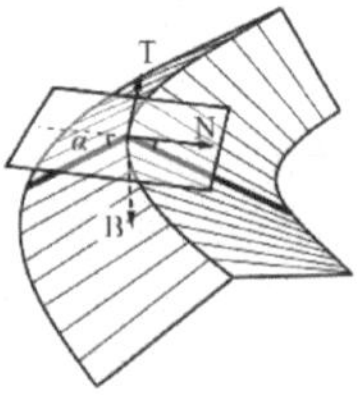

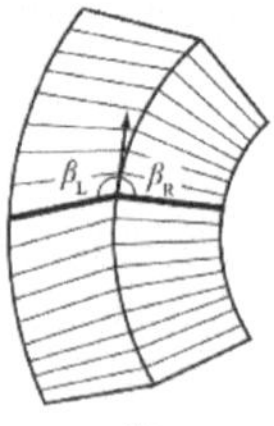

**Figure 2:** *Definition of tangent vector T, normal vector N, binormal vector B, folding angle $\alpha$ and ruling directions $\beta_L$, $\beta_R$. (a):3D shape. (b):2D crease curve.*

## 3.2   Discrete expression of T, N, B, curvature, and torsion

To apply the theory introduced in section 3.1 to the calculation of the curved crease as a polygon model, the curve is to be expressed in a discrete form $\{X_i\}$, or a set of sample points aligned along the curve $X(s)$. Similarly, other vectors and parameters on the point $X_i$ are denoted as $T_i$, $N_i$, $B_i$, $k_i$, $\tau_i$, $\alpha_i$, and $k_{2Di}$, instead of arc-length parametric functions $T(s)$, $N(s)$, $B(s)$, $k(s)$, $\tau(s)$, $\alpha(s)$, and $k_{2D}(s)$. The derivative of $X_i$, or $dX_i$, is calculated as follows, using three consecutive sample points $\{X_{i-1}, X_i, X_{i+1}\}$

$$dX_i = \frac{1}{2}\left(\frac{dX_{i,i-1}}{|dX_{i,i-1}|} + \frac{dX_{i+1,i}}{|dX_{i+1,i}|}\right), \tag{8}$$

where $dX_{i+1,i} = X_{i+1} - X_i$, is a vector connecting two sample points. Similarly, derivatives of the vectors $dT_i$, $dN_i$, and $dB_i$ are calculated as

$$dT_i = \frac{1}{2}\left(\frac{T_i - T_{i-1}}{|dX_{i,i-1}|} + \frac{T_{i+1} - T_i}{|dX_{i+1,i}|}\right)$$
$$dN_i = \frac{1}{2}\left(\frac{N_i - N_{i-1}}{|dX_{i,i-1}|} + \frac{N_{i+1} - N_i}{|dX_{i+1,i}|}\right), \tag{9}$$
$$dB_i = \frac{1}{2}\left(\frac{B_i - B_{i-1}}{|dX_{i,i-1}|} + \frac{B_{i+1} - B_i}{|dX_{i+1,i}|}\right)$$

which is an average of two differences between the consecutive points normalized by the distance. Figure 3 (a), (b) shows an example of discrete derivatives $dX_i$ and $dT_i$. As in the continuous form, vectors $T_i$, $N_i$, $B_i$, curvature $k_i$, and torsion $\tau_i$ are calculated from the 3D curve $\{X_i\}$, using the following equations. According to equation (1) and the fact that $T_i$, $N_i$, $B_i$ are unit vectors,

$$T_i = \frac{dX_i}{|dX_i|}$$
$$N_i = \frac{dT_i}{|dT_i|} \tag{10}$$
$$B_i = T_i \times N_i$$

Derived from equation (2), or $dT = kN$, $dN = -kT + \tau B$, $dB = -\tau N$ in discrete form, and that $N_i$ and $B_i$ are unit vectors, curvature $k_i$ and torsion $\tau_i$ are calculated by

$$k_i = |dT_i|, \tag{11}$$

$$\tau_i = |dN_i + k_i T_i| = |-dB_i|. \tag{12}$$

Conversely, the 3D curve $\{X_i\}$ can be obtained from a sequence of curvatures $\{k_i\}$ and torsions $\{\tau_i\}$, by first calculating the vectors $\{T_i\}$, $\{N_i\}$, $\{B_i\}$ sequentially from $\{k_i\}$ and $\{\tau_i\}$, and then $\{X_i\}$ from $\{T_i\}$. First, the vectors on one end of the curve are set as orthogonal unit vectors, for example,

$$[\boldsymbol{T}_0 \quad \boldsymbol{N}_0 \quad \boldsymbol{B}_0] = \begin{bmatrix} 1 & 0 & 0 \\ 0 & 1 & 0 \\ 0 & 0 & 1 \end{bmatrix}. \tag{13}$$

Then, using curvature, torsion, and the vectors on the first point, the vectors on the second point are calculated as

$$\begin{cases} \boldsymbol{T}_1 = \boldsymbol{T}_0 + d\boldsymbol{T}_0 = \boldsymbol{T}_0 + k_0 dx \boldsymbol{N}_0 \\ \boldsymbol{N}_1 = \boldsymbol{N}_0 + d\boldsymbol{N}_0 = \boldsymbol{N}_0 - k_0 dx \boldsymbol{T}_0 + \tau_0 dx \boldsymbol{B}_0, \\ \boldsymbol{B}_1 = \boldsymbol{T}_1 \times \boldsymbol{N}_1 \end{cases} \tag{14}$$

by substituting $\boldsymbol{T}'(s)$, $\boldsymbol{N}'(s)$, $\boldsymbol{B}'(s)$ with $d\boldsymbol{T}_i$, $d\boldsymbol{N}_i$, $d\boldsymbol{B}_i$ in equation (2). $dx$ is the distance between two consecutive sample points, which we set as constant in this work. The vectors on the following points are calculated as

$$\begin{cases} \boldsymbol{T}_i = \boldsymbol{T}_{i-2} + 2d\boldsymbol{T}_{i-1} = \boldsymbol{T}_{i-2} + 2k_{i-1} dx \boldsymbol{N}_{i-1} \\ \boldsymbol{N}_i = \boldsymbol{N}_{i-2} + 2d\boldsymbol{N}_{i-1} = \boldsymbol{N}_{i-2} - 2k_{i-1} dx \boldsymbol{T}_{i-1} + 2\tau_{i-1} dx \boldsymbol{B}_{i-1}, \\ \boldsymbol{B}_i = \boldsymbol{T}_i \times \boldsymbol{N}_i \end{cases} \tag{15}$$

considering that the derivatives $d\boldsymbol{T}_i$, $d\boldsymbol{N}_i$, $d\boldsymbol{B}_i$ are calculated using three points as in equation (9). Figure 3 (c) shows $\boldsymbol{T}_i$ derived from $\boldsymbol{T}_{i-2}$ and $d\boldsymbol{T}_{i-1} = k_{i-1} dx \boldsymbol{N}_{i-1}$, using equation (15). At last, sample points on the curve are calculated as follows.

$$\boldsymbol{X}_i = \boldsymbol{X}_{i-1} + dx \frac{\boldsymbol{T}_i + \boldsymbol{T}_{i-1}}{|\boldsymbol{T}_i + \boldsymbol{T}_{i-1}|}. \tag{16}$$

Similarly, sample points of the 2D curve $\{\boldsymbol{X}_{2D_i}\}$ are calculated from the sequence of the 2D curvatures $\{k_{2D_i}\}$ as follows.

$$[\boldsymbol{T}_{2D_0} \quad \boldsymbol{N}_{2D_0}] = \begin{bmatrix} 1 & 0 \\ 0 & 1 \end{bmatrix}, \tag{17}$$

$$\begin{cases} \boldsymbol{T}_{2D_1} = \boldsymbol{T}_{2D_0} + k_{2D_0} dx \boldsymbol{N}_{2D_0} \\ \boldsymbol{N}_{2D_1} = [-\boldsymbol{T}_{2D_1} y \quad \boldsymbol{T}_{2D_1} x]^{T'} \end{cases} \tag{18}$$

$$\begin{cases} \boldsymbol{T}_{2D_i} = \boldsymbol{T}_{2D_{i-2}} + 2k_{2D_{i-1}} dx \boldsymbol{N}_{2D_{i-1}} \\ \boldsymbol{N}_{2D_i} = [-\boldsymbol{T}_{2D_i} y \quad \boldsymbol{T}_{2D_i} x]^{T} \end{cases} \tag{19}$$

$$\boldsymbol{X}_{2D_i} = \boldsymbol{X}_{2D_{i-1}} + dx \frac{\boldsymbol{T}_{2D_i} + \boldsymbol{T}_{2D_{i-1}}}{|\boldsymbol{T}_{2D_i} + \boldsymbol{T}_{2D_{i-1}}|} \tag{20}$$

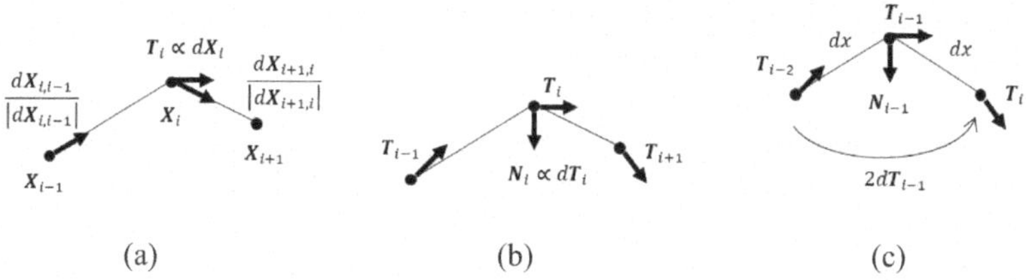

**Figure 3:** *Derivatives and cumulative vector calculation in discrete expression.*

## 3.3 Rectification of Torsion and Folding Angle

The mathematical framework explained in section 3.1 assumes the 2D and 3D folding curve to have non-zero curvature, as the curvature $k(s)$ is in the denominator of equations (4) and (5). In other words, the folding curve is not supposed to include a straight line. However, in practice, users may want to design a curve which includes a point with zero curvature. One example of such curve is shown in Figs. 4 and 5, where a curve changes from clockwise to counter-clockwise as we follow the curve in the direction of tangent vector. Since the curvature is positive in clockwise curve, and negative in counter-clockwise curve, as described in section 3.1, at the changing point the curvature becomes zero. At this point, as can be seen in equations (4) and (5), any non-zero value of torsion $\tau$ or difference of folding angle $\alpha'$ causes $\cot\beta_L$ and $\cot\beta_R$ to have infinite values, and $\beta_L$ and $\beta_R$ to be either 0 or $\pi$, which means that the ruling is parallel to the tangent vector, as can be seen in Fig. 4(a) and Fig. 5(a). On the other hand, observing paper in the real world, torsion is always zero and the folding angle is constant at zero curvature point. This is because a sheet of paper is a developable surface which can be flattened onto a plane without distortion. Non-zero torsion at the zero curvature point means that the surface twists along a straight fold line, which requires the surface to stretch at the point far from the line (Fig. 6(a)). A straight fold line with increasing or decreasing folding angle also has the same property (Fig. 6(b)). These surfaces are ruled surfaces but are not developable.

In our system, however, the user input is not restricted to meet this property, as it is not effective enough to impose constraints only on the control points. Instead, to keep this restriction, we propose a rectification process for torsion $\{\tau_i\}$ and folding angle $\{\alpha_i\}$ to restrict the set of parameters on all sample points on the curve. Figure 4 shows the 2D and 3D shapes of a curved fold and a graph indicating the value of the folding angle along the curve, before and after the rectification. The 3D shapes are constructed using the method described in section 4.1.3. They are composed of quads placed between the 3D rulings whose directions are obtained by equations (6) and (7) and their lengths are set to end at the paper edge. In case of rulings crossing, the polygon model is not developable as one point of a 2D plane corresponds to more than one point on the 3D polygon mesh. As shown in the graph, the folding angle is rectified to be constant in section (i), where the curvature is smaller than a threshold. In section (ii), which is adjacent to section (i), the graph is interpolated by Bezier spline curve to smoothly meet the original point of the graph on both ends. Section (ii) range from the end of section (i) to either a local maximum or minimum of the folding angle or a midpoint of two zero curvature points if there is any. Otherwise, section (ii) range to the end of the curve. This process shifts the change in folding angle from section (i) to section (ii).

Figure 5 shows the rectification of torsion. The graph in the top row indicates $\tau(s)/k(s)$ and in the bottom row shows $\tau(s)$. As given in equations (4) and (5), $\tau(s)/k(s)$ is proportional to $\cot \beta_L(s)$ , $\cot \beta_R(s)$ given constant folding angle $\alpha$. This means that making this graph smooth leads to the ruling direction changing gradually along the curve, which generates a smooth curved surface. In the rectification process, $\tau$ is set to zero in section (i), around zero curvature point, which also sets $\tau(s)/k(s)$ to be zero. Then, in section (ii), $\tau(s)/k(s)$ is interpolated by Bezier spline curve to keep the graph smooth. The ranges of section (i) and (ii) are set in the same method as in folding angle rectification, but using $\tau(s)/k(s)$ instead of the folding angle to obtain the end of section (ii). $\tau(s)$ is then normalized so that the total amount of torsion on the curve is the same before and after the rectification, expecting the curve end to be close to the original point before the rectification. These processes change the shape of the 3D curve and the adjacent curved surface to make the whole surface developable, as a sheet of paper in the real world. Also, as shown in Fig.7, the system eliminates the rulings around the zero curvature point because it is not adequate to define rulings on such planar area, and they cannot be calculated using equations (4) and (5).

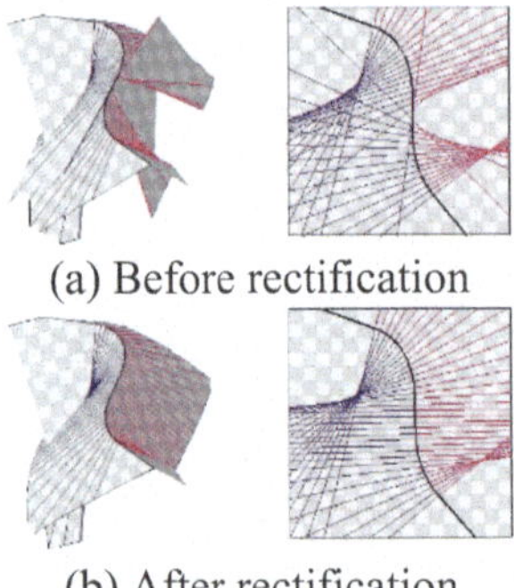

(a) Before rectification

(b) After rectification

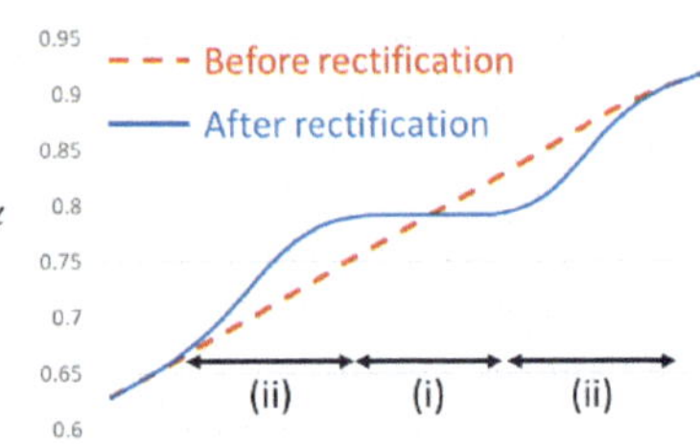


(c) Graph of folding angle before and after rectification.

**Figure 4:** *Folding angles before and after rectification. In section (i) of (c), folding angle is rectified to be constant, and in section (ii), it is interpolated by Bezier curve.*

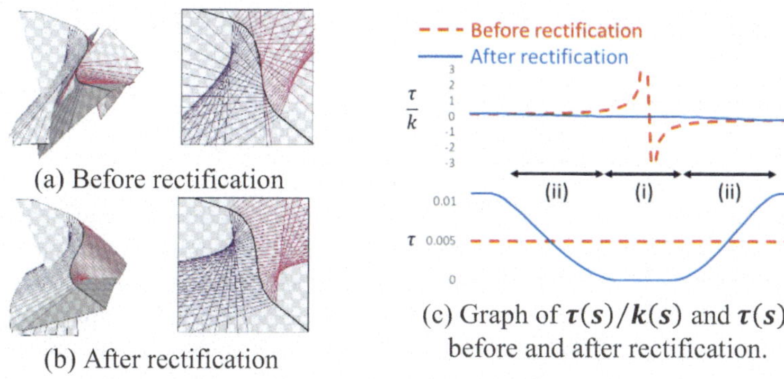



**Figure 5:** *Torsion before and after rectification. In section (i) of (c), torsion is set to be zero and in section (ii), it is interpolated by Bezier curve.*

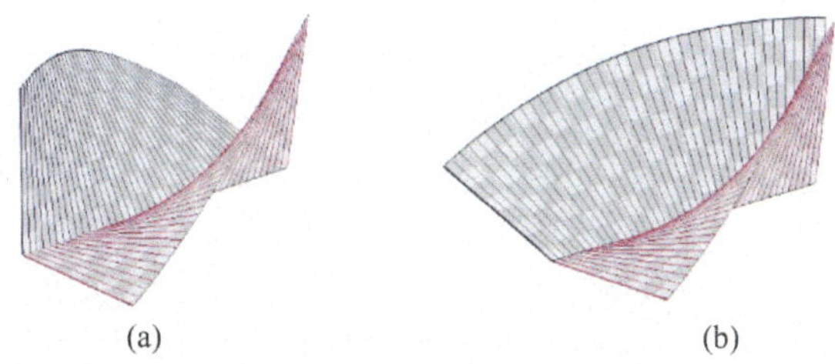



**Figure 6:** *Examples of ruled and non-developable surfaces. (a) Surface twisting along a straight fold line. (b) Surface with increasing folding angle at straight fold line.*

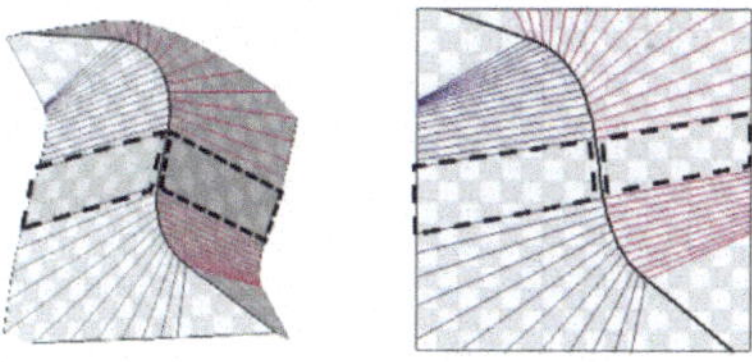

**Figure 7:** *Elimination of rulings. Rulings are eliminated in the dotted area where the curvature of the crease is close to zero.*

# 4   Shape and Motion Design of Developable Surface

## 4.1   Designing the Shape of a Curved Fold

As mentioned in section 3.1, the shape of developable surface with a curved fold may be derived given two of the three elements: 3D curve, 2D curve, and folding angle, using the mathematics in section 3. In our system, the shape of the curve is derived by curvatures and torsions of some control points on the curve. The parameter values on the control points are set and adjusted by the user through GUI, described in 4.1.1, which are then interpolated to all the sample points, as explained in 4.1.2. As the user changes the parameter value on a control point, the ruling directions and the shape of the surface are updated instantly, so that the user can adjust the parameters according to the feedback. The user may need to adjust different parameters on different control points alternately to create the desired shape that is developable and has no rulings crossing. The user may also trim the surface to eliminate the area where the crossings of the rulings occur. The following subsections describe the GUI and the process to update the surface shape according to the user input. The overall process is shown in Fig. 8.

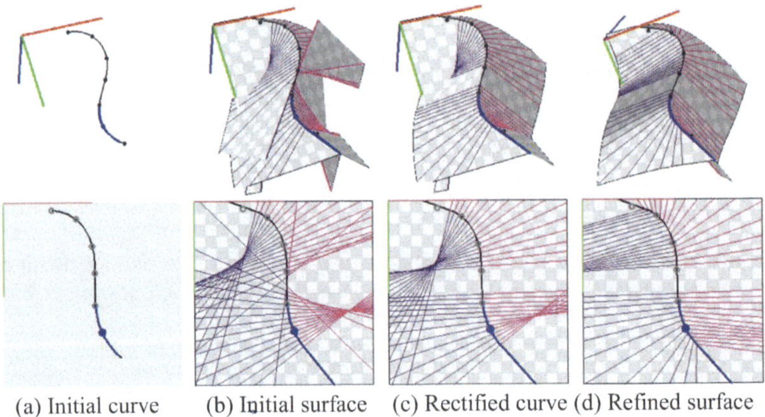

(a) Initial curve     (b) Initial surface     (c) Rectified curve (d) Refined surface

**Figure 8:** *The process of designing a curved fold.*

### 4.1.1 GUI System

Our GUI system shows the developable surface in 3D space, 2D crease pattern, and control panel with parameter values along the curve shown in the graphs (Fig. 9). Through the control panel and the keyboard input, the user sets and adjusts the parameter values on the control points on the crease curve. The types of parameters adjusted by the user are curvature $\{k_i\}$ and torsion $\{\tau_i\}$ of the 3D curve, curvature of 2D curve $\{k_{2D_i}\}$, and folding angle $\{\alpha_i\}$. The control points are placed on the curve at even intervals. On the control panel, the user specifies the "control mode" that decides which parameters to be calculated according to the other parameters in this operation. This mode corresponds to cases A, B, and C described in section 3.1. Then the user chooses which parameters to adjust and on which control point it should be changed. Other operations supported by the GUI system are activation and deactivation of the rectification process (section 3.3), folding and unfolding of the whole crease (section 4.2), trimming of the paper (section 4.1.5), and showing or hiding visual information. Through the mouse operation, the user may also change the location and orientation of the whole curve or change the paper size.

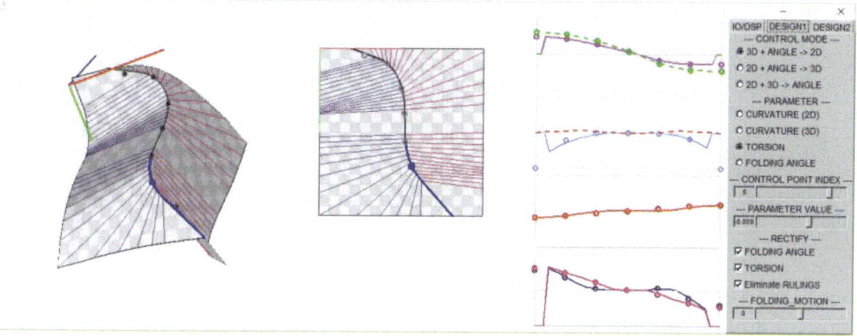

**Figure 9:** *Screenshot of the GUI in our system. From left to right, 3D shape, 2D crease, graphs of parameter values, and control panel. Parameters shown in the top row: curvature of 2D curve (pink) and 3D curve (dotted green); in the second row: torsion (blue) and derivative of folding angle (dotted red); in the third row: folding angle (red), and in the bottom row: ruling directions in 2D space (pink and purple, in the same colours as the rulings in 3D and 2D figures). The small circles on the graph and the figures indicate the control points.*

## 4.1.2 Derivation of curves and rulings

Given the parameter values on the control points set by the user, the shapes of the curves and ruling directions are derived. The steps in this process are explained below, and the parameters and elements processed in each step are summarized in Table 1 for cases A, B, and C described in section 3.1.

Step 1. Values of the given parameters on the control points are interpolated throughout the curve by spline interpolation. This makes the graph, or the arc-length parametric functions, of the parameters to be smooth, and the ruling directions to change gradually along the curve. Rectification of torsion and folding angle, described in section 3.3, is applied at this stage if needed.

Step 2. From the given parameters, 3D curve $\{X_i\}$ and/or 2D curve $\{X_{2D_i}\}$, are reconstructed, using equations (13) to (16) and equations (17) to (20). After the curve reconstruction, the parameters and vectors are recalculated to avoid the accumulation of numerical errors, using equations (9) to (12) or their 2D version.

Step 3. From given parameters on every sample point, the parameters to be derived are updated using equation (3).

Step 4. In cases A and B, 3D curve $\{X_i\}$ or 2D curve $\{X_{2D_i}\}$, are reconstructed using the derived parameters, followed by the recalculation of the parameters.

Step 5. The rulings are calculated by equations (4) to (7).

*Table 1: Parameters and elements processed in each step.*

| | | Case A | Case B | Case C |
|---|---|---|---|---|
| | Input elements | 3D curve<br>Folding angle | 2D curve<br>Folding angle | 2D curve<br>3D curve |
| | Derived element | 2D curve | 3D curve | Folding angle |
| Step 1 | Interpolated parameters | $\{k_i\},\{\tau_i\},\{\alpha_i\}$ | $\{k_{2D_i}\},\{\alpha_i\}$ | $\{k_i\},\{\tau_i\},\{k_{2D_i}\}$ |
| | Rectified parameters | $\{\tau_i\},\{\alpha_i\}$ | $\{\alpha_i\}$ | $\{k_i\}$ |
| Step 2 | Reconstructed curve | $\{X_i\}$ | $\{X_{2D_i}\}$ | $\{X_i\},\{X_{2D_i}\}$ |
| | Recalculated parameters and vectors | $\{k_i\},\{\tau_i\}$<br>$\{T_i\},\{N_i\},\{B_i\}$ | $\{k_{2D_i}\},$<br>$\{T_{2D_i}\},\{N_{2D_i}\}$ | $\{k_i\},\{\tau_i\},\{k_{2D_i}\}$<br>$\{T_i\},\{N_i\},\{B_i\},$<br>$\{T_{2D_i}\},\{N_{2D_i}\}$ |
| Step 3 | Derived parameters | $\{k_{2D_i}\}$ | $\{k_i\}$ | $\{\alpha_i\}$ |
| Step 4 | Reconstructed elements | $\{X_{2D_i}\}$ | $\{X_i\}$ | - |
| | Recalculated parameters and vectors | $\{k_{2D_i}\},$<br>$\{T_{2D_i}\},\{N_{2D_i}\}$ | $\{k_i\},\{\tau_i\}$<br>$\{T_i\},\{N_i\},\{B_i\}$ | - |
| Step 5 | Rulings calculated | $\{\beta_{L_i}\},\{\beta_{R_i}\}$ | $\{\beta_{L_i}\},\{\beta_{R_i}\}$ | $\{\beta_{L_i}\},\{\beta_{R_i}\}$ |

### 4.1.3 Making polygon shapes

After all curves and rulings are derived, developable surface is constructed as two quad strips adjacent to the fold curve. Each quad in the strips is composed of four edges: a crease, two rulings, and the boundary of the paper. If the fold curve lies across the paper edge, the part of the curve outside the paper is cut off. This is done by locating a new vertex on the intersection of the curve and the paper edge, and placing a new quad or a triangle on its sides. If the end of the curve is located within the paper boundaries, the curve is extended to the paper edge as a straight line, and flat polygons with no rulings are placed on its sides.

In our GUI system, the user can trim the paper on a free form curve input by the mouse drag (Fig. 10). This operation can eliminate the undesirable areas of the paper, such as crossings of the rulings.

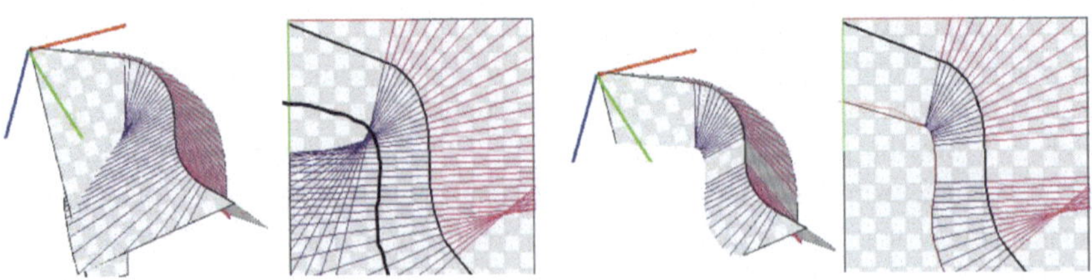

**Figure 10:** *Trimming of the paper.*

## 4.2 Simulation of Folding Motion

Now the user can simulate the folding and unfolding motion of the developable surface designed by the procedure in section 4.1, as shown in Fig. 11. To unfold the paper, folding angle $\{\alpha_i\}$ and torsion $\{\tau_i\}$ on all sample points are changed linearly from the designed values to zero, while 2D curve is fixed. The rest of the parameters are recalculated in the same method as section 4.1.2, and polygon shapes are updated as in section 4.1.3. To make the folding motion, folding angle $\{\alpha_i\}$ are linearly changed from the designed value to $\pi/2$, 2D curve being fixed, and other parameters and polygon shapes are updated. In both folding and unfolding, crossing of the rulings may occur at some point of the motion, as a result of ruling transition.

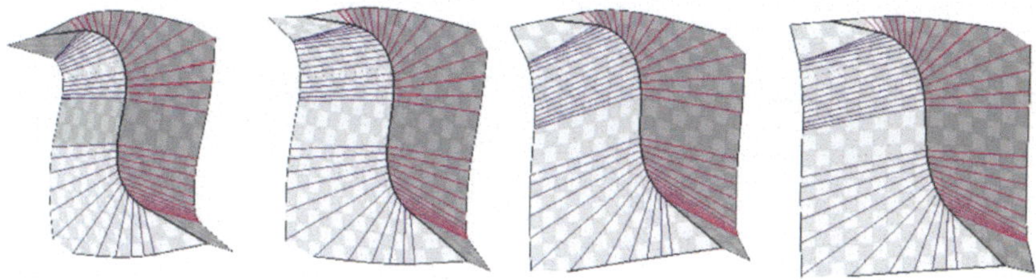

**Figure 11:** *Folding/unfolding motion of curved fold.*

# 5   Results and Discussion

As an evaluation of our system, visual comparison with real paper and the numerical check of its developability were performed. The results of visual comparison are shown in Fig. 12. For numerical check, (A) sum of the corner angles adjacent to a vertex and (B) the flatness of each quad were evaluated to check the developability of the curved surface. As to (A), the sum of the corner angles was expected to be $2\pi$ for all sample points, if the surface was perfectly developable. So we evaluated the error at sample point $X_i$ as

$$Error_i = 2\pi - \left\{\angle\left(T_i, r_{L_i}\right) + \angle\left(T_i, r_{R_i}\right) + \angle\left(-T_{i-1}, r_{L_i}\right) + \angle\left(-T_{i-1}, r_{R_i}\right)\right\},$$

where $r_{L_i}$ and $r_{R_i}$ are discrete version of $r_L$ and $r_R$ in equations (6) and (7). Table 2 shows the average and the maximum error of all sample points for the three examples of the curved folds in Fig. 12. As for (B), the flatness of a quad is calculated as the distance between one vertex of the quad and the plane passing through the other three vertices. This value should be zero for perfectly flat quads. Table 3 shows the average and maximum value for all quads placed between two rulings. From observation of the results, both types of errors are small enough to conclude that the 3D polygon model represents a developable surface. From the result of (A), the Gaussian curvature is close to zero on all the sample points on the crease curve, and with the result of (B) that all quads adjacent to the curve is flat, the Gaussian curvature is also close to zero on all points of the quads.

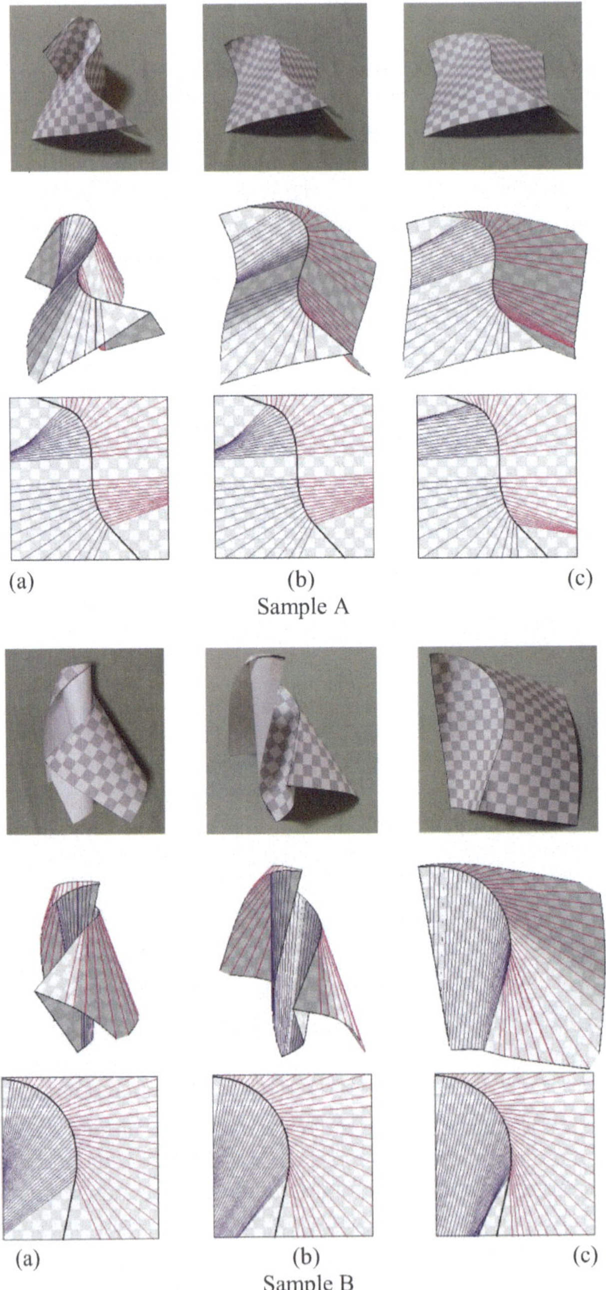


**Figure 12:** *Comparison with real paper.*

**Table 2:** *Errors in sum of corner angles. (Units: degrees).*

|  | Sample A | | | Sample B | | |
|---|---|---|---|---|---|---|
|  | (a) | (b) | (c) | (a) | (b) | (c) |
| Average | 0.0096 | 0.0028 | 0.00069 | 0.0053 | 0.0052 | 0.00067 |
| Maximum | 0.030 | 0.0079 | 0.0027 | 0.026 | 0.041 | 0.0053 |

**Table 3:** *Flatness of quads. (Units: mm. The length of the paper edge is 200mm).*

|  | Sample A | | | Sample B | | |
|---|---|---|---|---|---|---|
|  | (a) | (b) | (c) | (a) | (b) | (c) |
| Average | 0.036 | 0.017 | 0.0068 | 0.0079 | 0.0071 | 0.0033 |
| Maximum | 0.17 | 0.061 | 0.027 | 0.028 | 0.028 | 0.016 |

## 6 Conclusions

We have proposed a user interface system to design curved folding with a single curved crease on a sheet of paper and to simulate its folding motion. The system supports the transition of ruling by updating the ruling directions according to the geometric relationship between the curves, folding angle, and the ruling directions. It allows to create a curve with a zero curvature point, which the above relationship does not support, by rectifying the torsion to be zero and the folding angle to be constant around the point. The system was evaluated by visual comparison with real paper and by numerical verification of developability to show its usability and reliability.

As future work, we aim to develop a method to design and simulate multiple curved creases, with a framework to design and simulate creases affecting each other in the shape of the curve, folding angle, and ruling directions. As to our current system, the parameter rectification process may be improved by better interpolation rules or by refining Bezier interpolation. This part should be discussed in consideration with the behaviour of real paper. Also, in the simulation of the folding motion, the changing values of the folding angle and torsion on each point of the curved crease should be controlled, rather than just a linear change, to generate more complex and variated folding motions.

## References

[Demine et al. 14] Eric D. Demine, Martine L. Demine, David A. Huffman, Duks Koschitz, Tomohiro Tachi. "Characterization of Curved Creases and Rulings: Design and Analysis of Lenz Tessellations." Origami 6: The Sixth International Conference on Origami in Science, Mathematics, and Education (2014) 209–230.

[Fuchs and Tabachnikov. 99] Dmitry Fuchs, Serge Tabachnikov. "More on Paperfolding". The American Mathematical Monthly, Vol. 106, No.1, pp. 27-35, 1999..

[Kilian et al. 08] Martin Kilian, Simon Floery, Zhonggui Chen, Niloy J. Mitra, Alla Sheffer, Helmut Pottmann. "Curved Folding." ACM Transactions on Graphics 27:3 (2008), 75:1-9.

[Narain et al. 13] Rahul Narain, Tobias Pfaff, and James F. O'Brien. "Folding and Crumpling Adaptive Sheets". ACM Transactions on Graphics, 32(4):51:1–8, July 2013. Proceedings of ACM SIGGRAPH 2013, Anaheim.

[Tachi 09] Tomohiro Tachi. "Simulation of rigid origami." Origami 4: The Fourth International Conference on Origami in Science, Mathematics, and Education (2009) 175–187.

[Tachi 11] Tomohiro Tachi. "One-DOF Rigid Foldable Structures from Space Curves." in Proceedings of the IABSE-IASS Symposium 2011, London, UK, September 20-23, 2011.

[Tachi 13] Tomohiro Tachi. "Composite Rigid-Foldable Curved Origami Structure." in Proceedings of Transformables 2013, Seville, Spain, Sep.18-20, 2013.

[Tang et al. 16] Chengcheng Tang, Pengbo Bo, Johannes Wallner, Helmut Pottmann. "Interactive design of developable surfaces." ACM Transactions on Graphics, 35(2), 2016, 12:1-12.

Yuka Watanabe
Yuka Watanabe, University of Tsukuba, e-mail: yukakohno@hotmail.com

Jun Mitani
Jun Mitani, University of Tsukuba, e-mail: mitani@cs.tsukuba.ac.jp

# Fast, Interactive Origami Simulation using GPU Computation

*Amanda Ghassaei, Erik D. Demaine, Neil Gershenfeld*

**Abstract**:

*We present an explicit method for simulating origami that can be rapidly computed on a Graphics Processing Unit (GPU). Previous work on origami simulation methods model the geometric or structural behaviors of origami with a focus on physical realism; in this paper we introduce a compliant, explicit numerical simulation method that emphasizes computational speed and interactivity. We do this by reformulating existing techniques for simulating origami so that they can be computed on highly parallel GPU architectures. We implement this method in an open-source, GPU-accelerated WebGL app that runs in any modern browser. We evaluate our method's performance, stability, and scalability to existing methods (Freeform Origami, MERLIN) and demonstrate its capacity for real-time interaction through a traditional GUI and immersive virtual reality.*

## 1   Introduction

Physical simulations of origami allow designers to better understand how modifications to a crease pattern affect its folded state. Real-time simulation-based feedback can provide more intuitive ways to edit a folded pattern or to enforce desired geometric constraints, such as global developability. In particular, simulation provides an approach to inverse design, where computational tools generate valid designs according to high-level user-specified goals. An alternative to algorithmic inverse design (as in e.g. [Lang 96, Demaine and Tachi 17]) is to develop interactive design tools coupled with tightly integrated simulation capability [Tachi 10, Tang et al. 16]. This work aims to create a fast, interactive simulation environment that could serve as the backbone for new computational design tools for origami.

## 2   Simulation Methods

Numerical methods for simulating origami typically discretize an origami crease pattern into a rigid-body linkage or finite element mesh and implicitly solve a linear system of equations for small displacements. Through an iterative process, these methods compute large, non-linear deformations of origami from its initially flat or folded state. Rigid origami simulators [Tachi 06] model the rigid motions of origami with an emphasis on kinematic accuracy. Finite element methods (FEM) explore the structural properties of the origami's folded or partially folded states using realistic material models and higher-order elements such as plates/shells [Schenk et al. 14, Peraza Hernandez et al. 16], or volumetric elements [Ablat and Qattawi 18].

Bar-and-hinge models are an efficient modeling abstraction for approximating the mechanical behavior of origami [Tachi 10, Liu and Paulino 16]. Despite their simplicity, bar-and-hinge models can be used to capture realistic behaviors, such as out-of-plane bending [Tachi 13, Schenk and Guest 11] and in-plane shearing [Filipov et al. 17] of regions between creases.

The purpose of our solver is not to model physical behavior that has not been previously explored, but rather to rapidly compute the folded state of origami for applications with direct user interaction. Our solver reframes prior implicit methods into a dynamic, explicit form that is solved in a highly parallel process on a GPU. We use compliant constraints to guide the folding of the origami model, allowing the user to trade off computational speed for accuracy (or vice versa).

Sections 2.1–2.4 outline the setup for the method (based on prior work by [Schenk and Guest 11] and [Tachi 10], repeated here for clarity); Section 2.5 introduces our explicit integration method; and Section 3 discusses the parallelization of this method on the GPU.

## 2.1 Meshing

First we discretize the origami crease pattern into a triangulated mesh. We triangulate the crease pattern by adding extra edges across any polygonal faces with more than three sides (Figure 1). Following [Schenk and Guest 11], we call these edges "facet creases" and elastically constrain them to remain flat while folding.

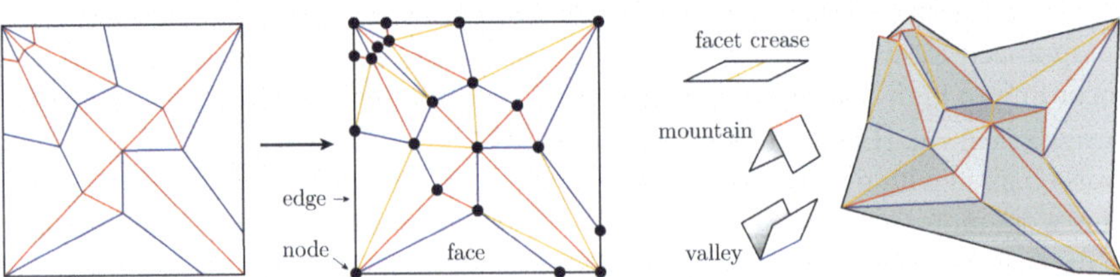


**Figure 1:** *Triangulation of an origami crease pattern (left) into a triangulated mesh (center). Polygonal faces may be triangulated in an automated process, or controlled directly by the user. (right) Unlike mountain and valley creases, facet creases formed by triangulation are driven flat by elastic constraints during folding.*

Each of the edges in the resulting mesh is modeled as a beam in a pin-jointed truss, subject to axial and angular constraints, which we now detail.

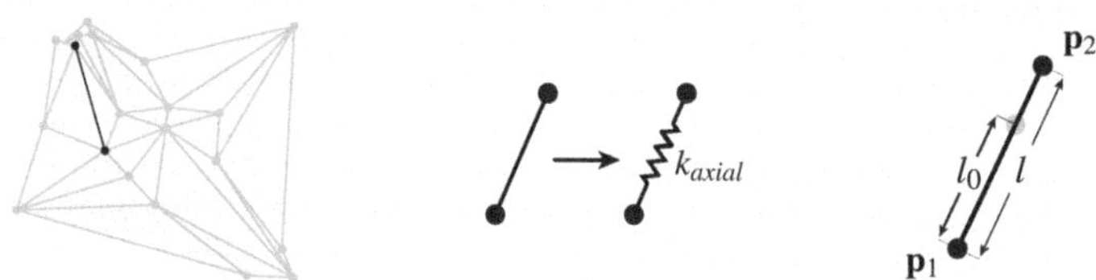

**Figure 2:** *(left) A crease pattern is triangulated and simulated as a pin-jointed truss network. (center) Beams in the truss are modeled as linear-elastic springs. (right) Formulation of axial constraints.*

## 2.2  Axial Constraints

Axial constraints prevent stretching and compression of the origami surface during folding. Each beam element behaves like a linear-elastic spring that applies forces only in its axial direction (Figure 2). At each step of the simulation, we calculate the length of each beam from the position of the nodes at its endpoints. We compute the force exerted on each node in the beam's local coordinate space by Hooke's Law:

$$F_l = -k_{axial}(l - l_0),$$

where $F_l$ is the force along the beam's axis, $k_{axial}$ is the axial stiffness of the beam element, $l$ is the length of the beam, and $l_0$ is the nominal length of the beam.

We must convert $F_l$ into a vector $\mathbf{F}_{axial}$ in global coordinate space before we apply it to the nodes. This axial force vector is related to node position $\mathbf{p}$ by

$$\mathbf{F}_{axial} = -\nabla V(\mathbf{p}) = -\frac{\partial V}{\partial \mathbf{p}},$$

where $V$ is the potential energy of the system. By the chain rule, we have

$$\mathbf{F}_{axial} = -\frac{\partial V}{\partial l}\frac{\partial l}{\partial \mathbf{p}} = F_l \frac{\partial l}{\partial \mathbf{p}},$$

$$\mathbf{F}_{axial} = -k_{axial}(l - l_0)\frac{\partial l}{\partial \mathbf{p}}. \tag{1}$$

For the two nodes attached to each beam, we have

$$\frac{\partial l}{\partial \mathbf{p}_1} = -\hat{\mathbf{l}}_{12}, \qquad \frac{\partial l}{\partial \mathbf{p}_2} = \hat{\mathbf{l}}_{12},$$

where $\hat{\mathbf{l}}_{12}$ is a unit vector from node 1 to node 2 and $\mathbf{p}_1$ is the 3D position of node 1 in global coordinate space.

Axial stiffness can be related to the material properties of the substrate by

$$k_{axial} = \frac{EA}{l_0},$$

where $E$ is Young's modulus and $A$ is the cross-sectional area of the beam. For this analysis, we consider only the relative stiffness of the axial and angular constraints and disregard their physical meaning. We choose constant $EA$ for our model and calculate $k_{axial}$ of each beam based on its $l_0$.

## 2.3 Crease Constraints

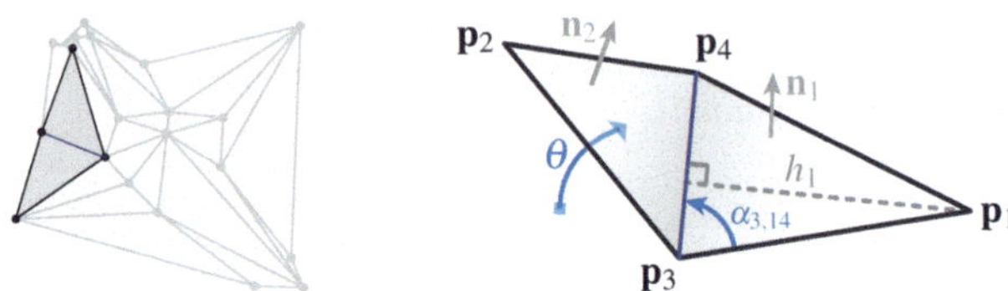

**Figure 3:** *Formulation of crease constraints.*

With distance constraints alone, we can model an arbitrary linkage connected by frictionless, spherical joints. We add constraints on the dihedral angle between triangulated faces of the mesh to drive and constrain folding.

The fold angle ($\theta$) of a crease is the supplement of the dihedral angle between two neighboring triangular faces. We model angular constraints as linear-elastic torsional springs that drive neighboring triangular faces toward some target fold angle. Similarly to Equation 1, angular constraints apply forces to neighboring nodes by

$$\mathbf{F}_{crease} = -k_{crease}(\theta - \theta_{target})\frac{\partial\theta}{\partial\mathbf{p}}, \tag{2}$$

where $\mathbf{F}_{crease}$ is a 3D force vector in global coordinate space, $\theta_{target}$ is the desired fold angle of the crease, $\mathbf{p}$ is the position of the node, and $k_{crease}$ is the stiffness of the constraint. Following prior work [Schenk and Guest 11], we choose $k_{crease}$ according to its type and scale it by its nominal length ($l_0$):

$$k_{crease} = \begin{cases} l_0 k_{fold} & \text{for a mountain or valley crease,} \\ l_0 k_{facet} & \text{for a facet crease (from Section 2.1),} \\ 0 & \text{for a boundary edge or undriven crease.} \end{cases}$$

We choose our stiffnesses so that $k_{axial} \gg k_{fold}$ and $k_{facet}$. Angle $\theta_{target}$ also depends on the type of crease:

$$\theta_{target} = \begin{cases} <0 & \text{for a mountain crease,} \\ >0 & \text{for a valley crease,} \\ 0 & \text{for a facet crease.} \end{cases}$$

The precise value of $\theta_{target}$ for mountain/valley creases is set by the user; see further Section 3.

Each angular constraint applies forces to four neighboring nodes, as indicated in Figure 3. The partial derivatives of $\theta$ with respect to $\mathbf{p}$ are given by

$$\frac{\partial \theta}{\partial \mathbf{p}_1} = \frac{\mathbf{n}_1}{h_1}, \tag{3}$$

$$\frac{\partial \theta}{\partial \mathbf{p}_2} = \frac{\mathbf{n}_2}{h_2}, \tag{4}$$

$$\frac{\partial \theta}{\partial \mathbf{p}_3} = \frac{-\cot \alpha_{4,31}}{\cot \alpha_{3,14} + \cot \alpha_{4,31}} \frac{\mathbf{n}_1}{h_1} + \frac{-\cot \alpha_{4,23}}{\cot \alpha_{3,42} + \cot \alpha_{4,23}} \frac{\mathbf{n}_2}{h_2}, \tag{5}$$

$$\frac{\partial \theta}{\partial \mathbf{p}_4} = \frac{-\cot \alpha_{3,14}}{\cot \alpha_{3,14} + \cot \alpha_{4,31}} \frac{\mathbf{n}_1}{h_1} + \frac{-\cot \alpha_{3,42}}{\cot \alpha_{3,42} + \cot \alpha_{4,23}} \frac{\mathbf{n}_2}{h_2}, \tag{6}$$

where $\mathbf{n}_1$ and $\mathbf{n}_2$ are face normals, $h_1$ and $h_2$ are lever arms from the crease to the outer nodes, and in-plane angles $\alpha$ are oriented according to Figure 3.

## 2.4 Face Constraints

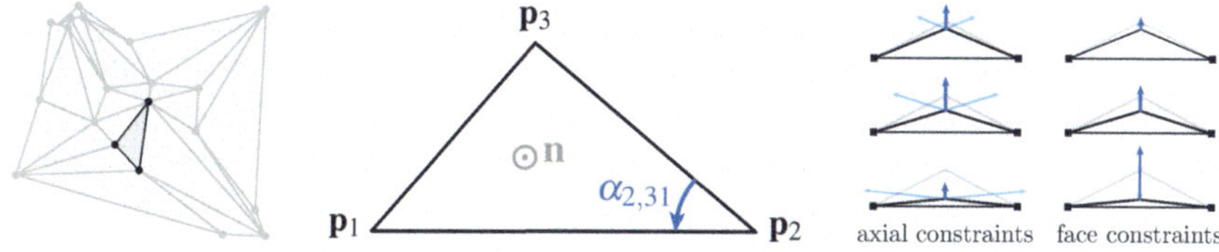


**Figure 4:** *(left and center) Formulation of face constraints. (right) As a triangular face (with two bottom nodes fixed) flattens, the restoring force applied to the center node by two neighboring axial constraints lessens the more the triangle is deformed. In contrast, three face constraints apply a greater force as the triangle deforms. Individual contributions shown in light blue, net force shown in dark blue, undeformed state shown in grey, fixed nodes indicated with squares.*

With beam and crease constraints alone, we have enough to simulate folding. Though not strictly necessary, the addition of constraints on each interior angle of the triangular faces of the mesh helps to prevent shearing of the folding surface, especially for high-aspect-ratio triangles (Figure 4). In practice, we have found that face constraints increase the stability of the simulation across a variety of input crease patterns. As with the previous constraints, we model this as a linear-elastic spring.

For each interior angle of a triangular face, we apply forces to its three neighboring nodes according to

$$\mathbf{F}_{face} = -k_{face}(\alpha - \alpha_0)\frac{\partial \alpha}{\partial \mathbf{p}},$$

where $\alpha$ is the current angle, $\alpha_0$ is its nominal angle in the flat state, $\mathbf{p}$ is the position of a neighboring node, and $k_{face}$ is the stiffness of the face constraint. The partial derivatives of $\mathbf{p}$ with respect to $\alpha$ are given by

$$\frac{\partial \mathbf{p}_1}{\partial \alpha_{2,31}} = \frac{\mathbf{n} \times (\mathbf{p}_1 - \mathbf{p}_2)}{\|\mathbf{p}_1 - \mathbf{p}_2\|^2},$$

$$\frac{\partial \mathbf{p}_2}{\partial \alpha_{2,31}} = -\frac{\mathbf{n} \times (\mathbf{p}_1 - \mathbf{p}_2)}{\|\mathbf{p}_1 - \mathbf{p}_2\|^2} + \frac{\mathbf{n} \times (\mathbf{p}_3 - \mathbf{p}_2)}{\|\mathbf{p}_3 - \mathbf{p}_2\|^2},$$

$$\frac{\partial \mathbf{p}_3}{\partial \alpha_{2,31}} = -\frac{\mathbf{n} \times (\mathbf{p}_3 - \mathbf{p}_2)}{\|\mathbf{p}_3 - \mathbf{p}_2\|^2},$$

where $\mathbf{n}$ is the normal vector of the triangular face and $\alpha$ is defined according to Figure 4.

## 2.5  Numerical Integration

So far, the governing equations we have described in the previous sections are a mix of the methods described by [Tachi 10] and [Schenk and Guest 11]. Our methods differ from prior work by calculating small displacements of the nodes under axial and angular constraints using an explicit method; these computations occur in parallel on a per-node basis, without a global stiffness/Jacobian matrix.

We calculate the total force on a node as the sum of the forces applied by neighboring beams, creases, and faces:

$$\mathbf{F}_{total} = \sum_{beams} \mathbf{F}_{beam} + \sum_{creases} \mathbf{F}_{crease} + \sum_{faces} \mathbf{F}_{face}.$$

We compute nodal acceleration by

$$\mathbf{a} = \frac{\mathbf{F}_{total}}{m},$$

where $\mathbf{a}$ is a 3D acceleration vector in global coordinate space and $m$ is the mass of the node. In this analysis, we assume the mass at each node is 1. A more accurate calculation of nodal mass would result in more realistic dynamics.

We calculate velocity and displacement at each node by forward Euler integration:

$$\mathbf{v}_{t+\Delta t} = \mathbf{v}_t + \mathbf{a}\Delta t,$$

$$\mathbf{p}_{t+\Delta t} = \mathbf{p}_t + \mathbf{v}_{t+\Delta t}\Delta t,$$

where $\Delta t$ is a small time step. In general, $\Delta t$ should be chosen so that it is small enough to keep the simulation numerically stable, but not too small that it slows down the simulation unnecessarily. We have chosen $\Delta t$ to satisfy

$$\Delta t < \frac{1}{2\pi \omega_{max}}, \tag{7}$$

where $\omega_{max}$ is the maximum natural frequency of any constraint in the model. Because we will always choose $k_{axial} \gg k_{fold}$, $k_{facet}$, and $k_{axial} > k_{face}$, we consider only axial constraints for this analysis:

$$\omega = \sqrt{\frac{k_{axial}}{m_{min}}}, \tag{8}$$

where $m_{min}$ is the minimum mass of the two nodes on either end of the beam. We assume the initial condition:

$$\mathbf{v}_0 = \mathbf{0}.$$

The above formulation may eventually converge to a static state due to numerical dissipation, but this may take more iterations than is practical. We introduce viscous damping between neighboring vertices in the mesh:

$$\mathbf{F}_{damping} = c(\mathbf{v}_{neighbor} - \mathbf{v}),$$

where $c$ is the viscous damping coefficient and $\mathbf{v}_{neighbor} - \mathbf{v}$ is the relative velocity between a node and its neighbor. We use this simplified approach to damping rather than calculating the damping force for every constraint in the system to minimize floating-point operations per simulation cycle. Additionally, we found that even with critical damping on all constraints, the structure was globally underdamped [Hiller and Lipson 14]. A more thorough approach to damping could result in more realistic dynamics.

We compute $c$ for each node according to the stiffest constraint in our system ($k_{axial}$):

$$c = 2\zeta \sqrt{k_{axial}m},$$

where $\zeta$ is the damping ratio and $m$ is the mass of the node (again, assumed to be 1 in this analysis). We've found that most patterns are stable with $0.01 \leq \zeta \leq 0.5$.

## 3  Implementation

We implemented this solver in an open-source,[1] interactive WebGL app that runs in any modern web browser.[2] We use the FOLD format of [Demaine et al. 16] for input/output and as an internal data structure to maximize interoperability with other origami software; we also support SVG input/output with opacity mapped to final fold angle and color mapped to crease type. We make few assumptions about the topology of the folded structure, allowing our solver to simulate origami, kirigami, discretized curved creases, underconstrained patterns with undriven hinges, patterns with holes, and 3D structures that do not have a flat state. We provide an interface to control simulation parameters in real time, such as stiffness and damping coefficients and the number of simulation steps per render cycle (Figure 6).

---

[1]https://github.com/amandaghassaei/OrigamiSimulator/
[2]http://apps.amandaghassaei.com/OrigamiSimulator/

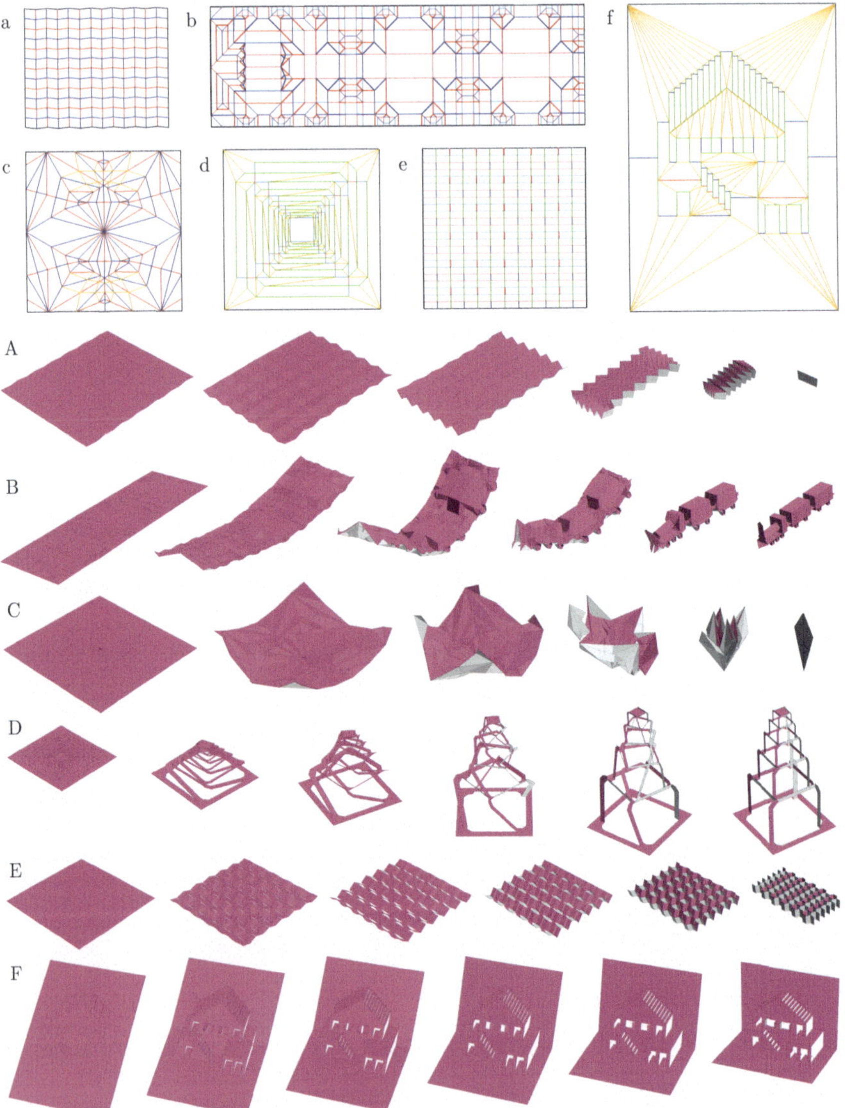

**Figure 5:** *Static solution of simulated origami at 0%, 20%, 40%, 60%, 80%, and 100% of target fold angle. Patterns depicted are (A) Miura-ori tessellation, (B) Mooser's Train (rigidly foldable design by William Gardner), (C) Robert Lang's Orchid Blossom, (D) Yoshinobu Miyamoto's RES Square Tower, (E) kirigami honeycomb [Saito et al. 14], and (F) pop-up house by Popupology. Corresponding crease patterns above with line opacity indicating final fold angle and color indicating crease type: mountain (red), valley (blue), border (black), cut (green), user-defined facet crease (orange). EA = 20, $k_{fold} = 0.7$, $k_{facet} = 0.7$, and $k_{face} = 0.2$.*

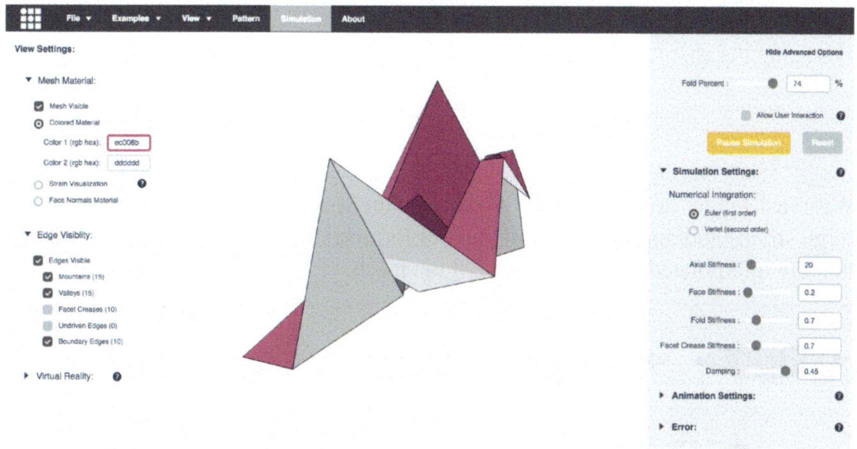

**Figure 6:** *Screenshot of Origami Simulator application.*

Before the solver begins, we precompute stiffness and damping coefficients as well as a lookup table of geometric relationships between elements. We run a series of GPU programs in WebGL fragment shaders to compute one step of the simulation:

1. Calculate face normals of all triangular faces in mesh (one face per thread).

2. Calculate current fold angle for all edges in mesh (one edge per thread).

3. Calculate coefficients of Equations 3–6 for all edges in mesh (one edge per thread).

4. Calculate forces and velocities for all nodes in mesh (one node per thread).

5. Calculate positions for all nodes in mesh (one node per thread).

In other GPU programming frameworks (e.g., NVIDIA CUDA), this solver could be implemented in fewer steps, but we chose to use WebGL shaders to maximize portability across operating systems and for easy browser support.

After a specified number of simulation steps, we render the geometry to the screen. Figure 5 shows a collection of simulated structures.

## 3.1   Strain Visualization

We implement a strain visualization tool to help users identify locally deformed regions in the mesh (Figure 7). We approximate the engineering strain across the surface of a folded origami structure by measuring the displacement of the axial constraints. For a given beam in the pin-jointed truss, we define engineering strain $\varepsilon$ by

$$\varepsilon = \frac{l - l_0}{l_0},$$

where $l$ is the length of the beam, and $l_0$ is the nominal length of the beam.

The magnitude of engineering strain at a node with $N$ beams is calculated by averaging the absolute value of the strain contribution from each beam:

$$\varepsilon_{node} = \frac{1}{N} \sum_{i=1}^{N} \frac{|\Delta l_i|}{l_i}.$$

We translate this strain into an RGB color on the spectrum of blue (no strain) to red (high strain) and apply it to the 3D model for visualization. The vertex colors are linearly interpolated across the faces of the mesh.

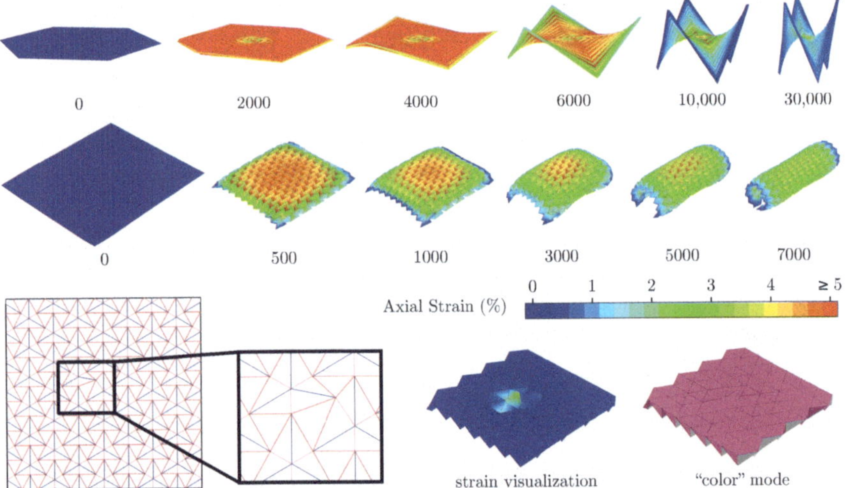


**Figure 7:** *Axial strain visualization of a hexagonal hypar variant (top, folded to $\pm 3\pi/4$ at all creases) and waterbomb tessellation (middle, folded to $\pm 3\pi/5$ at all creases). The sequences show our method solving for a fixed final fold angle in order of increasing time, with the number of simulation steps indicated below. Initially, both models move to a nearly planar state with high internal strain, which is slowly relaxed as they curl into the third dimension. (bottom) An irregular vertex in a Resch tessellation deforms the mesh slightly in material "color" rendering mode, but is clearly visible in the axial strain visualization. Simulation parameters for hypar variant: $EA = 100$, $k_{fold} = 0.7$, $k_{facet} = 0.7$, and $k_{face} = 1$, $\zeta = 0.45$. Simulation parameters for waterbomb and Resch tessellation: $EA = 20$, $k_{fold} = 0.7$, $k_{facet} = 0.7$, and $k_{face} = 0.2$, $\zeta = 0.45$.*

## 3.2   User Interaction

We introduce user interaction by modifying the boundary conditions of the simulation in real time. Compliance in the simulated structure allows users to toggle between stable states of bistable patterns (Figure 8A). Using the WebVR API, we are able to run our app on the HTC Vive and Oculus virtual reality (VR) headset and controllers (Figure 8B). In this mode, users can manipulate the origami with both hands and view the rendering in an immersive 3D environment. Future work will close the loop between design and simulation, allowing for interactive modifications to the folded structure.

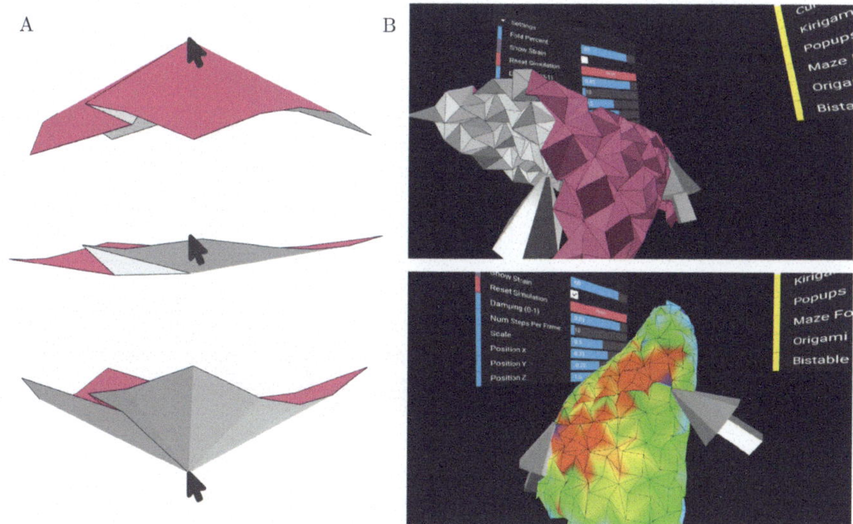

**Figure 8:** *(A) User interaction pushing bistable pleat model from one stable state to the other. (B) Virtual reality interface showing direct user interaction with folding Huffman waterbomb model. (B, bottom) With strain visualization turned on, users can pull and push on the mesh while visualizing strain in real time.*

## 4   Discussion

We performed a series of benchmarking tests to understand how the simulation speed of our solver scales with the number of nodes in the origami model (Figure 9). We ran these tests using the Chrome browser on two different GPUs: an Iris Graphics 6100 with 348 cores and a 300MHz clock, and two SLI-linked GeForce GTX 980's with 4096 combined cores and a 1.1GHz clock. We imported a series of $n \times n$ Miura-Ori tessellations of different dimensions into the solver and measured the simulation speed without rendering.

We found that the time to compute one cycle of the simulation was approximately constant while the number of nodes in the mesh ($N$) was less than the number of available GPU cores (indicated by the blue X's in Figure 9a). As $N$ grows beyond the available cores, we see approximately linear scaling of simulation time with $N$ as the GPU queues batches of threads and executes the batches in series.

### 4.1   Comparison with Existing Methods

We compare the speed of our solver to existing methods MERLIN [Liu and Paulino 17] and Freeform Origami [Tachi 10] (Figure 10). Both methods implicitly solve for small displacements of a triangulated origami mesh on the CPU.

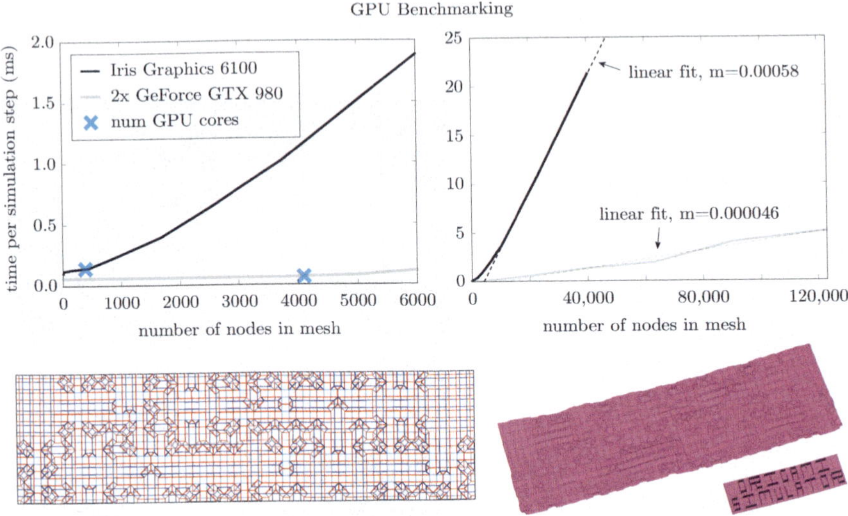


**Figure 9:** *Benchmarking time per simulation step as the number of nodes (N) in the mesh increases. (top left) When the number of nodes is smaller than the number of available GPU cores (indicated by the blue X's) the time to complete one simulation cycle is approximately constant. (top right) For very large meshes, simulation time scales approximately linearly with N. (bottom) Real-time simulation of a mesh with 2374 nodes.*

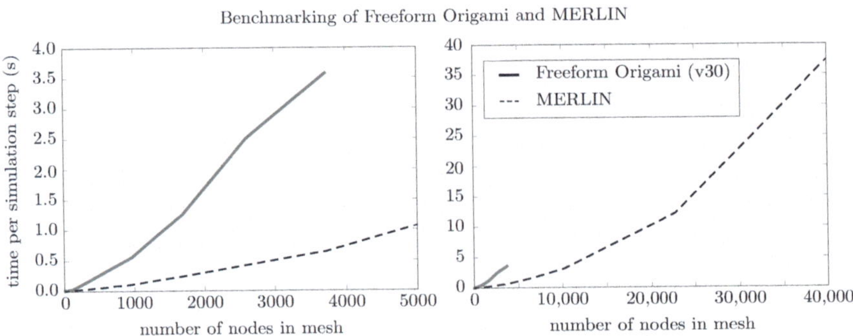


**Figure 10:** *Benchmarking solver iteration time versus mesh size in Freeform Origami and MERLIN.*

Each iteration of the **MERLIN** solver computes a global stiffness matrix and solves a system of equations that relates residual force to internal and external applied forces using the modified generalized displacement control method. Though the time to compute each of these steps is about 1000X longer than a single step in our method (Figure 10), MERLIN takes larger steps while maintaining stability for stiffer material settings. A $10 \times 10$ cell Miura-ori tessellation (121 nodes) reaches its final folded state in about 500 iterations of the MERLIN solver with material stiffnesses $k_{fold} = 0.1$, $k_{facet} = 10,000$, $EA = 100,000$, while about 150,000 iterations are needed using our method at these stiffer settings.

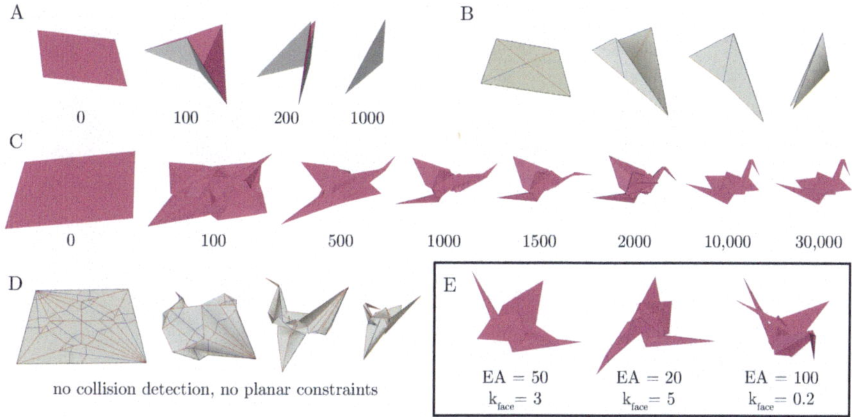

**Figure 11:** *Comparison of our method (magenta) and Freeform Origami (gray) [Tachi 10]. Unless otherwise indicated, simulation settings using our method are $EA = 20$, $k_{fold} = 0.7$, $k_{facet} = 0.7$, $k_{face} = 0.2$, and $\zeta = 0.45$, and the number of simulation steps is indicated below each image in the sequence. (A) Due to compliance in our axial and face constraints, some deformation of the mesh occurs as it tries to simultaneously fold a simply folded vertex; (B) by contrast, Freeform Origami more accurately captures the sequential nature of the folds. (C) The compliance of our method allows non-rigidly foldable designs like the crane to find their final folded state (allowing for some self-intersection). (D) Freeform Origami minimizes geometric error at each step of the simulation, effectively modeling infinitely stiff meshes. Even with collision detection and planar constraints (infinitely stiff facet creases) turned off, Freeform Origami is not able to correctly fold a crane. (E) Similarly, increasing the material stiffness in our solver prevents the crane from reaching the correct folded state.*

Freeform Origami is a rigid origami simulator (though it can also model elastic out-of-plane bending deformations of facets [Tachi 13]). The data points in Figure 10 were gathered from the FPS display in the app with face, edge, and vertex rending disabled and collision detection disabled. Each rendering cycle of Freeform Origami is comprised of many iterations of a conjugate gradient solver, which should be taken into consideration when comparing the benchmarking data. Figure 11 depicts the modeling differences between our method and Freeform Origami. Though Freeform captures the sequential folding of a simple vertex better than our method, Freeform's rigid constraints prevent it from folding non-rigidly foldable patterns like the crane; our method fails similarly when the axial and face stiffnesses are increased (Figure 11E).

In contrast with the quasi-static methods of Freeform Origami and MERLIN, the intermediate steps of our explicit solver depict the dynamics of the structural system. Though we have not taken much care to construct a realistic model of mass distribution and damping, our solver can be used to produce animations with plausible time-dependant behavior (Figure 12).

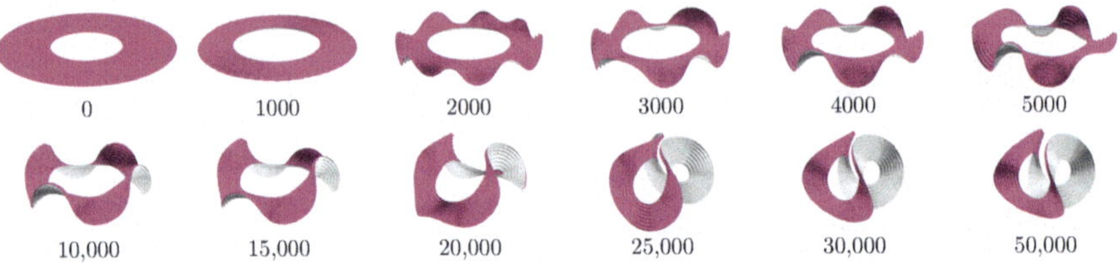

Figure 12: *Dynamic animation of a circular pleat pattern with iteration number indicated below. Simulation settings: target fold angle of all mountain/valley creases = $\pm 7\pi/10$, $EA = 20$, $k_{fold} = 0.7$, $k_{facet} = 0.7$, $k_{face} = 0.2$, and $\zeta = 0.2$.*

## 4.2   Limitations

According to Equations 7 and 8, there is a tradeoff between geometric accuracy (constraint stiffness) and simulation speed ($\Delta t$). Implicit formulations are advantageous in this regard because they can model very stiff systems while maintaining large step sizes, even though each step requires more computational effort to compute. To mitigate this, we provide easy access to stiffness parameters so that users may increase constraint stiffness when the model is near its static solution to avoid unnecessary computational time in the early stages of the simulation.

As is typical with other bar-and-hinge simulation methods, the behavior of the simulation is sensitive to the discretization of its mesh. This is especially prominent in the meshing of facets, where there may be more than one way to triangulate an arbitrary polygonal face (Section 2.1). If specific out-of-plane bending modes are desired, users may manually define facet creases before importing the designs into our app (Figure 13). Filipov et al. describe the effect of facet meshing on in-plane and out-of-plane deformations in bar-and-hinge models of origami in more detail [Filipov et al. 17].

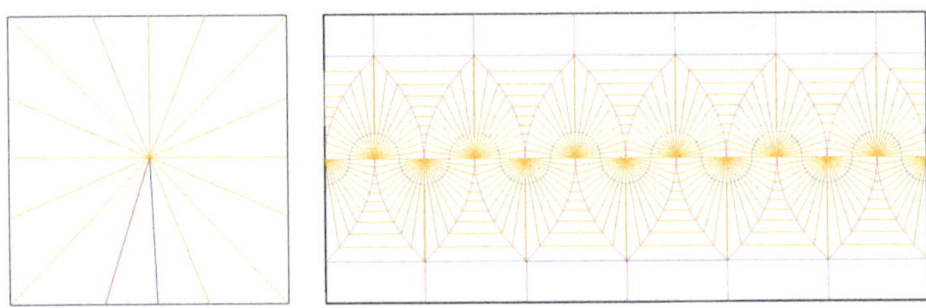

Figure 13: *User defined facet creases along rule lines of curved crease patterns guide expected out-of-plane bending of regions between mountain/valley creases.*

High-aspect-ratio triangles may pose a problem in simulation if they become so deformed that their face normal (defined by the cross product of two edges) flips orientation. This problem is not unique to our method, but may be exacerbated by the compliance of the constraints. This instability is mitigated by carefully chosen, user-defined facet creases or by increasing axial and face stiffnesses relative to fold stiffness.

Finally, our method requires a target fold angle for each crease as an input to the solver. Incorrect target fold angles may deform the final folded state, though this can usually be diagnosed with the strain visualization tool.

## 5  Future Work

Parallelization on the GPU is not unique to the explicit integration method introduced in Section 2.5 or to bar-and-hinge models in general. We chose to start with the methods outlined in this paper in part because of the simplicity of their implementation, but (as noted in Section 4.2) there are limitations to our current approach. In immediate future work, we plan to apply GPU parallelization to implicit formulations of bar-and-hinge models, so that we can model stiffer structural systems in a highly efficient manner. Through GPU acceleration, it may even be possible to perform more complex FEM in real time, opening up new possibilities for interactive origami design, optimization, and analysis tools.

Future work may also grow to include collision detection, adaptive time stepping, more robust mesh triangulation methods, and the incorporation of a design interface in our app.

## Acknowledgments

This work was supported by sponsors of the Center for Bits and Atoms.

## References

[Ablat and Qattawi 18]  Muhammad Ali Ablat and Ala Qattawi. "Finite Element Analysis of Origami-Based Sheet Metal Folding Process." *Journal of Engineering Materials and Technology* 140:April (2018), 1–7.

[Demaine and Tachi 17]  Erik D Demaine and Tomohiro Tachi. "Origamizer: A Practical Algorithm for Folding Any Polyhedron." *33rd International Symposium on Computational Geometry* :34 (2017), 1–15.

[Demaine et al. 16]  Erik D. Demaine, Jason S. Ku, and Robert J. Lang. "A New File Standard to Represent Folded Structures." In *Abstracts from the 26th Fall Workshop on Computational Geometry*, 2016. See https://github.com/edemaine/fold.

[Filipov et al. 17]  E T Filipov, K Liu, T Tachi, M Schenk, and G H Paulino. "Bar and hinge models for scalable analysis of origami." *International Journal of Solids and Structures* 124 (2017), 26–45.

[Hiller and Lipson 14]  Jonathan Hiller and Hod Lipson. "Dynamic Simulation of Soft Multimaterial 3D-Printed Objects." *Soft Robotics* 1:1 (2014), 88–101.

[Lang 96] Robert J Lang. "A Computational Algorithm for Origami Design." *Proceedings of the Twelfth Annual Symposium on Computational Geometry.*

[Liu and Paulino 16] Ke Liu and Glaucio H Paulino. "MERLIN: A MATLAB implementation to capture highly nonlinear behavior of non-rigid origami." *Proceedings of the IASS Annual Symposium.*

[Liu and Paulino 17] K Liu and G H Paulino. "Nonlinear mechanics of non-rigid origami: an efficient computational approach." *Proceedings of the Royal Society A.*

[Peraza Hernandez et al. 16] Edwin A Peraza Hernandez, Darren J Hartl, Ergun Akleman, and Dimitris C Lagoudas. "Computer-Aided Design Modeling and analysis of origami structures with smooth folds." *Computer-Aided Design* 78 (2016), 93–106.

[Saito et al. 14] Kazuya Saito, Sergio Pellegrino, and Taketoshi Nojima. "Manufacture of Arbitrary Cross-Section Composite Honeycomb Cores Based on Origami Techniques." *Journal of Mechanical Design* 136:May (2014), 1–9.

[Schenk and Guest 11] Mark Schenk and Simon D Guest. "Origami Folding: A Structural Engineering Approach." In *Origami⁵: Proceedings of the 5th International Meeting of Origami Science, Mathematics, and Education*, pp. 291–304. CRC Press, 2011.

[Schenk et al. 14] M Schenk, S D Guest, and G J Mcshane. "Novel stacked folded cores for blast-resistant sandwich beams." *International Journal of Solids and Structures* 51:25-26 (2014), 4196–4214.

[Tachi 06] Tomohiro Tachi. "Simulation of Rigid Origami." In *Origami⁴: Proceedings of the 4th International Meeting of Origami Science, Mathematics, and Education*, 2006.

[Tachi 10] Tomohiro Tachi. "Freeform Variations of Origami." *Journal for Geometry and Graphics* 14:2 (2010), 203–215.

[Tachi 13] Tomohiro Tachi. "Interactive Form-Finding of Elastic Origami." *Proceedings of the International Association for Shell and Spatial Structures (IASS) Symposium* :5 (2013), 7–10.

[Tang et al. 16] C C Tang, P B Bo, J Wallner, and H Pottmann. "Interactive Design of Developable Surfaces." *Acm Transactions on Graphics* 35:2 (2016), 12.


---


Amanda Ghassaei

Center for Bits and Atoms, Massachusetts Institute of Technology, Cambridge, MA, e-mail: amandaghassaei@gmail.com

Erik D. Demaine

Computer Science and Artificial Intelligence Laboratory, Massachusetts Institute of Technology, 32 Vassar Street, Cambridge, Massachusetts 02139, USA e-mail: edemaine@mit.edu

Neil Gershenfeld

Center for Bits and Atoms, Massachusetts Institute of Technology, Cambridge, MA, e-mail: neil.gershenfeld@cba.mit.edu

# Highly Efficient Nonlinear Structural Analysis of Origami Assemblages using the MERLIN2 Software

*Ke Liu, Glaucio H. Paulino*

**Abstract**: *The bar-and-hinge model is a highly efficient structural analysis approach for origami systems based on reduced-order modeling, which predicts their global mechanical behaviour surprisingly well. We implemented this method in MERLIN, a Matlab software for structural analysis of origami assemblages. Here we present MERLIN2, an extended version of the software, offering several new capabilities, such as implementation of a new triangulation scheme that allows for consideration of polygonal panels, convenient import/export capabilities, and displacement loading. The Matlab code and associated implementation are explained in detail, and a numerical example is presented to illustrate the MERLIN2 capabilities.*

## 1  Introduction

Emerging concepts associated to origami engineering [Schenk and Guest 13, Filipov et al. 15, Lang et al. 15] pose a challenge on our understanding of origami mechanics. In practical applications, engineers need to know how the origami structural system interacts with environment and responds to human control. Although origami folding is often idealized as rigid origami, concerning only geometry and kinematics; in practice, the folding process exhibits complicated behaviour, beyond simple folding, which cannot be explained by geometry alone, owing to the flexibility of panels. Modeling of origami structures by means of shell finite-elements (FE) provides high-resolution analysis, but also requires a time-consuming cycle for both modeling and computing, leading to unnecessary cost and effort especially in the preliminary design stage [Ramm and Wall 04, Ota et al. 16]. Thus, there is a need for a simple and effective analysis approach that fills the gap between the overly simplified rigid origami simulations and the detailed and expensive full-scale FE analyses, while shedding light on the essence of origami mechanics. This leads us to the bar-and-hinge approach [Filipov et al. 17, Liu and Paulino 17].

The essence of the bar-and-hinge approach [Schenk and Guest 11, Filipov et al. 17] is the simplification of the kinematics of origami assemblages, whose rationale lies with the fact that admissible deformations of origami structures are strongly confined by geometry. Deformed thin panels in origami are likely to display concentrated bending curvatures along diagonals due to the singular ridge effect [Witten 07]. The bar-and-hinge model represents the kinematic space of an origami

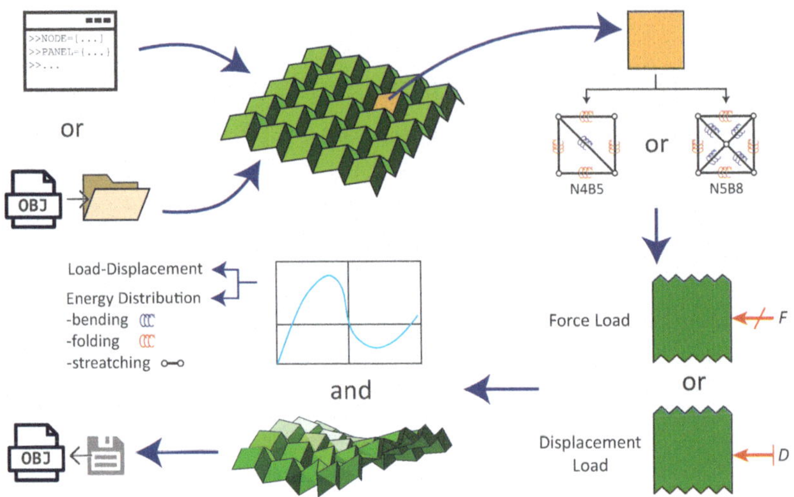



**Figure 1:** *A typical workflow of MERLIN2.*

structure with a bar (or truss) frame associated with constrained out-of-plane rotations, which successfully captures three fundamental deformation modes in origami: (in-plane) stretching, (out-of-plane) crease folding, and (out-of-plane) panel bending. Bars are placed along straight fold lines, and across panels for in-plane stiffness. The rotational hinges are along bars connecting panels to model folding of creases, and along bars across panels to model bending of panels. With only a few degrees of freedom, the reduced order bar-and-hinge model predicts surprisingly well the overall mechanical behaviour of origami structures.

Recently, we developed a nonlinear formulation for origami structural analysis using bar-and-hinge models to enable large deformation. We implemented the formulation in the software MERLIN, a dedicated open-source Matlab code for structural analysis of origami assemblages, which simulates the entire deformation process of an origami structure subject to prescribed applied forces. *Here we present MERLIN2, an extended version of MERLIN with several new capabilities, such as implementation of a new N5B8 super-element [Filipov et al. 17], consideration of polygonal panels, import/export capabilities with wavefront OBJ format, and displacement loading.*

## 2   Nonlinear elastic formulation for a general bar-and-hinge model

Here, we briefly describe the nonlinear elastic formulation of the bar-and-hinge method [Liu and Paulino 17]. We consider a discretized origami assemblage as an elastic system. The total potential energy ($\Pi$) of the system has contributions from the bars ($U_S$), bending hinges ($U_B$) and folding hinges ($U_F$). The total potential energy of the system can be written as:

$$\Pi(\mathbf{u}) = U_S(\mathbf{u}) + U_B(\mathbf{u}) + U_F(\mathbf{u}) - \mathbf{f}^T\mathbf{u}, \tag{1}$$

where $\mathbf{f}$ is the externally applied load, and all the other energy terms are nonlinear functions of the nodal displacements $\mathbf{u}$. Equilibrium is obtained when $\Pi$ is local stationary, and therefore the equilibrium equation and the finite element matrices can be derived as [Filipov et al. 17, Liu and Paulino 17]:

$$\mathbf{T}(\mathbf{u}) = \mathbf{T}_S(\mathbf{u}) + \mathbf{T}_B(\mathbf{u}) + \mathbf{T}_F(\mathbf{u}) - \mathbf{f} = \mathbf{0}, \tag{2}$$

$$\mathbf{K}(\mathbf{u}) = \mathbf{K}_S(\mathbf{u}) + \mathbf{K}_B(\mathbf{u}) + \mathbf{K}_F(\mathbf{u}), \tag{3}$$

where:

$$\mathbf{T}_S(\mathbf{u}) = \frac{\partial U_S(\mathbf{u})}{\partial \mathbf{u}}, \ \ \mathbf{T}_B(\mathbf{u}) = \frac{\partial U_B(\mathbf{u})}{\partial \mathbf{u}}, \ \ \mathbf{T}_F(\mathbf{u}) = \frac{\partial U_F(\mathbf{u})}{\partial \mathbf{u}}, \tag{4}$$

and

$$\mathbf{K}_S(\mathbf{u}) = \frac{\partial^2 U_S(\mathbf{u})}{\partial \mathbf{u}^2}, \ \ \mathbf{K}_B(\mathbf{u}) = \frac{\partial^2 U_B(\mathbf{u})}{\partial \mathbf{u}^2}, \ \ \mathbf{K}_F(\mathbf{u}) = \frac{\partial^2 U_F(\mathbf{u})}{\partial \mathbf{u}^2}. \tag{5}$$

The system equilibrium and tangent stiffness are summations of elemental contributions, which is defined through an elastic constitutive model for each element in the system.

## 2.1  Bar elements

For each bar element, we define its stored energy as:

$$U_S^i = ALW(E_{xx}) \tag{6}$$

where $A$ denotes the member area, $L$ denotes the member length, and $W$ is the energy density as a function of the one dimensional Green-Lagrange strain $E_{xx}$. The energy conjugate 2nd Piola-Kirchhoff (PK) stress and the tangent modulus are defined as

$$S_{xx} = \frac{\partial W}{\partial E_{xx}}, \quad \text{and} \quad C = \frac{\partial S_{xx}}{\partial E_{xx}}. \tag{7}$$

Thus for a single bar element, its contributions to the system equilibrium and tangent stiffness become

$$\mathbf{T}_S^i = ALS_{xx}\frac{\partial E_{xx}}{\partial \mathbf{u}}, \ \text{and} \ \mathbf{K}_S^i = AL\left(C\frac{\partial E_{xx}}{\partial \mathbf{u}} \otimes \frac{\partial E_{xx}}{\partial \mathbf{u}} + S_{xx}\frac{\partial^2 E_{xx}}{\partial \mathbf{u}^2}\right). \tag{8}$$

Given that the Green-Lagrange strain is a function of the nodal displacements [Wriggers 08, Liu and Paulino 17], we are able to complete the components contributed by a single bar element.

## 2.2  Special rotational spring elements

Folding and bending deformations are both represented by rotational springs in a bar-and-hinge model (see [Liu and Paulino 17]). Conceptually, they are kinematically the same, but may have different constitutive behaviors. Examples can be found in Fig. 2 and 3. A rotational spring element consists of 4 nodes, 5 bars, and 1 dihedral angle between two triangles. Borrowing ideas from standard nonlinear elasticity, we assume a stored energy function $\psi$ for each rotational spring to describe its constitutive behavior, which is a function of the rotation angle ($\theta$):

$$U_B = \psi_B(\theta), \quad \text{or} \ U_F = \psi_F(\theta), \tag{9}$$

where $\psi_B$ is the stored energy function of a bending hinge, while $\psi_F$ is the stored energy function of a folding hinge. Taking a bending hinge as example, we can define the resistant moment $M$ and tangent rotational modulus $k$ as:

$$M = \frac{d\psi_B(\theta)}{d\theta}, \quad k = \frac{dM}{d\theta}. \tag{10}$$

We define the rotation angle $\theta \in [0, 2\pi)$ using absolute measurements such that $\theta = \pi$ when the panel is flat. We call $\theta_0$ the neutral angle of a rotational spring if $M(\theta_0) = 0$. Following this convention, a linear elastic rotational spring can be defined by:

$$\psi_B(\theta) = \frac{1}{2}LK(\theta - \theta_0)^2, \quad M = LK(\theta - \theta_0), \quad k = LK, \tag{11}$$

where $L$ denotes the undeformed length of the rotational hinge and $K$ is the rotational modulus per unit length along the hinge. To differentiate bending and folding hinges, we can assign different values for $K$.

Thus for a bending rotational spring element, its contributions to the system equilibrium and tangent stiffness become

$$\mathbf{T}_B^i = M\frac{d\theta}{d\mathbf{u}}, \quad \text{and} \quad \mathbf{K}_B^i = k\frac{d\theta}{d\mathbf{u}} \otimes \frac{d\theta}{d\mathbf{u}} + M\frac{d^2\theta}{d\mathbf{u}^2}. \tag{12}$$

The same formulation also applies to folding rotational spring elements. Given that the the rotation angle is completely defined from the displacements and original coordinates of the 4 associated nodes, we are able to complete the components contributed by a single rotational spring element which either represents a bending hinge or folding hinge.

## 3  Discretization schemes

Bar-and-hinge approaches are mesh-dependent by design, and thus it is crucial to choose a representative triangulation scheme for origami structures. Currently, there are two types of triangulation schemes that differ by how they discretized quadrilateral panels. One is the N4B5 scheme, which simply divide a quadrilateral panel by one of its diagonals, discretizing it into two triangles. The other is the N5B8 scheme, which adds an extra interior node within each quadrilateral panel, known as a Steiner point in computational geometry [Bern and Eppstein 95], and divide it into four triangles. Triangular panels are not further discretized by both schemes. In this section, we review the basic concepts of these two schemes and generalize them to the triangulation of convex polygonal origami panels, as non-convex panels rarely appear in origami patterns.

### 3.1  The N4B5 scheme and its generalization to polygonal panels

The N4B5 discretization scheme is most commonly adopted for reduced-order modeling of origami structures with quadrilateral panels [Schenk and Guest 13, Filipov et al. 15], which divides each quadrilateral panel into two triangles by its shorter diagonal, as demonstrated in Fig. 2(a). If we assume that the panel bending

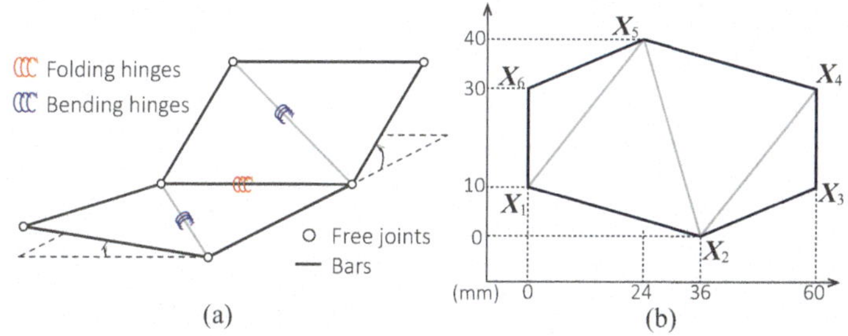


**Figure 2:** *(a) Illustration of a N4B5 model. (b) A hexagonal origami panel triangulated by the generalized N4B5 scheme.*

stiffness is the same per unit length along both diagonals, then shorter diagonals are easier to bend and thus require lower energy.

To extend this idea to the discretization of polygonal panels, we progressively bisect a polygon by the shortest diagonals until the original polygon is triangulated, which can be efficiently achieved using a divide and conquer algorithm [Heath 97]. In the example shown in Fig. 2(b), we first bisect hexagon 1-2-3-4-5-6 by its short-est diagonal 1-5 (could also be 2-4), dividing the hexagon into one triangle and one pentagon. Then we divide the pentagon 1-2-3-4-5 by its shortest diagonal 2-4. Finally, we dived the quadrilateral 1-2-4-5 by its shortest diagonal 2-5 to finish the triangulation.

The major advantage of the generalized N4B5 scheme is its simplicity. For certain origami patterns, such as the Miura-ori with only parallelogram panels, its accuracy is also satisfactory [Liu and Paulino 17]. However, ambiguity arises when multiple diagonals of a panel are of the same length.

## 3.2   The N5B8 scheme and its generalization to polygonal panels

As illustrated in Fig. 3(a), the N5B8 triangulation scheme adds a Steiner point at the intersection of the two diagonals of a quadrilateral panel, dividing it into four triangles, hence there are 5 nodes and 8 bars belong to each panel. The N5B8 scheme allows the discrete model to capture more realistic doubly curved out-of-plane deformations and isotropic in-plane behaviors of thin panels, potentially yielding higher resolution than the N4B5 scheme [Filipov et al. 17].

To extend the N5B8 discretization scheme to a polygon, a Steiner point that can simultaneously locate on all of the diagonals would be ideal. However, it is impossible for general polygonal panels to have such a Steiner point as specific quadrilaterals, as demonstrated in Fig. 3(b). Here we propose to pursue a point that has the shortest overall (weighted) distance to all diagonals of a convex polygon by solving the following optimization problem:

$$X^* = \arg\min_{X} \sum_{|j-i|>1} \frac{h_{ij}}{d_{ij}}, \tag{13}$$

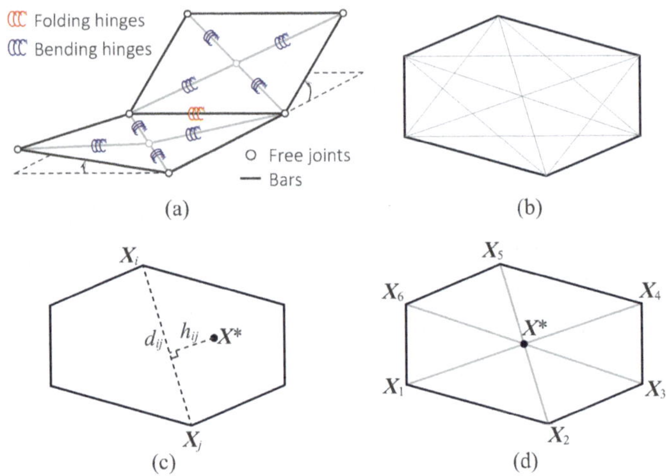



**Figure 3:** *(a) Illustration of a N5B8 model. (b) All diagonals of a hexagonal origami panel. (c) Illustration for $d_{ij}$ and $h_{ij}$. (d) Triangulation by the generalized N4B5 scheme.*

where

$$d_{ij} = \|X_i - X_j\|, \quad \text{and} \quad h_{ij} = \frac{\|(X_i - X^*) \times (X_j - X^*)\|}{d_{ij}}. \tag{14}$$

Geometrically, $d_{ij}$ is the length of the diagonal $i$-$j$, and $h_{ij}$ is the distance from the Steiner point (i.e. $X^*$) to the diagonal $i$-$j$ (see Fig. 3(c)). The measure of $(h_{ij}/d_{ij})$ is a weighted distance that favours shorter diagonals because they are more likely to bend than longer ones, judged by the scaled stiffness of bending ridges [Witten 07, Filipov et al. 17]. Notice that, for quadrilateral panels, the optimal solution for $X^*$ by formula (13) is simply the intersection point of the two diagonals, and thus the extended scheme is consistent with the previous N5B8 model. In the example shown in Fig. 3(d), the optimal Steiner point for the hexagon (same as in Fig. 2(b)) is at the common intersection of three diagonals.

To recover the in-plane Poisson's effect of origami panels, the original N5B8 model [Filipov et al. 17] provides formulas for assigning bar areas based on rectangular panels. However, we have not yet derived formula for polygonal and triangular panels. Realizing that it is difficult to account for the Poisson effect for those panels, we propose to assign the bar areas such that they preserves the (linear elastic) strain energy of a continuous panel (of any shape) under uniform in-plane dilation assuming plane stress state. For a polygonal panel of thickness $t$ and polygon area $S$, we assign a uniform area $A$ for all bars belonging to this panel by:

$$A = \frac{2St}{(1-v)\sum_i L_i}, \tag{15}$$

where $\sum_i L_i$ is the total length of bars, and $v$ denotes the Poisson's ratio. When applied to rectangular panels, the results are similar to those calculated by the existing formulas [Filipov et al. 17], which is satisfactory as an estimation tool.

## 3.3 Preliminary study on discretization of polygonal origami panels

Using the same hexagonal panel shown in Fig. 2(b) and Fig. 3(d) as an example, we apply load on it and compare the performance of two bar-and-hinge models triangulated by the generalized N4B5 scheme and N5B8 scheme. The applied load generates a torque on the polygonal panel (see Fig. 4(b)). Photograph of a twisted polygonal panel reveals that the surface (mean) curvature concentrates near a long diagonal, and there are two curved ridges in slight bending that spans two short diagonals, as illustrated in Fig. 4(a). We found that the generalized N5B8 model produces deformation that is closer to the physical model than the generalized N4B5 model. The deformation modes of the two discretized models are different, and the generalized N4B5 model is stiffer, as demonstrated in Fig. 4(c). We use the actual material properties of the Mylar sheet from which the polygon was made, i.e. modulus of elasticity $E = 5GPa$, Poisson's ratio $v = 0.35$, sheet thickness $t = 0.127mm$. Constitutive relations for bending hinges in both models implement a super-linear model [Filipov et al. 17]. We plan to make further comparison with experimental and FE results to verify and better calibrate the proposed discretization schemes.

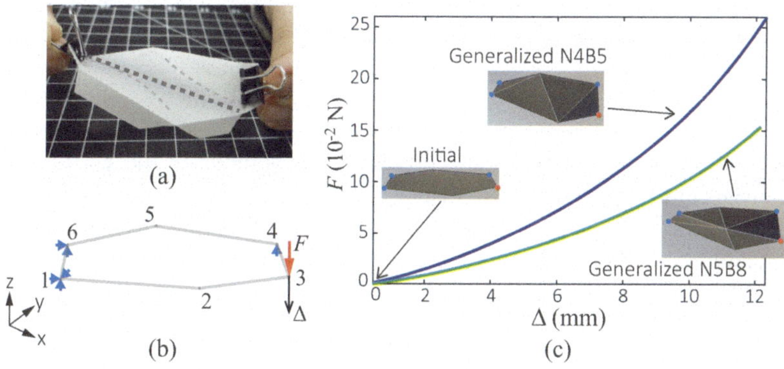


**Figure 4:** *(a) Photograph of a twisted hexagonal origami panel made with 0.127mm-thick Mylar sheet. Flanges are used to imitate surrounding panels. (b) Boundary conditions: blue arrows indicate fixed degrees of freedom, and red arrow indicates the applied force. (c) Load-displacement curve for both generalized N4B5 model and generalized N5B8 model. The insets show the deformed shapes of the polygonal panel using two different discretization models. The blue dots and red dots implies supporting nodes and loading node(s), respectively.*

# 4   Matlab implementation

In this section, we explain the details of the MERLIN2 code for origami structural analysis. We begin by describing the structure of the code, and the input/output parameters. We also make some comments on the supporting functions in the Appendix.

The kernel of MERLIN2 has two parts: discretization and nonlinear analysis. The function `PrepareData` discretizes the input origami model and formats necessary data for the structural analysis. The function `PathAnalysis` solves the nonlinear equilibrium problem defined by the formulation described in Sec. 2.

## 4.1 The `PrepareData` function

The following variables are inputs to the `PrepareData` function:

```
[truss,angles,AnalyInputOpt]=
   PrepareData(Node,Panel,Supp,Load,AnalyInputOpt).
```

`Node`   This is a three-column array of size $N_{node} \times 3$ whose $i$th row represents the coordinates of the $i$-th node (or vertex), where $N_{node}$ is the number of nodes in the original origami model (before discretization).

`Panel`   The is a Matlab cell array containing the topology information about an origami structure. The $j$th entry of the cell array contains the indices of the nodes incident on the $j$th panel in counter-clockwise order. Each entry can vary in size, which allows for freedom on defining origami patterns that contain polygonal panels with different number of vertices.

`Supp`   This is a four-column array containing information for boundary conditions of the structure. The first column holds the nodal index, and the second, third and fourth columns give support conditions for that node in the $x$-, $y$-, and $z$-direction, respectively. Value of 0 means that the node is free, and value of 1 specifies a fixed node.

`Load`   This is structured in a similar way as `Supp`, except for the values in the second to fourth columns that represent the magnitude of the $x$-, $y$-, and $z$-components of the applied force or displacement.

`AnalyInputOpt`   This is a Matlab structure array that contains miscellaneous information needed for the analysis. The following fields can be specified:

1. `ModelType` - This parameter specifies the discretization scheme. The user chooses between `N4B5` and `N5B8`, which refer to the generalized versions as explained in Sec. 3.

2. `MaterCalib` - This parameter specifies the calibration method for elemental constitutive behaviour. The user chooses between `manual` and `auto` mode. In the `manual` mode, the constitutive models involved in a bar-and-hinge model need to be specified explicitly using the input fields: `BarCM`, `RotSprBend`, `RotSprFold`, `Abar`, `Kb`, and `Kf`. In the `auto` mode, the constitutive models are specified implicitly based on actual material properties using the formulas defined in [Filipov et al. 17]. The following input fields are required: `ModElastic`, `Poisson`, `Thickness`, and `LScaleFactor`. The `auto` mode only applies to `N5B8` models.

3. `BarCM` This is a function that defines the constitutive model for bar elements in the following format: `[Sx,Ct,W]=BarCM(Ex)`.
   The only input is the Green-Lagrange strain $E_{xx}$ (`Ex`), and the three outputs are the 2nd PK stress $S_{xx}$ (`Sx`), tangent modulus $C$ (`Ct`), and strain energy density $W$

(W). The default function is a hyper elastic model (@Ogden) as reported in [**Liu and Paulino 17**]. To customize, one can write a separate function and pass a function handle to this parameter.

4. RotSprBend This is a function that defines the constitutive model for bending rotational spring elements in the following format:

[M,k,P]=RotSprBend(he,h0,Kp,L0).

The four inputs are: deformed bending angles $\theta$ (he), neutral angles $\theta_0$ (h0), stiffness parameter(s) that may differ for each element (Kp), and hinge lengths $L$ (Lo). The three outputs shall be resistant moment $M$ (M), tangent rotational modulus $k$ (k), and stored energy $\psi$ (P). Input and output values (except for Kp) are all $N_{bend} \times 1$ arrays, where $N_{bend}$ denotes the total number of bending hinges. The default function implements an enhanced linear elastic model [**Liu and Paulino 17**] (@EnhancedLinear) in the manual mode, and a super-linear model [**Filipov et al. 17**] (@SuperLinearBend) in the auto mode.

5. RotSprFold This function specifies the constitutive model for folding rotational spring elements, following the same format as RotSprBend, except that the size of input and output arrays shall be $N_{fold} \times 1$, where $N_{fold}$ denotes the total number of folding hinges. The default function is the enhanced linear elastic model (@EnhancedLinear) in both manual and auto modes.

6. Abar - Areas of bar elements, which could either be a scalar (for uniform area) or a $N_{bar} \times 1$ array where $N_{bar}$ denotes the total number of bar elements in the bar-and-hinge model.

7. Kb - Stiffness parameters for bending rotational spring elements, which could either be a $1 \times m$ array (for uniform property) or a $N_{bend} \times m$ array, where $m$ is the number of required stiffness parameters, which typically equals 1. This is the input Kp for RotSprBend.

8. Kf - Stiffness parameters for folding rotational spring elements, following the same format as Kb. This is the input Kp for RotSprFold.

9. ModElastic - Modulus of elasticity $E$ (or Young's modulus) of the material of the origami structure. This parameter is used in the auto mode.

10. Poisson - Poisson's ratio $v$ of the material, used in the auto mode.

11. Thickness - Thickness $t$ of the origami panels, used in the auto mode.

12. LScaleFactor - Ratio of length scale factor $L^*/L_F$, where $L_F$ is the length of a folding hinge and $L^*$ is the length scale factor defined in [**Filipov et al. 17**]. This parameter is used to determine the rotational modulus of folding rotational springs, used in the auto mode. Typical values of $L^*/L_F$ is between 1 to 5. The default value is 3.

13. ZeroBend - This parameter specifies the neutral angles of bending hinges. User can choose between Flat and AsIs, or specify user-defined values. This is useful when the initial configuration involves bended panels. The option AsIs uses the bended shapes of panels as their undeformed states; while the option Flat always assumes that the neutral angles of bending hinges are $\pi$ (referring to flat panels). When specifying user-defined values, we use a scalar for uniform neutral angles and a $N_{bend} \times 1$ array for non-uniform neutral angles.

14. `LoadType` - User can choose between `Force` and `Displacement`.

15. `Load` - In `Force` mode, the actual applied loads are multiples of the specified load, the scalar multipliers (known as load factor) is determined automatically by a numerical continuation algorithm [Leon et al. 14]. In `Displacement` mode, the solver attempts the specified amount of displacements in an incremental manner, and stops when the specified amount of displacements is achieved.

16. `AdaptiveLoad` - This is a function that specifies an adaptive load in the following format: `[F] = LoadFun(Node,U,icrm)`.
    The inputs are initial nodal coordinates (`Node`), displacement vector (`U`), and incremental number (`icrm`), which can be regarded as pseudo-time. The output is the load vector, which can be either `Force` or `Displacement`, but each entry must corresponds to the same degree of freedom as in `U`. In `Displacement` mode, the size of each incremental step must be considered within the function. Once specified, the `AdaptiveLoad` overrides `Load`.

17. `InitialLoadFactor` - Initial load factor for the numerical continuation algorithm [Liu and Paulino 17, Leon et al. 14], used in `Force` mode.

18. `MaxIcr` - Maximum number of increments ($N_{icrm}$) for the numerical continuation algorithm, used in `Force` mode.

19. `DispStep` - Number of incremental steps to achieve a `Displacement` load.

20. `StopCriterion` - A function that specifies a stopping criterion for the numerical simulation based on configurational changes of the origami structure. Use the following format: `[Flag] = StopCriterion(Node,U,icrm)`. The function returns 1 when the stopping criterion is met; otherwise the function should return 0.

The outputs of the `PrepareData` function are:

`truss`  This is a Matlab structure array that contains information about the bar elements. It has the following fields:

1. `Node` - This is a three-column array of size $N'_{node} \times 3$ in the same format as `Node`, but contains additional Steiner points for generalized N5B8 models.

2. `Bars` - Connectivity information of bar elements stored in a $N_{bar} \times 2$ array. The two entries in a row refer to the indices of the two end nodes of a bar element.

3. `Trigl` - Triangulation information stored in a $N_{tri} \times 3$ array, where $N_{tri}$ equals the total number of triangles in the bar-and-hinge model after discretization. The three entries in a row are vertex indices of each triangle.

4. `B` - Compatibility matrix of the bar frame [Filipov et al. 17] of size $N_{bar} \times N_{dof}$, where $N_{dof}$ is the total number of degrees of freedom in the system ($= 3N'_{node}$).

5. `L` - Initial lengths of bar elements stored in a $N_{bar} \times 1$ array.

6. `FixedDofs` - Indices of fixed degrees of freedom specified by `Supp`.

7. `CM` - Constitutive model for bar elements, passed from `AnalyInputOpt.BarCM`.

8. `A` - Member areas of bar elements stored in a $N_{bar} \times 1$ array.

9. `U0` - Initial displacement referring to the unformed configuration. Most of the time, the initial configuration of a structure in a simulation is the undeformed

configuration, thus U0 is a $N_{dof} \times 1$ array of zeros. This parameter is specified outside the `PrepareData` function, but before the `PathAnalysis` function is executed. If not specified, the `PathAnalysis` function assigns `truss.U0` a vector of zeros.

`angles` This is a Matlab structure array that contains information about the rotational spring elements. It has the following fields:

1. `Panel` - Same as the aforementioned input parameter `Panel`.

2. `fold` - A $N_{fold} \times 4$ array that contains indices of associated nodes of folding rotational springs. The first two entries define the rotation axis, and the later two refer to off-axis points of the two adjacent triangles in a rotational spring element. See [Liu and Paulino 16] for examples.

3. `bend` - A $N_{bend} \times 4$ array that contains indices of associated nodes of bending rotational springs, in the same format as `angles.fold`.

4. `pf0` - Neutral angles of folding rotational springs stored in a $N_{fold} \times 1$ array.

5. `pb0` - Neutral angles of bending rotational springs stored in a $N_{bend} \times 1$ array.

6. `CMbend` - Same as `AnalyInputOpt.RotSprBend`.

7. `CMfold` - Same as `AnalyInputOpt.RotSprFold`.

8. `Kb` - Stiffness parameters for bending rotational springs specified for each element in an array of $N_{bend}$ rows.

9. `Kf` - Stiffness parameters for folding rotational springs specified for each element in an array of $N_{fold}$ rows.

## 4.2 The `PathAnalysis` function

The `PathAnalysis` function takes in the outputs of the `PrepareData` function and conducts the nonlinear structural analysis:

```
[Uhis,Fhis]=PathAnalysis(truss,angles,AnalyInputOpt).
```

In `Force` mode, the function implements a numerical continuation algorithm called Modified Generalized Displacement Control Method (MGDCM) [Leon et al. 14]. In addition, to improve convergence performance, the function heuristically adjusts the constraint radius, which determines the incremental load factors. In each increment, when the number of iterations to convergence is large, the constraint radius is reduced to make a more conservative subsequent increment; when the number of iterations is very small, the constraint radius is increased, yielding a more aggressive step for the next increment. In `Displacement` mode, the function divides the `Displacement` load into multiple small increments. Each increment is solved by a damped Newton-Raphson algorithm [Wriggers 08]. A standard incremental step of displacement load is the prescribed amount divided by the number of `DispStep`. The implemented algorithm makes attempts to reduce or increase the incremental movements, and adjusts damping factors based on the convergence performance of the current increment. Therefore, the actual number of increments $N_{icrm}$ performed to achieve the prescribed displacement load may not equal to `DispStep`.

The outputs of the function `PathAnalysis` are:

1. `Uhis` - History of nodal displacements at the end of each increment, stored in a $N_{dof} \times N_{icrm}$ array.

2. `Fhis` - In `Force` mode, this is a $N_{icrm} \times 1$ array that stores the values of load factors at the end of each increment. In `Displacement` mode, this is a $N_{icrm} \times N_{disp}$ array, whose columns store the negative values of the resistant forces in the degrees of freedom that are imposed with displacement loads, where $N_{disp}$ is the number of imposed degrees of freedom.

## 5 An example using MERLIN2

In this example, we demonstrate how MERLIN2 can be used to conduct structural analysis of a custom designed origami pattern. A generalized Miura-ori is created using the Freeform Origami software by Tomohiro Tachi [Tachi 10] and imported to MERLIN2 as an OBJ file. Simulation results are shown in Fig. 5. We choose the `N5B8` model under `auto` mode by specifying modulus of elasticity $E = 1GPa$, Poisson's ratio $v = 0.3$, material thickness $t = 0.25mm$, and ratio of length scale factor $L^*/L_F = 2$. The structure is loaded by displacement.

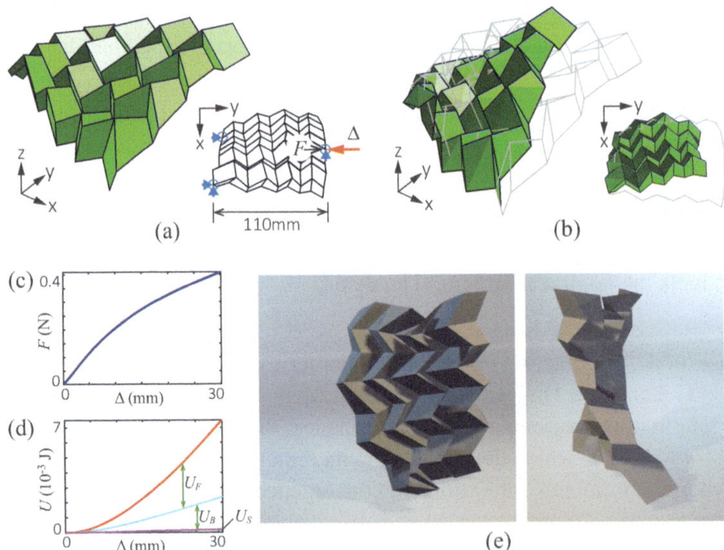


**Figure 5:** *(a) Initial configuration of a generalized Miura-ori. The plan view shows boundary conditions for numerical simulation. Blue arrows indicates support in x and y-direction, while blue circles indicates support in the z-direction. the red arrow shows the applied displacement load. Black arrow marks the balance force along the specific direction of loading. (b) The deformed configuration. The grey wire frame refers to initial configuration. (c) Load-displacement plot. (d) Energy-displacement plot. Contributions from three deformation modes are differentiated: folding ($U_F$), bending ($U_B$) and stretching ($U_S$). (e) Two views of the rendering of the deformed origami, generated using the exported OBJ file from MERLIN2.*

## 6   Conclusions

Sharing and publication of open-source and educational software has been a good tradition in the origami community, such as "TreeMaker" [Lang 11], "Freeform Origami" [Tachi 10], and "Rigid Origami Toolbox" [Gattas et al. 13]. In this paper, we present the MERLIN2 software as a powerful and easy-to-use tool for nonlinear structural analysis of origami assemblages, aiming at speeding origami design cycles and educating origami engineering. The MERLIN2 software is available at the following url: http://paulino.ce.gatech.edu/software.html.

**Acknowledgements**

This paper has been awarded the 7OSME Gabriella & Paul Rosenbaum Foundation Travel Award. We also thank the support from the US National Science Foundation (NSF) through grant no.1538830, the China Scholarship Council (CSC), and the Raymond Allen Jones Chair at the Georgia Institute of Technology.

## Appendix: Some comments on the supporting functions

In addition to the kernel functions, the MERLIN2 software provides supporting functions for data communication and interpretation.

ReadOBJ and Write2OBJ   The wavefront OBJ file format is a standard format for sharing 3D geometry, supported by many design software programs including the origami design tool "Freeform Origami" [Tachi 10]. The ReadOBJ function directly imports an OBJ coded 3D origami assemblage into MERLIN2. The Write2OBJ function writes the geometry of a deformed origami simulated by MERLIN2 to OBJ file. Necessary inputs to this function are: file name, nodal coordinates (Node), and triangulated mesh (truss.Trigl).

PostProcess   This function computes physical measures for the bar-and-hinge model based on the deformation history stored in Uhis, such as strains of bar elements and system energies. All measures are stored in a Matlab structure array (named as STAT). A complete list of outputs are shown below:

1. bar - Information about bar elements at every increment: Green-Lagrange strain (bar.Ex), 2nd PK stress (bar.Sx), stored energy of each bar element (bar.USi), and total stored energy of bar elements (bar.US). Attributes bar.Ex, bar.Sx, and bar.USi are of size $N_{bar} \times N_{icrm}$. Attribute bar.US is a $1 \times N_{icrm}$ array.

2. fold - Information about folding rotational springs at every increment: folding angle (fold.Angle), resistant moment (fold.RM), stored energy of each folding hinge (fold.UFi), and total stored energy of folding hinges (fold.UF). Attributes fold.Angle, fold.RM, and fold.UFi are of size $N_{fold} \times N_{icrm}$. Attribute fold.UF is a $1 \times N_{icrm}$ array.

3. bend - Same as STAT.fold, but for bending rotational springs, which also has 4 attributes: bending angle (bend.Angle), resistant moment (bend.RM), stored energy of each bending hinge (bend.UBi), and total stored energy (bend.UB).

4. PE - Total potential energy of the origami structure, stored in a $1 \times N_{icrm}$ array.

PlotOri   This function generates origami renderings with various options. The format of this function reads:

```
PlotOri(Node,Panel,Trigl,Name,Value).
```

Among the inputs, `Node` and `Panel` are necessary. If `Trigl` is provided, the function will draw the triangulated origami model, which should be used for plotting deformed origami structures with bent panels. We need to feed an empty array `[]` to the 3rd input of the function when `Trigl` is not needed, e.g. `PlotOri(Node,Panel,[])`. Users may use one or more `Name, Value` pair arguments to specify rendering properties. The following options are available:

1. `PanelColor` - Specify a uniform colour for panels. For example, to plot the origami in blue colour, specify `PlotOri(__,__,__, 'PanelColor',[0,0,1])`. The default colour is green.

2. `FoldEdgeStyle` - Line style for the folding hinges. To plot the folding hinges in dashed lines, specify `PlotOri(__,__,__,'FoldEdgeStyle','--')`. The default style is solid line.

3. `BendEdgeStyle` - Line style for the bending hinges. The default style is `none`, so that the origami plot does not display bending hinges. This property is only used when `Trigl` is provided.

4. `EdgeShade` - A scalar between 0 and 1 that specifies the greyscale of edges in the plot. Use value 1 for black (default) and 0 for transparent.

5. `ShowNumber` - The `PlotOri` function shows the indices of nodes in the origami plot if this property is set to `on`. This is very helpful for specifying boundary conditions (i.e. `Supp` and `Load`). When this option is `on`, the panels are plotted without colour (transparent). The default value is `off`.

6. `FaceVertexColor` - Face and vertex colours, specified as one colour per face, or one colour per vertex for interpolated face colour. For one colour per face, use an $N_{face} \times 3$ array of RGB triplets, where $N_{face}$ is the number of origami panels (`Panel`) when `Trigl` is not provided, or the number of triangles when `Trigl` is provided. For interpolated face colour based on vertex values, provide an $N'_{node} \times 3$ array of RGB triplets.

7. `EdgeColor` - Colours for all bars in the bar-and-hinge model, specified as one colour per bar, including bending hinges, folding hinges, and boundary edges. Use an $N_{bar} \times 3$ array of RGB triplets. To enable this property, connectivity of bar elements and $N_{bend}$ needs to be specified. For example, if the bar colours are stored in the array `CData`, then use `PlotOri` as:
   ```
   PlotOri(__,__,__,'EdgeColor',CData,'Bars',truss.Bars,...
           'NumBendHinge',size(angles.bend,1)).
   ```

`VisualFold` This function animates the numerical simulation. The format of this function reads:

```
VisualFold(Uhis,truss,angles,Fhis,instdof,Name,Value).
```

The input `Uhis` is the output of `PathAnalysis` function. The input `Fhis` is optional, which allows the `VisualFold` function to animate the load-displacement digram in accordance with the animation of deformation history. The parameter `instdof` is a $2 \times 1$ array that specifies the degree of freedom for displacement measure. The first entry stores the node being tracked, and the second entry is a signed integer defining the direction of nodal displacement to be monitored, for example, -3 indicates the minus $z$-direction and 2 indicates the positive $y$-direction. Use one or more `Name, Value` pair arguments to specify particular display options. The following options are available:

1. `Showinitial` - If set to `on`, the `VisualFold` function displays a grey-coloured wire frame of the initial configuration during the entire animation(s) of deformation history.

2. `Recordtype` - The `VisualFold` function can record the animation in video format (MP4) or animated image format (GIF). Specify `video` or `imggif` for video recording or image recording, respectively. The default option is `none`.

3. `Filename` - A file name for the generated video or animated image.

4. `IntensityMap` - This option enables the function to plot faces or edges in varying colours during the animation based on a physical measure specified in `IntensityData`. There are three choices: `Vertex`, `Edge`, and `Face`. This is useful for visualizing intensity of forces acting on nodes, strains in bars, or area change of triangulated panels.

5. `IntensityData` - A $N_{icrm}$-column array. Each column specifies a physical measure for all entities of a group (e.g. nodes, bars) in the origami model at one increment of simulation. Use a $N'_{node} \times N_{icrm}$ array for `IntensityMap` of `Vertex`, a $N_{bar} \times N_{icrm}$ array for `IntensityMap` of `Edge`, and a $N_{face} \times N_{icrm}$ array for `IntensityMap` of `Face`.

6. `Viewangle` - View angle for the animation of origami structure, specified in [azimuth angle, elevation angle] format.

7. `Axislim` - Axes limits for the animation of load-displacement curve. The default option automatically adjusts the axes sizes.

# References

[Bern and Eppstein 95] M. Bern and D. Eppstein. "Mesh generation and optimal triagulation." In *Computing in Euclidean Geometry*, edited by D. Z. Du and Hwang F. K., Second edition, pp. 47–123. World Scientific, 1995.

[Filipov et al. 15] Evgueni T. Filipov, Tomohiro Tachi, and Glaucio H. Paulino. "Origami tubes assembled into stiff, yet reconfigurable structures and metamaterials." *Proceedings of the National Academy of Sciences* 112:40 (2015), 12321–12326.

[Filipov et al. 17] E. T. Filipov, K. Liu, T. Tachi, M. Schenk, and G. H. Paulino. "Bar and hinge models for scalable analysis of origami." *International Journal of Solids and Structures* 124 (2017), 26–45.

[Gattas et al. 13] Joseph M. Gattas, Weina Wu, and Zhong You. "Miura-base rigid origami: parameterizations of first-level derivative and piecewise geometries." *Journal of Mechanical Design* 135 (2013), 111011.

[Heath 97] M.T. Heath. *Scientific Computing: An Introductory Survey*. Second Edition, McGraw-Hill, 1997.

[Lang et al. 15] Robert J. Lang, Spencer Magleby, and Larry L. Howell. "Single-degree-of-freedom rigidly foldable cut origami flashers." *Journal of Mechanisms and Robotics* 8 (2015), 031005.

[Lang 11] Robert J. Lang. *Origami Design Secrets*, Second edition. CRC Press, 2011.

[Leon et al. 14] Sofie E. Leon, Eduardo N. Lages, Catarina N. de Araújo, and Glaucio H. Paulino. "On the effect of constraint parameters on the generalized displacement control method." *Mechanics Research Communications* 56 (2014), 123–129.

[Liu and Paulino 16] Ke Liu and Glaucio H Paulino. "MERLIN : A MATLAB implementation to capture highly nonlinear behavior of non-rigid origami." In *Proceedings of the IASS Annual Symposium*, edited by K Kawaguchi, M. Ohsaki, and T Takeuchi. Tokyo, Japan, 2016.

[Liu and Paulino 17] K. Liu and G. H. Paulino. "Nonlinear mechanics of non-rigid origami: an efficient computational approach." *Proceedings of the Royal Society A* 473 (2017), 20170348.

[Ota et al. 16] N. S. N. Ota, L. Wilson, A. Gay Neto, S. Pellegrino, and P. M. Pimenta. "Nonlinear dynamic analysis of creased shells." *Finite Elements in Analysis and Design* 121 (2016), 64–74.

[Ramm and Wall 04] E. Ramm and W. A. Wall. "Shell structures - A sensitive interrelation between physics and numerics." *International Journal for Numerical Methods in Engineering* 60:1 (2004), 381–427.

[Schenk and Guest 11] Mark Schenk and Simon D Guest. "Origami folding: A structural engineering approach." In *Origami 5*, edited by Patsy Wang-Iverson, Robert J Lang, and Mark Yim, pp. 293–305. CRC Press, 2011.

[Schenk and Guest 13] Mark Schenk and Simon D. Guest. "Geometry of Miura-folded metamaterials." *Proceedings of the National Academy of Sciences* 110:9 (2013), 3276–3281.

[Tachi 10] Tomohiro Tachi. "Freeform Variations of Origami." *Journal for Geometry and Graphics* 14:2 (2010), 203–215.

[Witten 07] T. A. Witten. "Stress focusing in elastic sheets." *Reviews of Modern Physics* 79:2 (2007), 643–675.

[Wriggers 08] Peter Wriggers. *Nonlinear Finite Element Methods*. Springer, 2008.

---

K. Liu
School of Civil and Environmental Engineering, Georgia Institute of Technology, Atlanta, Georgia, USA. e-mail: ke.liu@gatech.edu

G. H. Paulino
School of Civil and Environmental Engineering, Georgia Institute of Technology, Atlanta, Georgia, USA. e-mail: paulino@gatech.edu

# Rigid-facet Kinematics Coupled with Finite Bending Rotation along Crease Lines

*R. Chudoba, K. H. Brakhage*

**Abstract**: *The paper describes a modeling approach that can combine kinematic and energetic criteria to simulate the behavior of a crease pattern exposed to force action due to gravity field or due to applied loads. The model is used to analyze the intermediate distribution of bending moments in the crease lines during the folding and loading process. The nonlinear kinematics affecting the force flow through the structure is presented and applied in the implementation of the energy potential governing the simulation. The feasibility of the model is examined by showing the simulated deformation behavior and the evaluted bending capacity of oricrete shell prototypes.*

## 1   Introduction

The simulation method presented in this paper is focused on the analysis of distribution of bending moment along the fold lines of folded-plate structures in combination with the rigid folding kinematics. The model development is motivated by the need to provide an intermediate feedback on mechanical performance of the folded structure both during the folding process and in response to loading ac-tions in interim and target configurations. The present paper is focused on model components that are used to describe both the manufacturing process of oricrete shells [Chudoba et al. 14] in combination with the intermediate stress states during the folding and in the target shape. During this process, the change of kinematic conditions is considered to seamlessly switch between the production and service conditions.

To illustrate the context in which the simulation framework has been developed let us first present a prototype of oricrete shell depicted in Fig. 1. The shells with the thickness of 10 mm were reinforced with one layer of carbon fabrics and have been produced by folding the parameterized waterbomb crease pattern. The shape design of the shells has been guided by the aim to achieve a high ultimate load in the displayed configuration. To indicate the level of structural performance, let us mention, that the ultimate vertical load applied at the middle of the span of 2.2 m measured in the conducted tests was 13.2 kN. This is 43 times larger than the self weight of the shell of 0.3 kN. It should be noted that besides the high ultimate load, other characteristics of the structural behavior need to be considered in the design, including the high initial stiffness and structural ductility ensuring large deformation at the level of ultimate load prior to failure.

Besides the structural performance, the shape design of the folded plate structure is conditioned by technological questions included in realization of the folding process. The shown oricrete shells were concreted in plane and folded to the final shape in fresh state within one hour after the pour using a foldable formwork. To achieve the goal of high structural performance in the target shape, the kinematics of the crease pattern must be locked at the end of the folding process. This can be done either locally, by introducing rotational stiffness around the crease lines, or using external kinematic constraints, stiffeners and linkages that prevent a relative displacement of vertexes.

In case of oricrete the development of cross-sectional stiffness of the crease lines is introduced automatically due to the hardening of concrete matrix and is thus implicitly included in the manufacturing method [van der Woerd et al. 17]. After the hardening, the shell can be lifted from the formwork and transported to the construction site where it is mounted to the supports and linked to other structural components.

The development of the present modeling framework has been motivated by the need to support the whole technological process including folding, transport and load carrying behavior of the thin light-weight shell elements [van der Woerd et al. 15, van der Woerd et al. 16]. In the present paper we focus on two steps within this process, namely the folding and the transport. By focusing on these two aspects we can provide an example of the modeling approach that couples the criteria related to folding process and to the interim stress distribution and, thus, document the feasibility of the approach to the model and software development. Other mentioned questions, i.e. the folding technology using a foldable formwork, process parameters related to the concrete hardening in combination with folding, or the analysis of the physically non-linear structural behavior in the final stage and their validation go beyond the scope of the present paper and will be published elsewhere.

The question of transport must be considered due to the fact that, in spite of high performance in the service state, the folded shells can be exposed to loading conditions during the manipulation that induce relatively high bending moments along the crease lines that can lead to cracking along the fold lines. Therefore, an efficient stress analysis of the shell element arbitrarily positioned in space is needed to identify the critical configurations causing damage during the manipulation. Fig. 2 shows examples of interim manipulation states.

The developed computational framework provides the possibility to configure and activate the numerical model components of the optimization framework contributing to the goal functions or to the equality and inequality constraints [Chudoba et al. 15]. In the present paper, focused on the simulation of folding process and stress state analysis at transport states due to self weight considered here, the model components representing the facet rigidity, instantaneous potential energy of gravity and stored bending deformation energy have been applied.

**Figure 1:** *Oricrete shells produced using foldable formwork in target configuration.*

The modeling approach applied here is related to the methods developed in the last decade with the ambition to analyze the mechanical performance of folded structures *on-the-fly*, without an impedance mismatch in the model of folding and the model of structural analysis. The pioneering work in this field has been presented by [Resch and Christiansen 70] introducing a framework for kinematic and elastic structural analysis of folded plate structures based on the principle of virtual work with three assumed deformation modes: normal deformation of fold lines, in-plane deformation of facets and bending of fold lines along the creases. Bending stiffness has been introduced into the simulation of origami using the bar-and-hinge discretization by [Schenk and Guest 11] and extended by [Tachi 13] to interactively analyze stiffness properties of folded structures. This method has been used by [Filipov et al. 15] to study the mechanical properties of the so called zipper origami tubes. These versions of the model used infinitesimal definition of bending rotation around the bars and hinges and could not account for large displacements and rotations. The nonlinear kinematics for the bar-and-hinge crease pattern representation has been derived recently by [Liu and Paulino 17].

The present paper is organized as follows: The nonlinear kinematics of rotation around the crease lines based on the facet rigidity condition is summarized in Sec. 2 in a form suitable for a vectorized implementation. The facet rigidity is combined with the energetic potentials that selectively introduce design relevant stress resultants into the model as exemplified here on the case of the bending moments in the crease lines are described in Sec. 3. This choice is relevant for the case of oricrete shells considered here and for the stress analysis upon manipulation conditions. To indicate the intended direction of development, the interesting properties of the waterbomb tesellation with six-crease base for design of shells with adjusted mechanical properties are exemplified in Sec. 4. Finally, in Sec. 5 the simulation of structural response under self-weight for different orientations and boundary conditions is exemplified and discussed.

## 2 Intermediate configuration of a crease pattern

### 2.1 Vertex, line and facet parameters

The geometrical configuration of a crease pattern during the folding process or due to a loading action is defined by the vertex positions

$$\mathbf{x}_N(t) \in \mathbb{R}^3, N \in \mathcal{N},\tag{1}$$

**Figure 2:** *Examples of transport states with non-optimal force flow.*

where $\mathscr{N}$ denotes the set of all vertices. The difference between the current and initial position is introduced as

$$\mathbf{u}_L^t = \mathbf{x}_L(t) - \mathbf{x}_L(0). \tag{2}$$

The crease pattern topology is defined using triples of vertices constituting triangular facets as

$$N(F,n) \to N; \quad F \in \mathscr{F}, \, n \in (0,1,2), \, N \in \mathscr{N}, \tag{3}$$

where $\mathscr{F}$ denots the set of all facets and $n$ enumerates the lines within a facet $F$. Based on this mapping, the oricreate package delivers three derived mappings:

$$N(L,n) \to N; \quad L \in \mathscr{L}, \, n \in (0,1), \, N \in \mathscr{N}, \tag{4}$$

$$L(F,\ell) \to L; \quad F \in \mathscr{F}, \, \ell \in (0,1,2), \, L \in \mathscr{L}, \tag{5}$$

$$\Phi(L,f) \to F; \quad L \in \mathscr{L}, \, f \in (0,1), \, F \in \mathscr{F}. \tag{6}$$

Here $N(L,n)$ returns the global index of a line $\ell$ in a facet $F$, $L(F,\ell)$ the global index of a node $n$ of the line $L$, and $\Phi(L,f)$ the global index of a facet $f$ attached to line $L$. The introduced mappings are used in the computational framewok as index functions to remap the geometrical parameters of lines and facets into a form that reflects the topology of the crease pattern. They establish the infrastructure for an efficient vectorized evaluation of energetic potentials and constraints.

Two geometrical parameters represent the starting point of evaluation in the simulation tasks considered here: direction vectors of the interior crease lines $\mathbf{v}_L$, and normal vectors of the facets $\mathbf{n}_F$. Both have been provided within the core functionality of the package oricreate. The array of vectors along all crease lines is defined as

$$\mathbf{v}_L^t = \mathbf{x}_{N(L,1)}^t - \mathbf{x}_{N(L,0)}^t. \tag{7}$$

The calculation of the facet normals $\mathbf{n}_F$ is done using the linear shape functions. Any point within a triangular facet can be addressed using barycentric coordinates $\eta_1, \eta_2 \in (0,1) \wedge \eta_1 + \eta_2 = 1$. Thus, for a facet $F$ the global coordinates of a point within the facets are obtained as

$$\mathbf{r}_F^t(\eta) = \eta_1 \mathbf{x}_{N(F,1)}^t + \eta_2 \mathbf{x}_{N(F,2)}^t + (1 - \eta_1 - \eta_2)\mathbf{x}_{N(F,3)}^t. \tag{8}$$

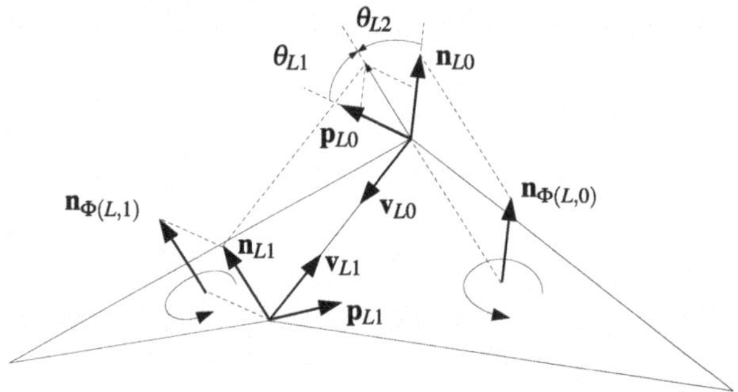



**Figure 3:** *Calculation of instantaneous dihedral angle.*

The normal vector of the facet $F$ is then obtained as a vector product of the derivatives of the reference vector $\mathbf{r}$ with respect to the parametric coordinates, i.e.

$$\mathbf{n}_F^t = \partial \mathbf{r}_F^t / \partial \eta_1 \times \partial \mathbf{r}_F^t / \partial \eta_2. \tag{9}$$

The surface area of the facet then reads

$$a_F^t = \frac{1}{2} \sqrt{\mathbf{n}_F^t \cdot \mathbf{n}_F^t}. \tag{10}$$

For later use, let us finally introduce the normalized version of the crease line vector and of the normal to the facet as

$$\hat{\mathbf{v}}_L^t = \mathbf{v}_L^t / \|\mathbf{v}_L^t\| \quad \text{and} \quad \hat{\mathbf{n}}_F^t = \mathbf{n}_F^t / \|\mathbf{n}_F^t\|. \tag{11}$$

### 2.2   Rotation around a crease line

Consider two adjacent facets shown in Fig. 3. The following conventions have been introduced:

- In a planar, unfolded configuration of the crease pattern, all normal unit vectors have the value $\hat{\mathbf{n}}_F = \{0, 0, 1\}^T$.

- Unit line vectors $\hat{\mathbf{v}}_L$ are mapped into an array $\hat{\mathbf{v}}_{F\ell}$, $\ell = (0, 1, 2)$ that makes the line vectors addressable via the facets and the local line number within the facet. The line vectors are ordered and directed counter-clockwise around the normal vector $\hat{\mathbf{n}}_F$.

- Starting from an interior line, its unit line vector can be accessed via two adjacent facets using the index function $\Phi(L, f)$

$$\hat{\mathbf{v}}_{Lf} = \hat{\mathbf{v}}_{\Phi(L,f)\ell(\Phi(L,f),L)}.$$

The additional index function $\ell(F, L)$ delivers the local index (0,1,2) of the line $L$ within the facet $F$. Due to the counter-clockwise enumeration and orientation of line vectors within a facet, the line vectors $\hat{\mathbf{v}}_{L0}$ and $\hat{\mathbf{v}}_{L1}$ have opposite directions.

- The normal unit vectors of two facets $f = 0, 1$ connected via the interior line $L$ are further abbreviated as

$$\hat{\mathbf{n}}_{Lf} = \hat{\mathbf{n}}_{\Phi(L,f)}.$$

The angle between these two vectors defines the dihedral angle between the neighboring facets.

To get the array of dihedral angles around all interior lines and their derivatives with respect to the unknown nodal coordinates we apply the following procedure: Find the in-plane outward vector of the facet $\Phi(L, f)$ by calculating the vector product

$$\hat{\mathbf{p}}_{Lf} = \hat{\mathbf{v}}_{Lf} \times \hat{\mathbf{n}}_{Lf} \tag{12}$$

and assemble an orthonormal basis for the line $L$ attached to the facet $f = 0$ by stacking the base vectors along the new dimension $j$

$$T_{L0ij} = \begin{bmatrix} v_{L0i} & n_{L0i} & p_{L0i} \end{bmatrix}. \tag{13}$$

Then, transform the normal vector of the second facet $n_{L1}$ into this basis

$$\hat{n}_{L1i} = T_{L0ji} n_{L1j}. \tag{14}$$

As indicated in Fig. 3, the first component of the vector $\hat{n}_{L10}$ is zero. The second and third components $\hat{n}_{L11}$ and $\hat{n}_{L12}$ represent the cosines and sines of the dihedral angles, respectively. To calculate the dihedral angle in the whole range of possible values $\psi_L \in (-\pi, \pi)$ we first calculate the angles with respect to the axes of the orthonormal base $\hat{n}_{L1(1,2)}$ as

$$\theta_{L(1,2)} = \arcsin\left(\hat{n}_{L1(1,2)}\right). \tag{15}$$

The dihedral angle can then be obtained by combining the two angles as

$$\psi_L = \theta_{L2} + \mathrm{sign}(\theta_{L2})(\mathrm{sign}(\theta_{L1}) - 1)\theta_{L1}. \tag{16}$$

The dihedral angles is a function of the vector of nodal positions changing during the folding or loading process controled by the parameter $t$, i.e. $\mathbf{x}_N(t)$. To simplify notation we shall denote the dihedral angle at staet of folding $t$ given by the vertex positions $\mathbf{x}_N(t)$ as $\psi_{t,L} = \psi_L(\mathbf{x}_N(t))$.

The described evaluation of dihedral angles serves as a basis for simulations of both the folding process controlled by a fold angle, and for the analysis of deformation behavior of a crease pattern with crease lines with finite bending stiffness. The implemented plugin module also evaluates the derivatives of $\psi$ with respect to vertex displacements. With the introduced index functions, the derived equations can be directly transformed to the implementation using the using the Einstein summation rule provided in the used numpy package.

## 3  Rigidity, potential energy and bending deformation energy

The geometrical and physical criteria defining the model are introduced by specifying the relations between the geometrical parameters of the crease pattern, e.g. line vectors, facet normals, dihedral angles, sector angles, etc. In the present study, the functions delivering the scalar value of the total potential energy enter the goal function

$$\Pi(\mathbf{u}_N) = U_\phi(\mathbf{u}_N) - W_g(\mathbf{u}_N) - W_P(\mathbf{u}_N). \tag{17}$$

The state that renders the minimum of this function while satisfying the condition of facet rigidity $G(\mathbf{u}_N)$ represents the sought physically and geometrically admissible configuration:

$$\min_{\mathbf{u}^l} \Pi(\mathbf{u}_N^l) \quad \text{subject to} \quad \boldsymbol{G}(\mathbf{u}_N^l, t) = 0 \tag{18}$$

The components of this optimization problem are briefly described in the sequel.

### 3.1  Facet rigidity

The rigidity constraint requires that the length of crease lines does not change during the folding or during the loading process. In the initial configuration the length of a crease line $L$ is evaluated as the norm of the vector introduced in Eq. (7)

$$\|\mathbf{v}_L(0)\| = \|\mathbf{v}_L(t)\|. \tag{19}$$

Using Eq. (2) with $t$ parameter omitted the constant-length constraint (Eq. 19) for a single crease line $L$ can be written as

$$G_L := \|\mathbf{v}_L(0) + \mathbf{u}_{L1}(t) - \mathbf{u}_{L0}(t)\| - \|\mathbf{v}_L(0)\| \tag{20}$$

$$= 2\mathbf{v}_L^0 \cdot \mathbf{u}_{L1} - 2\mathbf{v}_L^0 \cdot \mathbf{u}_{L0} + \mathbf{u}_{L1} \cdot \mathbf{u}_{L1} - 2\mathbf{u}_{L1} \cdot \mathbf{u}_{L0} + \mathbf{u}_{L0} \cdot \mathbf{u}_{L0} = 0. \tag{21}$$

By applying the length-constraint to all crease lines and boundary edges of the crease pattern a system of $\text{len}(\mathcal{L})$ quadratic equations with unknown $3 \cdot \text{len}(\mathcal{N})$ nodal displacements $\mathbf{u}_N$ is obtained in the form

$$G_L(\mathbf{u}_N) = 0. \tag{22}$$

To support gradient-based optimization strategies, each criterion implementation includes both the value and its derivatives with respect to the displacement vector $\mathbf{u}$. The contributions to the constraint derivatives of Eq. (20) evaluated for a line $L$ are

$$\partial G_L^{(\ell)}/\partial \mathbf{u}_{L0} = -2\mathbf{v}_L^{0T} - 2\mathbf{u}_{L1}^T + 2\mathbf{u}_{L0}^T, \quad \partial G_l^{(\ell)}/\partial \mathbf{u}_{L1} = 2\mathbf{v}_L^{0T} - 2\mathbf{u}_{L0}^T + 2\mathbf{u}_{L1}^T.$$

These terms are assembled in an algorithmic system matrix required for the calculation of the predictor step. By applying this constraint, we regard the crease pattern as a pin-jointed framework of the crease lines. The constraint delivers a set of quadratic equations capturing the kinematics of folding that can be included as equality constraints in the optimization problem (Eq. 18).

Additional kinematic constraints can be introduced to control the free degrees of freedom of the crease pattern and to solve the problem using the Newton-Raphson method. In both cases, the derivatives of the constraint with respect to the nodal displacements $\mathbf{u}_N^t$ are applied as a predictor for the next trial solution during the iterative solution procedure. For crease patterns with triangular facets, this assumption provides the necessary and sufficient condition for a rigid foldable crease pattern. For quadrilateral and polygonal facets additional constraint equations preserving the planarity of facets must be included.

## 3.2 External energy

The potential energy of gravity of a generally shaped shell $S$ with parametric coordinates $\mathbf{s} = [\eta_1, \eta_2]$ is given by a surface integral of its global vertical coordinate $r_3(\mathbf{s})$ multiplied by the thickness, material density $\rho$ and gravity acceleration $g$

$$W_{gS} = \int_S \rho g h \, r_3(\mathbf{s}) \, \mathrm{d}\mathbf{s}. \tag{23}$$

In the present case of a triangulated crease pattern, we can express the potential energy of gravity in a single facet $F$ using the barycentric coordinates introduced in Eq. (8) as

$$W_{gF} = \int_0^1 \int_0^{1-\eta_2} \rho g h \, r_{[F,3]}(\eta, \mathbf{x}) \, a_F \, \mathrm{d}\eta_1 \mathrm{d}\eta_2, \tag{24}$$

where $a_F$ represents the surface area given in Eq. (10). By applying a numerical quadrature over a triangular shape we can evaluate this integral expression in a closed form. The parametric coordinates of the quadrature points are $\eta_1 = \eta_2 = 1/3$ and a weighting factor of $w = 1/2$ lead to the following closed form expression for the overall potential energy of gravity within the crease pattern

$$W_g^t = \frac{1}{6} \rho g h \sum_{F \in \mathscr{F}} a_F^t \sum_{n=0}^2 \mathbf{x}_{\mathrm{N}(F,n)}^t. \tag{25}$$

Based on this formulation, analytical derivatives $\partial W_g / \partial \mathbf{u}_N$ have been derived using the chain rule. Let us note, that in the described case of gravity potential an explicit expression can be derived without using barycentric coordinates and numerical quadrature. However, we have used this description as it directly corresponds to the extensible implementation in the code to indicate that other energetic potentials, that are defined as integrals of more complex functions over the facet domain can be easily implemented within the framework. Examples of such potentials are elastic bending energy of facets or energy stored in in-plane facet deformation.

The second considered type of loading is represented by discrete forces $P_N$ at individual vertices $N$. The work exerted by these forces is given simply as

$$W_P = \sum_{N \in \mathscr{N}} \mathbf{P}_N \mathbf{u}_N.$$

### 3.3  Stored bending deformation energy

Focusing on the analysis of the bending deformation around the crease lines let us extend the model with a potential expressing the bending deformation energy around the crease lines. Assuming that there is an energetically admissible configuration of the crease pattern at the initial state of folding or loading $t = 0$ with the nodal coordinates $\mathbf{x}_N(0)$ and the corresponding dihedral angles $\psi_L^i$, the bending deformation corresponding to the control parameter $t$ is expressed by the rotation

$$\varphi_L^t = \psi_L^t - \psi_L^i.$$

Further assume that this rotation will result in a bending moment of the magnitude

$$M_L = \kappa_L \Lambda_L \varphi_L,$$

where $\kappa$ and $\Lambda$ are the bending stiffness of the crease line of a unit length and the length of the crease line, respectively. For simplicity, we assume linear elastic behavior. However, an extension to non-linear relation between the rotation and bending moment is straightforward.

To introduce the force flow into the global model of the crease pattern we evaluate the elastic bending energy stored in all interior crease lines as

$$U_\phi = \sum_{L \in \mathscr{L}} \frac{1}{2} M_L \varphi_L = \frac{1}{2} \kappa_L \Lambda_L \varphi_L^2.$$

Again, the derivatives of the potential with respect to the nodal displacements $\mathbf{u}_N$ have been derived and implemented in the vectorized way as a fundamental assumption for an efficient solution algorithm. Extension of the potential expressing the shear force along the crease line or a normal force accounting for tension and compression between two adjacent facets can be done analogically.

## 4  Simulation of the folding process

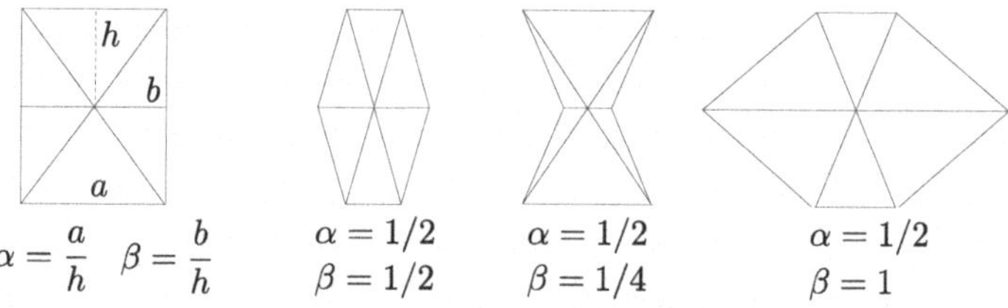


**Figure 4:** *Parameterized waterbomb base.*

The waterbomb tessellation with a six-crease base provides interesting properties for the design of structural elements with required high mechanical performance. By adjusting the ratio of crease line lengths within the hexagonal base,

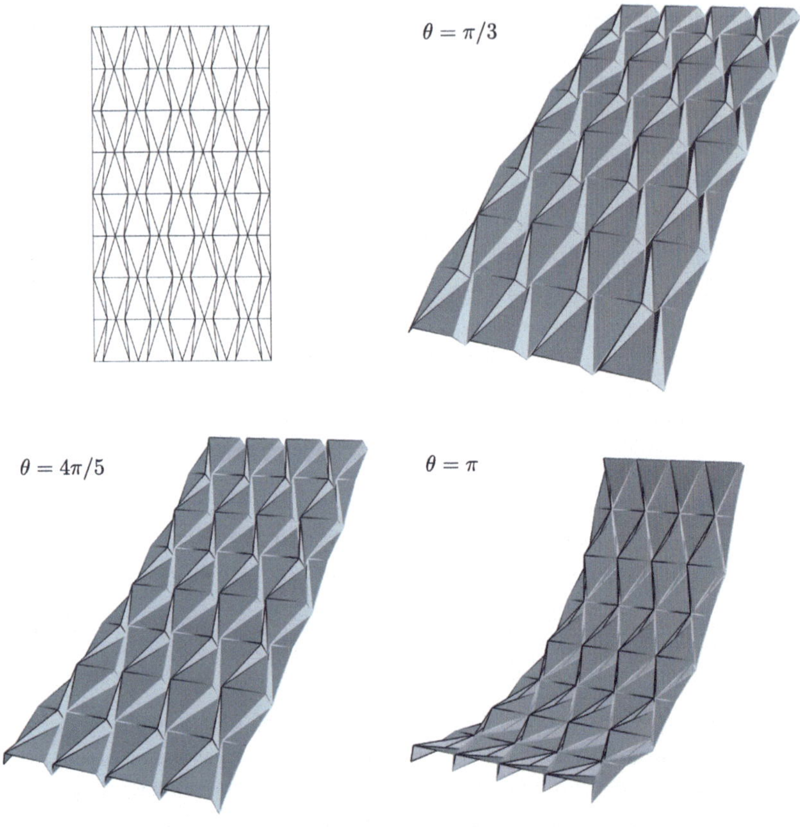



**Figure 5:** *Example of parameterized waterbomb crease tessellation with six-crease base for three different values of the control dihedral angle.*

the global curvature of the folded shell can be adjusted independently from the resulting local height of the folded shell cross section. The versatility of the pattern has been described in [Chen et al. 16]. Examples of studied tessellation bases are shown in Fig. 4.

The correspondence between the parameters of the tessellation base and the global shape is exemplified in Fig. 5 for the values $\alpha = 0.218$ and $\beta = 0.312$. These values have been selected to show the ability of the crease pattern to achieve both positive and negative global curvature. The folding process is shown for three values of the control dihedral angle. The simulation was purely kinematic and the system of non-linear equations were solved using the Newton-Raphson iteration scheme for each increment of the control angle $\eta$. It should be noted that the geometrical mapping between the parameters $\alpha, \beta$ and $\eta$ and the final shape can

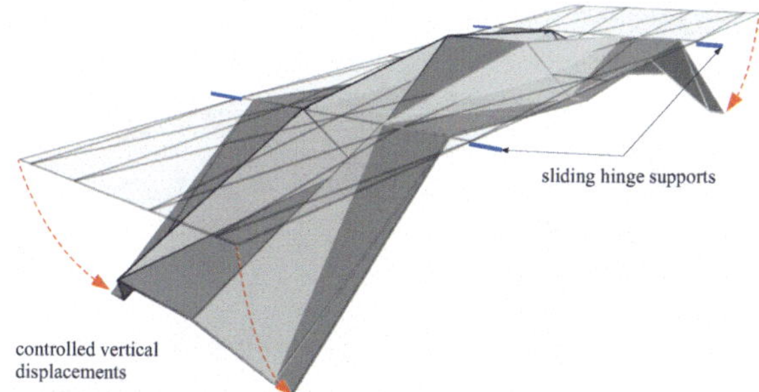


**Figure 6:** *Simulation of the folding process considering the self-balancing forces of self weight and the folding kinematics of the waterbomb crease pattern.*

be expressed analytically. The purpose of the performed studies was to verify the model component implementing the rigidity constraints described in Sec. 3.1, and to identify the crease pattern suitable for the production of the oricrete shells shown in Fig. 1.

Once the geometry for the production of oricrete shell prototypes has been selected, The simulation task has been extended with the model component representing the gravity load to solve the technological questions included in the process of folding. These questions were related to the construction of the folding formwork, positions of the supports that are suitable for the distribution of the self weight during the folding or the question of the maximum slope of a facet occurring during the folding process to verify that fresh concrete would not flow down along the facet during the folding. Based on these simulations a foldable formwork has been designed that allowed for a simple manual control of the folding process. The sketch of the folding process is shown in Fig. 6.

The simulations exemplified in Figs. 5 and 6 have been performed using different solution strategies. In the former case of rigid-facet folding simulation it was possible to define the same number of constraints as the number of degrees of freedom. The derivative of the facet rigidity condition than delivered square system matrix, so that the Newton-Raphson scheme could be applied, resulting in very fast convergence, of maximum three iteration steps in an increment. The shown model has 363 degrees of freedom, 320 of them are constrained by the constant length condition specified in Sec. 3.1 and 6 nodal displacement are fixed to prevent rigid body motions. Thus, the model possesses 37 free degrees of freedom that have been constrained by introducing symmetry conditions. This is done by linking the 36 dihedral angles across the tessellation. The remaining single degree of freedom was used to control the calculation.

On the other hand, for problems including the minimization of energy potential the sequential least squares programming optimization algorithm was applied [Kraft 88]. In the simulation of the folding process of the shell prototype shown in Fig. 6, the gravity force has been included so that the simulation was controlled by the optimization algorithm. Such type of simulation needs more iterations to find a solution. However, the computational of the models shown here is still in an acceptable range of milliseconds on a standard computer. The advan-

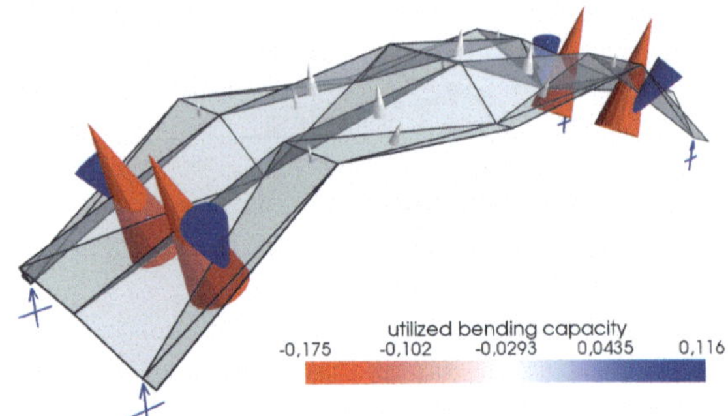



**Figure 7:** *Distribution of bending moments normalized with respect to the bending capacity of a cross section due to self-weight load.*

tage of this model configuration over the purely kinematic simulation described before is a higher flexibility in the definition of boundary conditions and more "intuitive" behavior of the simulated shell within a gravity field. For example, folding path leading to a symmetric shape is entered without the need to specify additional kinematic links.

## 5   Deformation behavior of the shell during transport

The folded configuration of the structure is used in the next simulation step to study its resistance to self-weight loading conditions that can occur during the transport from the formwork to the position within the structure.

The result of the simulation is shown Fig. 7. The size of the cones displayed at the middle of each crease line has the meaning of the magnitude of the bending moment. The orientation of the cone indicates the orientation of the bending moment with respect to the cross sectional height. The base of the cone is at the side of the shell with tensile bending stress and the top of the cone on the side with compressive bending stress.

The studied case corresponds to the transport state of the oricrete shell when lifted from the formwork as shown in the left picture of Fig. 2. The magnitude shown in the legend of the figure represents the utilization of the bending capacity of the cross section. It has been calculated as

$$\mu_L = m_L/m_u \tag{26}$$

where $m_L$ denotes the bending moment per unit length in a crease line $L$ and $m_R$ represents the ultimate bending moment per unit crease line length. The studied shell was $h = 0.01$ m, the Young's modulus of concrete was $E = 21$ GPa so that resulting bending stiffness was $\kappa = Ebh^3/12 = 1.75$ kN/m$^4$/m. With the assumed concrete strength of $\sigma_u = 3$ MPa the ultimate cross-sectional bending moment was $m_u = \sigma_u bh^2/6 = \pm 0.05$ kNm/m. With these characteristics, the simulated maximum utilization $\mu_L$ of a fold line was 17%. The corresponding deflection of the shell was 0.6 mm.

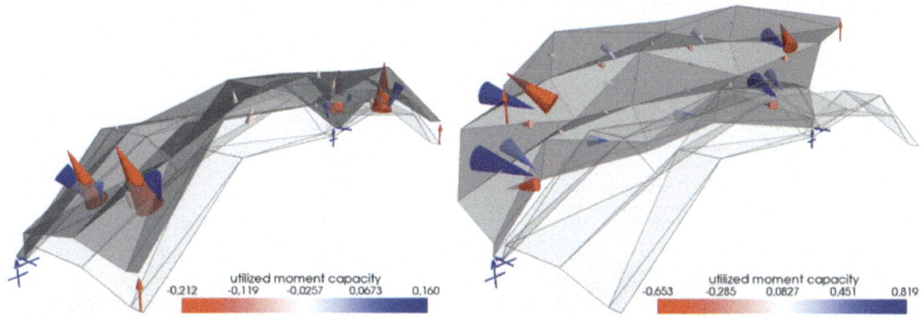

**Figure 8:** *Simulation of the oricrete shell rotation into an upright position with instantaneous evaluation of bending moments due to self weight.*

The study in Fig. 8 shows the distribution of the bending moments in an upright configuration of the shell due to self weight. This study corresponds to the situation in Fig. 2b. The results show that the crease lines with the largest cones would reach up to 82% of their bending moment capacity. This result is qualitatively plausible as the self weight in this configuration activates the folding kinematics of the waterbomb crease pattern.

The simulation shown in Fig. 9 demonstrates that the model accounts for large rotations in combination with the analysis of the stress resultants. In the studied example, a vertical load has been applied at the top two nodes of the shell. The three shown stages represent the true, non-scaled deformation.

The real behavior of the shell exhibits physical nonlinearity owing to the development of finely distributed cracks over the reinforced shell facets. The model component that accounts for this kind of strain-hardening behavior is prepared to be integrated into the framework to provide realistic prediction of the structural performance in terms of the ultimate load and ductility [Chudoba et al. 16].

# 6   Conclusions

The paper describes the concept of modeling approach with governing geometrical and physical criteria that can be combined depending on the simulation problem at hand. The modeling approach has been motivated by questions posed during the development of technology for the production of thin folded plate structures out of textile reinforced concrete.

In particular, the model was used to study the intermediate states of folding of the oricrete shell produced using the green folding method, i.e. folding of the fresh state of concrete matrix. The foldable formwork has been designed with regard to the distribution self weight of the folded structure so that minimal forces needed to be used to initiate and control the folding process.

The waterbomb crease pattern with six-crease base was used to produce a vault with adjustable cross sectional height. The resulting spatial structural shape is ide-

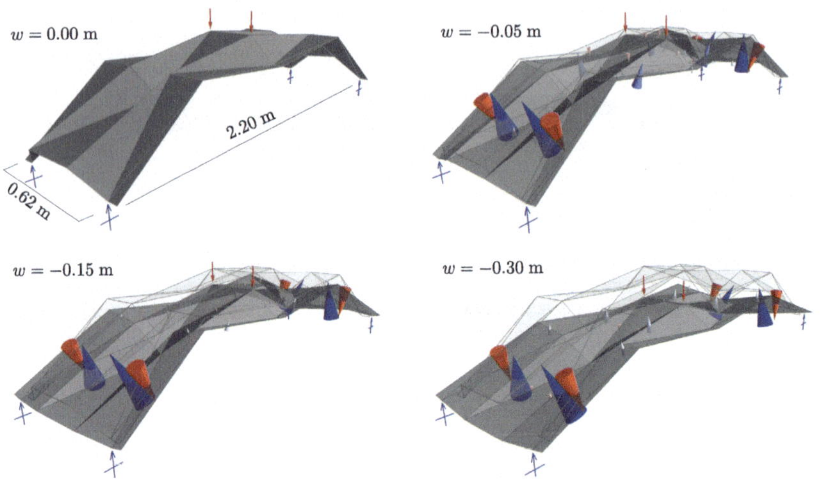



**Figure 9:** *Vertical displacement loading applied at the top of the shell to invoke large deflections and large rotations.*

ally suited for the production of light weight highly stiff and highly ductile elements made of carbon concrete composite in the target configuration.

However, the structure may be highly vulnerable to loading conditions occurring during the transport. The present model configuration provides an efficient simulation methods to study the behavior of the shell within a gravity field. Even though the model was developed with a focus on the oricrete technology, the modeling approach is general and can be enhanced and applied for a wider range of questions included in the design of folded structures.

# References


[Chen et al. 16] Yan Chen, Huijuan Feng, Jiayao Ma, Rui Peng, and Zhong You. "Symmetric waterbomb origami." *Proceedings of the Royal Society A: Mathematical, Physical and Engineering Science* 472:2190 (2016), 20150846. doi:10.1098/rspa.2015.0846. Available online (http://rspa.royalsocietypublishing.org/lookup/doi/10.1098/rspa.2015.0846).

[Chudoba et al. 14] R Chudoba, J van der Woerd, M Schmerl, and J Hegger. "ORICRETE: Modeling support for design and manufacturing of folded concrete structures." *Advances in engineering software* 72 (2014), 119–127.

[Chudoba et al. 15] Rostislav Chudoba, Jan Dirk van der Woerd, and Josef Hegger. "ORICREATE: Modeling framework for design and manufacturing of folded plate structures." In *Origami6 - Part II: Technology, Art, Education*. Koryo Miura, Toshikazu Kawasaki, Tomohiro Tachi, Ryuhei Uehara, Robert J. Lang and Patsy Wang-Iverson, American Mathematical Society, 2015.

[Chudoba et al. 16]  R Chudoba, E Sharei, and A Scholzen. "A strain-hardening microplane damage model for thin-walled textile-reinforced concrete shells, calibration procedure, and experimental validation." *Composite Structures* 152 (2016), 913–928.

[Filipov et al. 15]  Evgueni T. Filipov, Tomohiro Tachi, and Glaucio H. Paulino. "Origami tubes assembled into stiff, yet reconfigurable s tructures a nd m  etamaterials." *Proceedings of the National Academy of Sciences* 112:40 (2015), 12321–12326. doi:10.1073/pnas.1509465112.    Available online (http://www.pnas.org/content/112/40/12321).

[Kraft 88]  D. Kraft. "A software package for sequential quadratic programming." Tech. rep. DFVLR-FB 88-28, DLR German Aerospace Center  Institute for Flight Mechanics, Köln, Germany, 1988.

[Liu and Paulino 17]  K Liu and GH Paulino.  "Nonlinear mechanics of non-rigid origami: an efficient computational approach." *Proc. R. Soc. A* 473:2206 (2017), 20170348.

[Resch and Christiansen 70]  R. D. Resch and H. Christiansen. "The Design and Analysis of Kinematic Folded Plate Systems." *Proceedings of IASS Symposium on Folded Plates and Prismatic Structures.*

[Schenk and Guest 11]  Mark Schenk and Simon D Guest. "Origami folding: A structural engineering approach." In *Origami 5: Fifth International Meeting of Origami Science, Mathematics, and Education*, pp. 291–304. CRC Press, Boca Raton, FL, 2011.

[Tachi 13]  Tomohiro Tachi. "Interactive form-finding of elastic origami." In *International Association for Shell and Spatial Structures Symposium (IASS), Wrocław, Poland, Sept*, pp. 23–27, 2013.

[van der Woerd et al. 15]  Jan Dirk van der Woerd, Rostislav Chudoba, and Josef Hegger. "Design and construction of a thin barrel vault by folding." In *Future Visions. Proceedings of the IASS 2015 Symposium.* Amsterdam, 2015.

[van der Woerd et al. 16]  Jan Dirk van der Woerd, Rostislav Chudoba, and Josef Hegger. "Folded bike shell-ter: Application of oricrete design and manufacturing method." In *Proceedings of the IASS Annual Symposium 2016.* Tokyo, Japan, 2016.

[van der Woerd et al. 17]  J. D. van der Woerd, C. Bonfig, J. Hegger, and R. Chudoba. "Construction of a vault using folded segments made out of textile reinforced concrete by fold-in-fresh." In *Interfaces: architecture.engineering.science (IASS) Symposium. published on USB stick.* Hamburg, 2017.

---

Rostislav Chudoba
RWTH Aachen University, Germany,
e-mail: rostislav.chudoba@rwth-aachen.de

Karl-Heinz Brakhage
RWTH Aachen University, Germany,
e-mail: brakhage@igpm.rwth-aachen.de

# Quasi-static Crushing Behaviours of Folded Open-top Truncated Pyramid Structures with Interconnected Side Walls

*Zhejian Li, Wensu Chen, Hong Hao*

**Abstract:** *Three open-top truncated pyramid folded structures with different base shapes i.e. triangle, square and pentagon, are proposed in this study. Their crushing behaviours are investigated under quasi-static loading condition both experimentally and numerically. Superior performances of truncated triangle pyramid (TTP) are demonstrated, with higher average crushing stress, larger densification and lower initial peak stress over the conventional Miura-type foldcore. This indicates higher capacity of energy absorption and more consistent deformation for the proposed folded structure. These foldcores can be potentially used in the application of energy absorption such as cladding or crush box.*

## 1 Introduction

Origami and kirigami structures have been recently studied extensively. As these structures are formed by folding two-dimensional sheet material into complex three-dimensional objects, they possess several preferable characteristics such as deployable, continuous manufacture, consistent crushing resistance etc. Because of these advantages, folded structures have been used in many applications, such as spaceship solar panel [Miura. 1985], origami-inspired clothing that grows with kids [Smithers. 2017], folded plate shelters [De Temmerman et al. 2007, Gioia et al. 2012], light weight bellows [Butler et al. 2017], deep sea pipelines [Albermani et al. 2011], crush boxes [Zhou et al. 2017, Zhou et al. 2017]and sacrificial cladding for blast loading protection [Li et al. 2017, Hao et al. 2018].

Miura-origami, as one of the mostly studied folded structure, was firstly proposed by Miura [Miura. 1972]. It is a tessellated zig-zag pattern folded from a single un-broken sheet material along the creases without stretching or twisting of the faces. The creases are consisted of segments of straight lines. It was introduced as core of sandwich structures with the advantage of continuous manufacturing and open channel design to reduce accumulated heat and moisture inside the core of the panel [Liu et al. 2015]. The crushing resistance of Miura-origami, however, is not as comparable as to honeycomb core of same density [Gattas and You. 2014]. It also has non-uniform collapsing with a high peak force under out-of-plane impact caused by plate buckling failure mode of the core [Heimbs. 2013].

Curved-crease foldcores were then developed. Instead of straight segment of creases as in Miura-type, the creases of this type of foldcore are curved. Different variations of curved-crease foldcores were studied for their crushing behaviours [Gattas and You. 2014, 2014, 2015]. A higher and more consistent crushing resistance is shown comparing to the standard Miura-type foldcore. Different from Miura-type and curved-crease foldcore, the sheet material can be cut, stamped or punched prior to the folding process for kirigami foldcores [Nojima and Saito. 2006]. Pre-cut of sheet material allows the object with more complicated geometry to be folded and potentially further increases the crushing resistance of the folded structures. As reported from the studies, the cube strip and diamond strip kirigami foldcores have greatly outperformed the standard Miura-type and curved-crease foldcores with higher average crushing stress and more uniform collapsing [Fathers et al. 2015].

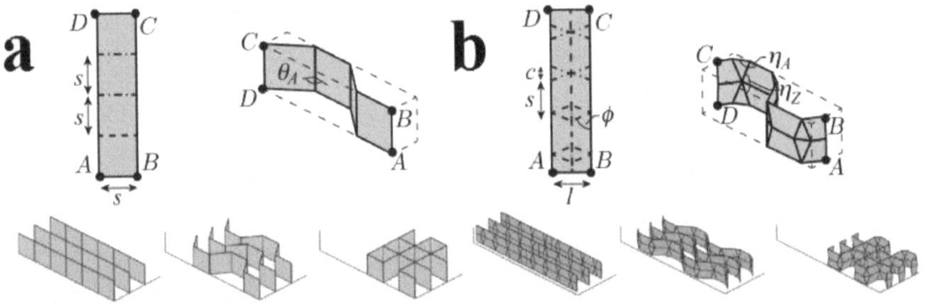

**Figure 1:** *(a) Cube strip; (b) Diamond strip kirigami foldcore [Fathers et al. 2015]*

However, both of the cube and diamond strip kirigami structures are folded individually by strip and placed next to other rows. Therefore, continuous fabrication cannot be achieved. Furthermore, the adjacent vertical faces in many kirigami structures are not connected. For instance, vertical faces between each row are placed individually and not connected for both cube and diamond strips, as shown in Figure 1. Improvement in crushing resistance is expected if vertical faces of the foldcore is connected to provide constraints to resist out-of-plane loading, though the connected sidewalls could lead to the increase in the initial peak crushing force and sensitivity to strain rate because of the inertial stabilization effect provided by the fully connected vertical sidewalls, similar to honeycomb structure [Xue and Hutchinson. 2006]. An open-top truncated square pyramid (TSP) folded structure was proposed [Li et al. 2017, 2018]. To increase the crushing resistance while remain the uniform collapsing and strain rate insensitivity, the proposed TSP foldcore has inclined sidewalls connected via triangular interconnections, as presented in Figure 2. Single sheet fabrication can be achieved in this pattern as well. Its structural behaviour under out-of-plane quasi-static and dynamic crushing was investigated and compared with the cube strip foldcore. Good performances including low initial peak stress, high average crushing stress and low strain rate sensitivity were obtained, indicating the potential application as energy absorber. Its blast mitigation capability as cladding core was also numerically studied and compared with honeycomb and aluminium foam [Li et al. 2017, Hao et al. 2018].

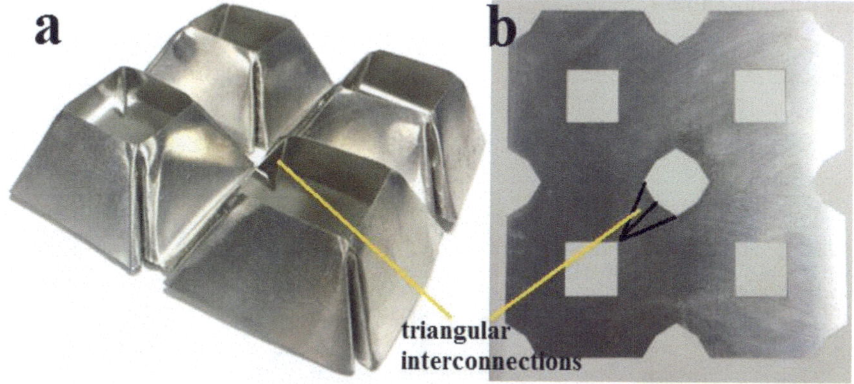


**Figure 2:** *(a) isometric view of hand-folded sample of truncated square pyramid with four unit cells; (b) patterned Aluminium sheet used for folding*

The out-of-plane crushing behaviours of truncated pyramid folded structures with different base shapes including triangle, square and pentagon are investigated in this paper, as listed in Figure 3. Samples of the foldcores are folded by hand using aluminium sheet and crushed under quasi-static loading. Numerical model is calibrated based on the experiment results and then used to simulate their crushing behaviour under dynamic crushing. Key criteria including initial peak stress, average stress and densification strain are used to evaluate and compare the performance of these foldcores under different loading conditions.

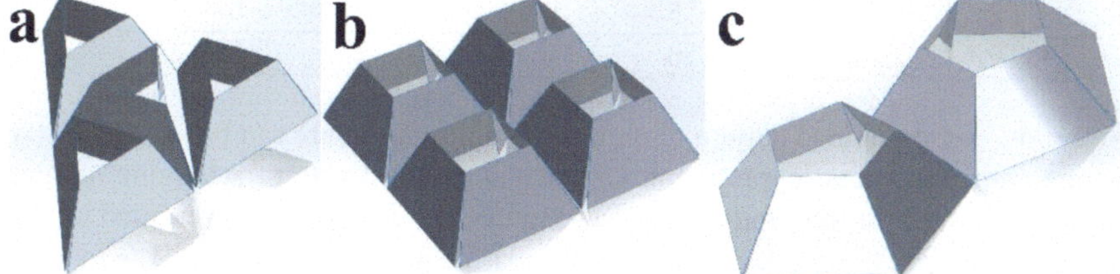


**Figure 3**: *open-top truncated pyramid structures (a) triangular; (b) square; (c) pentagonal*

## 2   Geometry parameters

Folding configuration and sample patterns of these three truncated pyramid kirigami structures can be seen in Figure 4. Three truncated pyramid structures are denoted as truncated triangular, square and pentagonal pyramid (named as TTP, TSP and TPP, respectively). Inclined sidewalls are connected with adjacent faces via triangular interconnections. A small folding gap of 0.5 mm at the corners of unit cell is considered in the numerical model as well. The heights of the foldcores are set to be equal to 20 mm with the top (*b*) and bottom edge (*a*) length of 20 and 40 mm, respectively. Other angles of the folds and inclination angle of sidewalls are listed in **Table 1**.

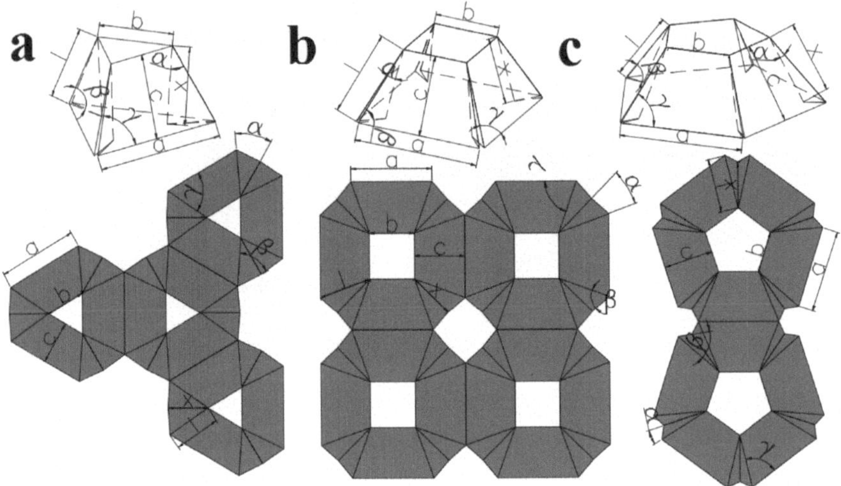

**Figure 4:** *Folding configurations with geometric parameters and folding creases for (a) TTP; (b) TSP; (c) TPP*

**Table 1:** *Folding angles of three foldcores*

| Foldcore | $\alpha$ (degree) | $\beta$ (degree) | $\gamma$ (degree) | Inclination angle (degree) |
|---|---|---|---|---|
| TTP | 34 | 60 | 64 | 74 |
| TSP | 22 | 55 | 67 | 63 |
| TPP | 14 | 50 | 68 | 55 |

Fully covered tessellated pattern of regular triangle and square can be formed easily without any gap between each unit. One of the simplest way of arranging regular pentagon is used for this study, shown in Figure 5. Small gaps are presented in this pattern, since it is impossible to arrange regular pentagons connected edge to edge without gap [Caspar and Fontano. 1996]. Therefore, the gap area between two pentagons is considered when calculating the crushing stress using crushing force and base area.

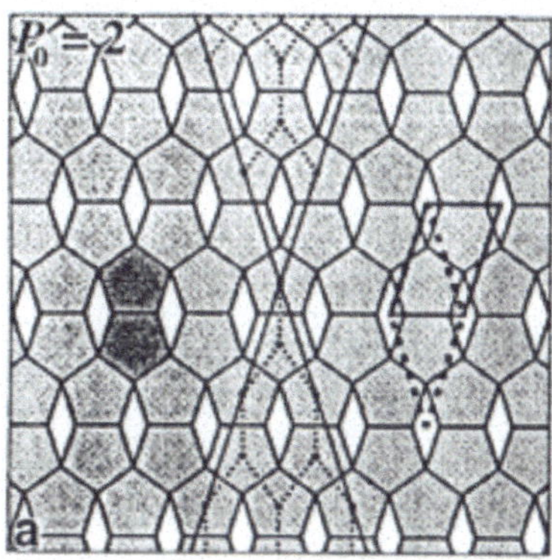

**Figure 5**: *One of the simplest tessellated pattern for regular pentagons [Caspar and Fontano. 1996]*

## 3   Model validation

### 3.1 Quasi-static crushing

Samples of three types are folded by hand and placed on base plates with a 2-mm high boundary to constrain the horizontal movements of bottom edges of the foldcores, as shown in Figure 6. No glue or any fixing is placed between the foldcores and the base plate. The samples are then crushed under quasi-static loading rate of 1mm/min. As given in Table 2, the thickness of aluminium sheet material and the relative density of each foldcore are listed. It is noted that the relative density of the TPP foldcore is smaller than TTP and TSP. As mentioned previously, the length of the top and bottom edges of each unit cell and their height are kept the same for all three types of folded structures (TPP, TSP and TTP).

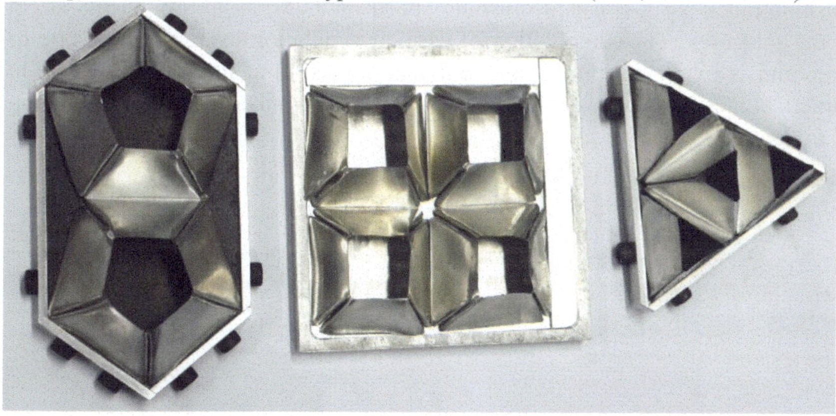

**Figure 6.** *Hand-folded samples of TPP, TSP and TTP (left to right) and their base plates with boundary*

Slight bent of the sidewalls and some uneven gaps between foldcore and base plate can be observed. These imperfections are caused by hand folding process and are unlikely to be avoided. Stamping or other advanced manufacturing method can be applied in the future to improve the quality and cost of the foldcore.

*Table 2: Parameters for hand-folded samples*

| Foldcore | a (mm) | b (mm) | H (mm) | t (mm) | $\rho_v$ % |
|----------|--------|--------|--------|--------|-----------|
| TTP | 40 | 20 | 20 | 0.15 | 2.7 |
| TSP | 40 | 20 | 20 | 0.26 | 2.7 |
| TPP | 40 | 20 | 20 | 0.26 | 1.7 |

Tensile test of aluminium sheet with thickness of 0.26mm used in the sample is also carried out according to the ASTM-E8M-04 [Astm. 2004]. The displacement and strain of the specimens are measured using Direct Image Correlation technique. Their true stress-strain data obtained from the test are listed in the Table 3.

*Table 3: True plastic stress-strain data of Aluminium 1060 sheet*

| Strain | 0 | 0.002 | 0.005 | 0.013 | 0.063 | 0.121 |
|--------|---|-------|-------|-------|-------|-------|
| Stress (MPa) | 0 | 66.7 | 112.3 | 120.1 | 125.8 | 130.6 |

## 3.2 Numerical simulation

LS-DYNA 971 is used for numerical simulation and Solidworks is used for model construction in this study. Folded structures are constructed using shell elements and placed between the base plate and the top crushing block. Same as in crushing test, the base plates with outer boundary are modelled, as shown in Figure 7. The bottom base plate is set to be fixed rigid solid block and the top crushing block has a constant crushing speed of 0.05 m/s to around 0.8 strain of the foldcores. The computation of explicit simulation using crushing rate of 1mm/min is extremely time consuming and the crushing speed used (2 m/s) in numerical model is found sufficient to accurately simulate quasi-static loading of foldcores [Fathers et al. 2015].

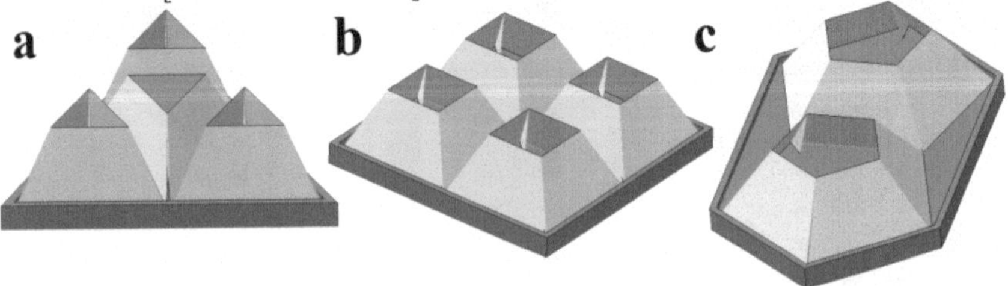


**Figure 7:** *Numerical models of (a) TTP; (b) TSP; (c) TPP; with base plate*

Material model *MAT024 PIECEWISE LINEAR PLASTICITY is used for aluminium sheet 1060 and its strain rate effect is not considered, and the material parameters are listed in Table 3 and Table 4. Self-contact of the foldcore itself during crushing is modelled using *CONTACT AUTOMATIC SINGLE SURFACE and the contact between foldcore to base plate is described by *CONTACT AUTOMATIC NODES TO SURFACE with friction considered, the friction coefficient is set to be 0.25 [Fathers et al. 2015].

*Table 4: Material properties of Aluminium 1060 sheet*

| Parameter | Young's modulus (GPa) | Poisson's ratio | Yield stress (MPa) | Density (kg/m$^3$) |
|---|---|---|---|---|
| Value | 69 | 0.33 | 66.7 | 2710 |

## 3.3 Model validation

Crushing resistance of three types of truncated foldcore (TTP, TSP and TPP) under quasi-static crushing test and numerical simulation are shown in Figure 8. Two key parameters, i.e., plateau stress and densification strain are quite well matched for all three types. The plateau stress is the average crushing resistance before the densification of the core, where the densification is shown as a sharp increase in the crushing resistance near the end of crushing caused by the compacted core. The deformation modes of the foldcores between numerical and experimental results are in good agreement as well. The deformation in numerical results are symmetrical whereas in experiments are not exactly symmetrical owing to imperfect specimens caused by hand folding. No tearing is noticed in both numerical and experiment results.

Some discrepancies of initial peak stress can be observed between numerical and experimental results. The higher value of initial peak stress and sharper rise of crushing resistance at early stages are demonstrated for all three types of truncated pyramid structures, especially for TTP and TSP. As mentioned in section 3.1, the hand folding induced imperfections such as bent sidewalls and uneven levelling of the foldcore, lead to the uneven loading and easier buckling process in the early stage of the crushing. Similar reduction of initial peak crushing resistance in numerical simulation can be observed in other folded structures as well [Fathers et al. 2015], due to the folding induced imperfections. Other key parameters for evaluating energy absorption including plateau stress and densification strain are in good agreement between numerical and experimental results. Therefore, the numerical models of these open-top truncated pyramid structure are acceptable for the study of their crushing behaviour and energy absorption.

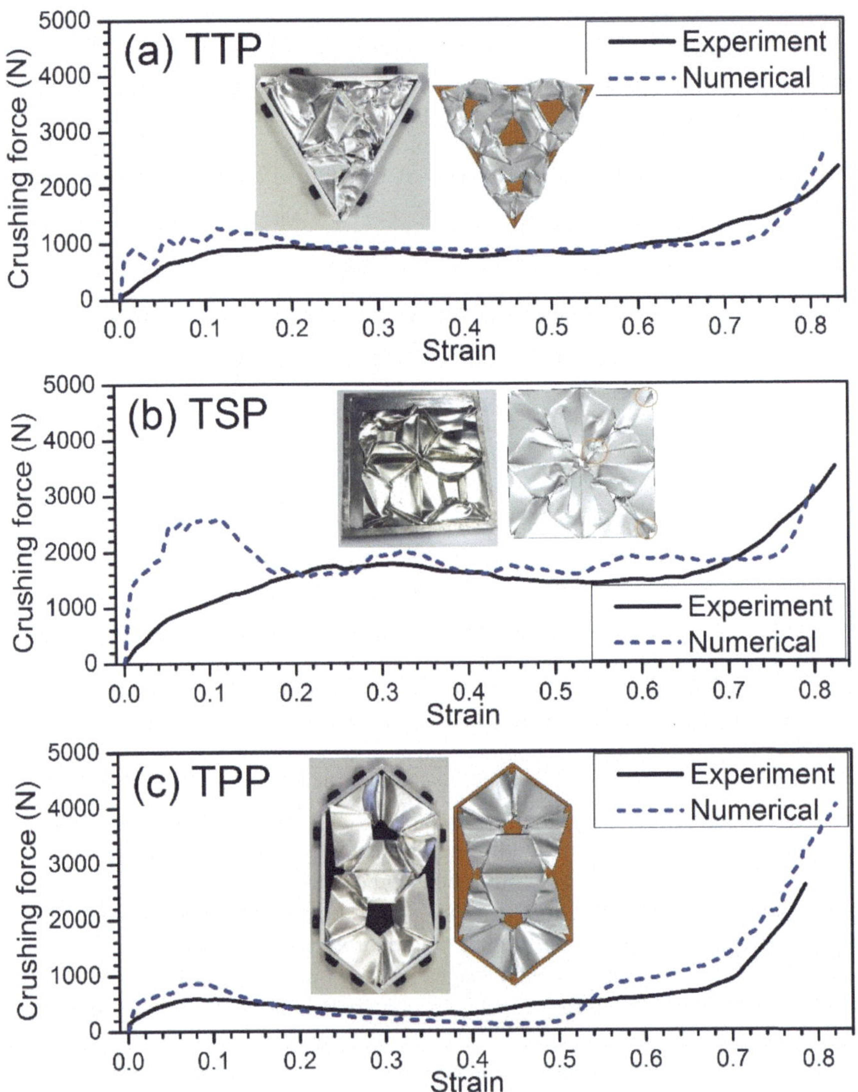



**Figure 8:** *Comparison of stress-strain curves and deformation mode of (a) TTP; (b) TSP; (c) TPP between numerical simulation and experiment data*

## 4   Comparison with Miura-origami foldcore

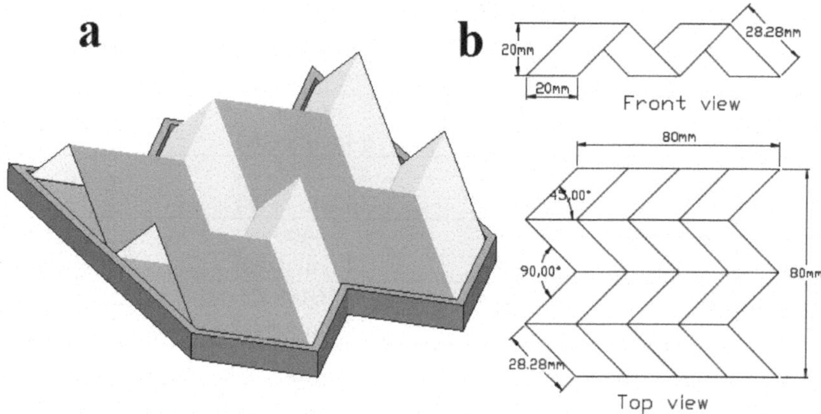



**Figure 9**: (*a*) *numerical model of Miura-type foldcore with four unit cells and base plate; (b) dimensions of Miura-type foldcore*

Structural responses of these three folded structures i.e. TTP, TSP and TPP are compared with one of the most common folded structure i.e. Miura-type origami. Numerical model of Miura-type foldcore is constructed with the same keywords in the finite element software. It has the base dimension of 80 x80 mm and a height of 20 mm. The unit cell of Miura-type foldcore has the same value (40 mm) of width, length and height as the unit cell of TSP folded structure, as shown in Figure 9. The same base plate with a 2 mm high boundary strip is used to constrain horizontal movements of the sidewalls under crushing.

In this section, the volumetric density is kept the same (2.7%) for Miura-type and three truncated pyramid folded structures. The thickness of the sidewalls is calculated as follows,

$$t = \frac{\rho_v \cdot A_{base} \cdot H}{A_{surf}} \tag{1}$$

where $t$ is the thickness of foldcore wall, $\rho_v$ is the volumetric density, $A_{base}$ and $A_{surf}$ are the base area and surface area of the foldcore, $H$ is the foldcore height. These parameters are listed in Table 5.

**Table 5:** *Parameters of four types of foldcores*

| Foldcore $\rho_v = 2.7\%$ | Wall thickness (mm) | Model base area (mm$^2$) | Model surface area (mm$^2$) |
|---|---|---|---|
| Miura | 0.31 | 6400 (4 unit cells) | 11081 |
| TTP | 0.147 | 2771 (4 unit cells) | 10207 |
| TSP | 0.26 | 6400 (4 unit cells) | 13337 |
| TPP | 0.43 | 6449 (2 unit cells) | 8539 |

**Table 6:** *Peak and average stress, densification strain of four types of foldcores under quasi-static loading condition*

| Foldcore | Miura | TTP | TSP | TPP |
|---|---|---|---|---|
| $\sigma_{peak}$ (MPa) | 0.486 | 0.458 | 0.405 | 0.326 |
| $\sigma_{ave}$ (MPa) | 0.268 | 0.340 | 0.286 | 0.262 |
| $\varepsilon_D$ | 0.66 | 0.74 | 0.76 | 0.66 |

These four types of foldcores of the same density are crushed under quasi-static loading in the numerical simulation. Same boundary conditions are applied as in Section 3.2 where foldcores are placed on top of base plates with 2mm high boundary to constrain in-plane movements. Stress-strain curves are shown in Figure 10. The peak, average crushing resistance and densification strain of these foldcore are listed in Table **6**. Truncated triangular pyramid structure (TTP) shows the best performance among these foldcores, with a low initial peak stress, the highest average stress and a large densification strain. As shown the Miura-type foldcore has the highest initial peak stress among these folded structures. The lower initial stress of the other three truncated pyramid structures indicates the easier initiation of the plastic deformation under crushing. Both TTP and TSP foldcores outperform Miura-type foldcore in all three key parameters, indicating superior performance as energy absorber. Comparing with conventional Miura-type folded structure, the average crushing stress of TTP foldcore is 27% higher and its densification strain is 12% higher, which lead to a 42% increase in energy absorption over Miura-type foldcore before it reaches densification.

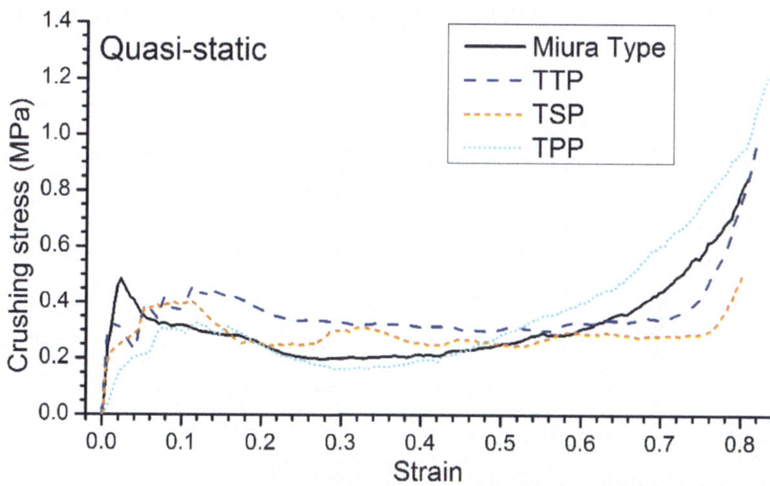


**Figure 10:** *Stress-strain curves of four types of foldcores under quasi-static crushing*

Deformation modes of four types of foldcores at strain of 0.4 under quasi-static loading are shown in Figure 11. Completely different deformations of the foldcore unit cell sidewalls are shown. For Miura-type foldcore, only buckling deformation is observed where the buckling occurs near the middle plane horizontally. This buckling initiation in the early stage causes the sharp rise of the initial peak crushing stress. For truncated triangular pyramid (TTP) folded structure, the sidewalls are bent towards centre of each unit cells which occurs at the early stage of the crushing. Some buckling can also be observed near the corners of sidewall intersection line. These connected sidewalls provide extra constraint while crushing therefore achieving higher average crushing stress. For TSP foldcore, local buckling near the intersection line of the sidewalls is presented similar to TTP, while other faces are bent vertically toward centre. For TPP foldcore, no local buckling can be observed which leads to the relatively low average crushing resistance. Some faces are vertically bent towards centre. Slight rise of the corner can be observed for all four samples due to the lack of constraints in vertical direction. The corners slide upwards as the bottom edges of foldcore sidewalls spread out under crushing. No tearing is observed along the folding lines or on the sidewalls which is consistent with the quasi-static test results.

This change of deformation mode from TTP to TPP is caused by the reduction of the sidewall inclination angle of the unit cell. With lower inclination angle, the foldcore sidewalls tend to spread out like TPP rather than buckling which provides higher crushing resistance. Moreover, the size of vertical triangular interconnections is on decreasing from TTP to TSP to TPP. Less lateral crushing resistance is provided by the decreasing size of interconnections. This is consistent with the stress-strain curves for these three truncated pyramid foldcores, where the average stress decreases with the increasing number of sides i.e. from triangle to square to pentagon.

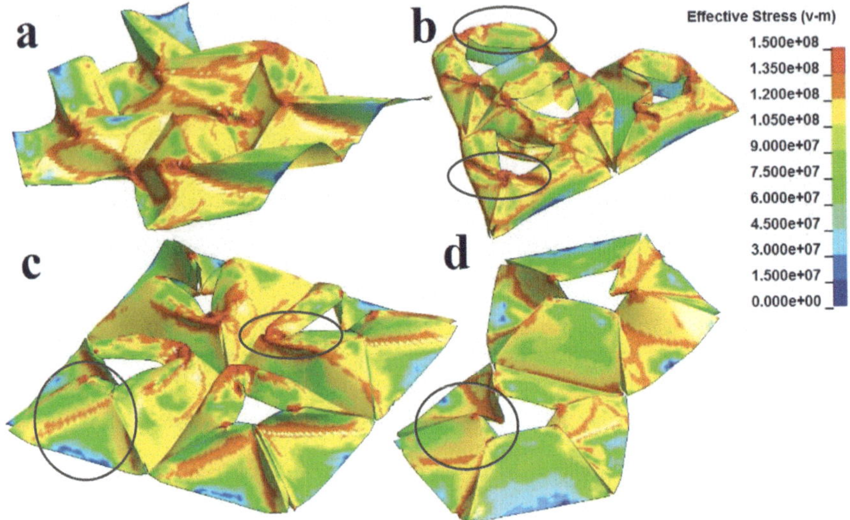



**Figure 11:** *Deformation modes of four types of foldcores under quasi-static loading condition at the strain of 0.4; (a) Miura type; (b) TTP; (c) TSP; (d) TPP*

## 5 Conclusions

Open-top truncated pyramid folded structures with three different base shapes, i.e. triangle, square and pentagon (i.e. TTP, TSP, TPP) are proposed in this study. Their crushing behaviours under quasi-static loading condition are investigated. The numerical models are calibrated and used for the investigation and comparison of these foldcores with the conventional Miura-type foldcore of the same relative density. All three types of the proposed foldcores have lower initial peak stress, and higher or comparable densification strain than Miura-type foldcore. Among them, TTP foldcore has a 27% increase in the average stress with a 12% increase in the densification strain, outperforming TSP, TPP and Miura-type foldcores. TTP foldcore also has a smaller initial peak stress and a consistent deformation under crushing, indicating excellent characteristics as energy absorber.

## 6 Acknowledgement

The authors acknowledge the support from Australian Research Council via Discovery Early Career Researcher Award (DE160101116). This paper has been awarded the 7OSME Gabriella & Paul Rosenbaum Foundation Travel Award.

# References

[Miura. 1985] Koryo Miura. "Method of packaging and deployment of large membranes in    space." title The Institute of Space and Astronautical Science report 618 (1985). 1

[Smithers. 2017] Rebecca Smithers. "Origami-inspired clothing range that grows with your child wins Dyson award." The Guardian (2017).

[De Temmerman et al. 2007] Ir arch Niels De Temmerman, Marijke Mollaert, Ir arch Tom Van Mele and Ir arch Lars De Laet. "Design and analysis of a foldable mobile shelter system." International Journal of Space Structures 22 (2007). 161-168

[Gioia et al. 2012] Francesco Gioia, David Dureisseix, René Motro and Bernard Maurin. "Design and analysis of a foldable/unfoldable corrugated architectural curved envelop." Journal of Mechanical Design 134 (2012). 031003

[Butler et al. 2017] Jared Butler, Spencer Magleby, Larry Howell, Stefano Mancini and Aaron Parness. "Highly Compressible Origami Bellows for Microgravity Drilling-Debris Containment." (2017).

[Albermani et al. 2011] F. Albermani, H. Khalilpasha and H. Karampour. "Propagation buckling in deep sub-sea pipelines." Engineering Structures 33 (2011). 2547-2553

[Zhou et al. 2017] Caihua Zhou, Yan Zhou and Bo Wang. "Crashworthiness design for trapezoid origami crash boxes." Thin-Walled Structures 117 (2017). 257-267

[Zhou et al. 2017] Caihua Zhou, Liangliang Jiang, Kuo Tian, Xiangjun Bi and Bo Wang. "Origami Crash Boxes Subjected to Dynamic Oblique Loading." Journal of Applied Mechanics 84 (2017). 091006

[Li et al. 2017] Zhejian Li, Wensu Chen and Hong Hao. "Blast resistant performance of multi-layer square dome shape kirigami folded structure,." 6th International Conference on Design and Analysis of Protective Structures (2017)

[Hao et al. 2018] Hong Hao, Zhejian Li and Wensu Chen. "Performance of sandwich panel with square dome shape folded kirigami core under blast loading." 13th International Conference on Steel, Space and Composite Structures (2018) Keynote

[Miura. 1972] Koryo Miura. "Zeta-core sandwich-its concept and realization." title ISAS report/Institute of Space and Aeronautical Science, University of Tokyo 37 (1972). 137

[Liu et al. 2015] Sicong Liu, Guoxing Lu, Yan Chen and Yew Wei Leong. "Deformation of the Miura-ori patterned sheet." International Journal of Mechanical Sciences 99 (2015). 130-142

[Gattas and You. 2014] J. M Gattas and Z You. "Quasi-static impact of indented foldcores." International Journal of Impact Engineering 73 (2014). 15-29

[Heimbs. 2013] Sebastian Heimbs. "Foldcore sandwich structures and their impact behaviour: an overview." Dynamic failure of composite and sandwich structures (2013). 491-544

[Gattas and You. 2014] J. M Gattas and Z You. "Miura-Base Rigid Origami: Parametrizations of Curved-Crease Geometries." Journal of Mechanical Design 136 (2014). 121404-121404-121410

[Gattas and You. 2015] J. M Gattas and Z You. "The behaviour of curved-crease foldcores under low-velocity impact loads." International Journal of Solids and Structures 53 (2015). 80-91

[Nojima and Saito. 2006] Taketoshi Nojima and Kazuya Saito. "Development of newly designed ultra-light core structures." JSME International Journal Series A Solid Mechanics and Material Engineering 49 (2006). 38-42

[Fathers et al. 2015] R. K Fathers, J. M Gattas and Z You. "Quasi-static crushing of eggbox, cube, and modified cube foldcore sandwich structures." International Journal of Mechanical Sciences 101-102 (2015). 421-428

[Xue and Hutchinson. 2006] Zhenyu Xue and John W. Hutchinson. "Crush dynamics of square honeycomb sandwich cores." International Journal for Numerical Methods in Engineering 65 (2006). 2221-2245

[Li et al. 2017] Zhejian Li, Wensu Chen and Hong Hao. "Numerical study of folded dome shape aluminium structure against flatwise crushing." 12th International Conference on Shock & Impact Loads on Structures (2017)

[Li et al. 2018] Zhejian Li, Wensu Chen and Hong Hao. "Crushing behaviours of folded kirigami structure with square dome shape." International Journal of Impact Engineering 115 (2018). 94-105

[Caspar and Fontano. 1996] Donald LD Caspar and Eric Fontano. "Five-fold symmetry in crystalline quasicrystal lattices." Proceedings of the National Academy of Sciences 93 (1996). 14271-14278

[ASTM. 2004] ASTM. "E8M-04 Standard Test Methods for Tension Testing of Metallic Materials (Metric) 1." (2004).

---

Zhejian Li
Centre for Infrastructural Monitoring and Protection, Curtin University, Kent Street, Bentley, WA 6102 Australia, e-mail: zhejian.li@postgrad.curtin.edu.au

Wensu Chen
Centre for Infrastructural Monitoring and Protection, Curtin University, Kent Street, Bentley, WA 6102 Australia, e-mail: wensu.chen@curtin.edu.au

Hong Hao
Centre for Infrastructural Monitoring and Protection, Curtin University, Kent Street, Bentley, WA 6102 Australia, e-mail: hong.hao@curtin.edu.au

# Blast Resistant Performance of Cladding with Folded Open-top Truncated Pyramid Structures as Core

*Zhejian Li, Wensu Chen, Hong Hao*

***Abstract:*** *Two open-top truncated pyramid structures with base shape of triangle and pentagon are developed, similar to the previously studied truncated square pyramid structure. The sidewalls of proposed structures are interconnected via triangular interconnections which provide extra constraints and inertia stabilization effect during the crushing. Their performances as sacrificial cladding are evaluated and compared with Miura-type foldcore. Superior blast mitigation capacity of cladding with truncated triangle and pentagon pyramid structure as core are demonstrated by yielding a further reduction of peak transmitted pressure to protected structure and a higher specific energy absorption.*

## 1   Introduction

Sandwich structure usually consists of a cellular core sandwiched by two thin skin layers. It has been increasingly adopted in many applications during the last decades, due to its excellent strength to weight ratio [Heimbs et al. 2010]. Different topologies of the core were developed, including lattices [Mcshane et al. 2006, Mines. 2008], polymeric foam [Ousji et al. 2017], honeycomb [Côtéet al. 2004, Radford et al. 2007], metallic foam [Radford et al. 2006], auxetic core [Hou et al. 2014, Imbalzano et al. 2016] and load-self-cancelling core [Chen and Hao. 2012, 2013, Li et al. 2018]. Recently, the applications of sandwich structure as energy absorber, such as protective cladding or crush box have been investigated extensively. The crushable core of sandwich structure may undergo large plastic deformation with a constant stress under dynamic loading conditions such as blast or impact. Due to the plastic deformation of the core, large amounts of energy are absorbed by the core and the force transmitted to protected structure behind the cladding is greatly reduced [Langdon et al. 2010].

As one of the developed cores, Miura-type folded structure has drawn the attention in the field of sandwich structures. It was firstly proposed as packaging of solar panel for space deployment [Miura. 1985]. The Miura-type origami structure was firstly folded from an un-broken sheet along the creases, no stretching or twisting on the faces. It is now used in many applications like commercial aviation [Herrmann et al. 2005], sacrificial cladding [Pydah and Batra. 2017], as the core of sandwich structure, due to its advantages such as continuous manufacturing, deployable and open channel design to allow heat and moist to escape. Different variations based on original Miura-type origami foldcore were then developed such as indented [Gattas and You. 2014], curved creased foldcore [Gattas and You. 2015], and kirigami foldcore [Fathers et al. 2015]. These modifications were aimed to increase the crushing resistance or enhance the crushing behaviour by reducing the initial peak stress.

However, the structural responses of Miura-type foldcore [Pydah and Batra. 2017] and some kirigami foldcores, like cube strip [Fathers et al. 2015] have a high initial peak stress and are strain rate dependent, i.e., the crushing resistance increases sharply for these foldcores under dynamic loading condition. This strain-rate sensitivity and the high initial crushing resistance are less ideal for many energy absorption applications such as impact attenuator [Van Oorschot. 2017], sacrificial cladding [Hanssen et al. 2002], and vehicle crush box [Zhou et al. 2016], etc. A new form of open-top truncated square pyramid structure was developed as shown in Figure 1[Li et al. 2017, 2018]. With adjacent sidewalls connected via triangular interconnections, this type of foldcore demonstrates superior crushing resistance and it shows a strain rate insensitivity comparing with other foldcores. Due to these ideal characteristics, their blast mitigation capacities were investigated numerically and compared with square honeycomb and aluminium foam of the same density [Li et al. 2017, Hao et al. 2018].

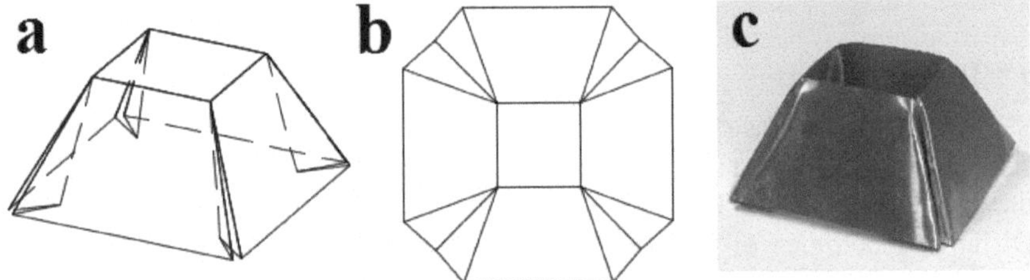


**Figure 1**: *(a) Isometric view of a single unit cell of TSP foldcore; (b) sheet pattern for folding; (c) isometric view of a hand-folded sample of TSP single unit cell*

In this paper, the blast mitigation capacities of truncated triangular pyramid (TTP), truncated square pyramid (TSP) and truncated pentagonal pyramid (TPP) folded structures are investigated. Hand-fold samples of the foldcores are crushed under quasi-static loading conditions and the aluminium sheet material is tested to obtain the mechanical properties. Numerical models are then constructed and calibrated using experiment data. Blast mitigation capabilities of these foldcore are then simulated and compared with Miura-type foldcore of the same density. Key parameters including peak stress transmitted to protected structure, energy absorption by core are used as indicators to evaluate their performances.

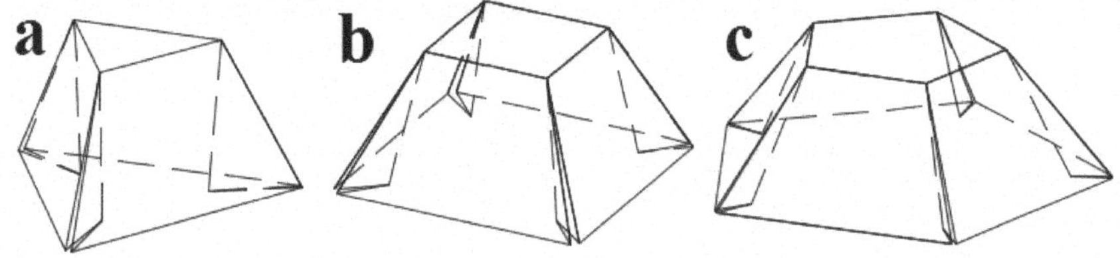

**Figure 2:** *Open-top truncated pyramid structures; (a) truncated triangular pyramid (TTP); (b) truncated square pyramid (TSP); (c) truncated pentagonal pyramid (TPP)*

## 2   Quasi-static crushing and material tensile test

### 2.1 Sample geometry

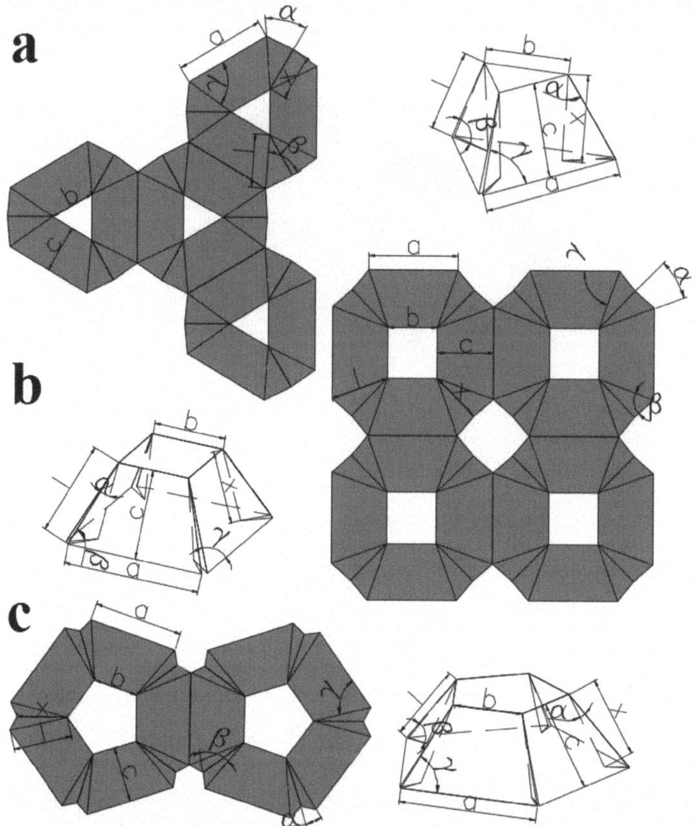

**Figure 3**: *Folding configuration and geometric parameters for (a) TTP; (b) TSP; (c) TPP*

Sheet sample patterns and the folding configurations of these truncated pyramid foldcore are shown in Figure 3. As shown, the adjacent sidewalls are connected via two triangular interconnections. Other geometric parameters shown in Figure 3 are determined by the top and bottom edge length, $b$, $a$, respectively and the core height, $H$. For the hand folded samples, these three parameters i.e., top edge length $b$, bottom edge length $a$ and the core height $H$, are kept the same for all three foldcores.

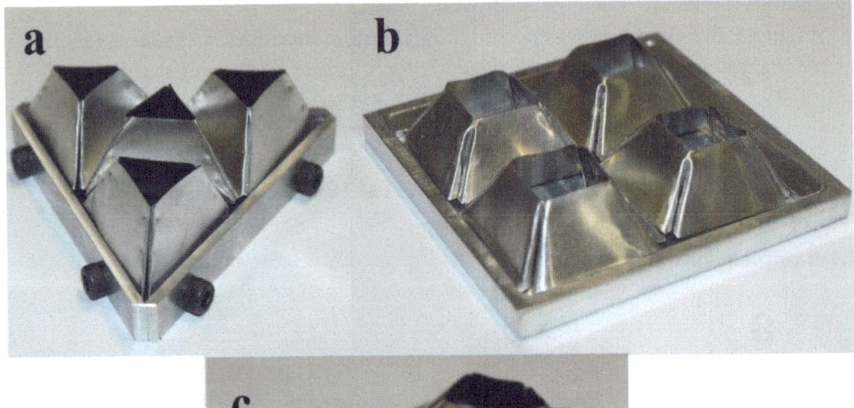

**Figure 4:** *Hand-folded samples of (a) TTP; (b) TSP; (c) TPP and the base plates with boundary*

**Table 1:** *wall thickness and volumetric density of the foldcores*

| Foldcore | t (mm) | $\rho_v$ % |
|---|---|---|
| TTP | 0.15 | 2.7 |
| TSP | 0.26 | 2.7 |
| TPP | 0.26 | 1.7 |

Samples of these three types of foldcores are shown in Figure 4. They all have the same top edge length of 20 mm, bottom edge of 40 mm and the height of 20 mm. The base plates have a 2 mm high boundary strip to constrain the horizontal movements of bottom edges of the foldcore under deformation. Samples are simply placed on the base plate without any glue or fixing and crushed under quasi-static loading of 1mm/min till around strain of 0.8. Aluminium 1060 sheet is used for sample folding. The aluminium sheet thickness, $t$ and the volumetric density, $\rho_v$ of the three foldcores are given in Table 1. The aluminium 1060 sheet material with thickness of 0.26mm is tested for its tensile properties according to the ASTM-E8M testing standard [ASTM. 2004].

## 2.2 Numerical model

LS-DYNA 971 is used for model construction. The foldcores are constructed by using Belytschko-Tsay type shell element and placed between two rigid solid block. The base plate with boundary is built to support foldcores, and the top block is moving towards base plate at a constant rate till around 0.8 strain. The material model *MAT024 PIECEWISE LINEAR PLASTICITY is applied in the finite element model with material properties and stress-strain data obtained from tensile test, as listed in Table 2 and Table 3. Friction and sliding of foldcore are considered for the contact setting in the numerical model.

***Table 2:*** *Material properties of Aluminium 1060 sheet*

| Parameter | Young's modulus (GPa) | Poisson's ratio | Yield stress (MPa) | Density (kg/m$^3$) |
|---|---|---|---|---|
| Value | 69 | 0.33 | 66.7 | 2710 |

***Table 3****: True plastic stress-strain data of Aluminium 1060 sheet*

| Strain | 0 | 0.002 | 0.005 | 0.013 | 0.063 | 0.121 |
|---|---|---|---|---|---|---|
| Stress (MPa) | 0 | 66.7 | 112.3 | 120.1 | 125.8 | 130.6 |

## 2.3 Model validation

The stress-strain curves of the foldcores under quasi-static loading from both numerical simulation and experiments are shown in Figure 5. There are some discrepancies between numerical and experimental results at the early stage of the deformation. The numerical results have a sharper rise and a higher initial peak force for these foldcores as opposed to the experimental results. This is caused by the imperfections induced in the hand folding process. As shown in Figure 4, slight bent of the sample sidewalls can be spotted, some gaps are presented between foldcore and bottom base plate as well. These folding induced imperfections are unlikely to be avoided by hand folding of the specimens. Similar difference has been observed in the previous studies of folded structure as well [Fathers et al. 2015].

Overall structural responses of these foldcores, however, are in good agreement. Two key parameters for energy absorption evaluation including average stress and densification are well matched for all three foldcores. Densification strain is defined by the sharp rise at the later stage of the crushing due to the compaction of the cellular core, and the average stress or plateau stress is the average crushing stress of the core before full densification. Therefore, the numerical models and experimental results are well matched and it is acceptable for the investigation on their blast mitigation performance as cladding cores.

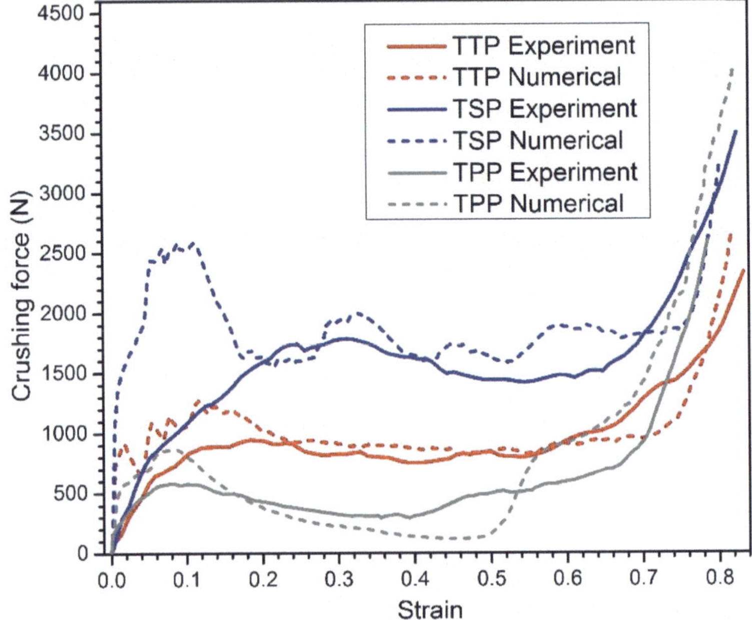



**Figure 5:** *Comparisons between numerical simulation and experimental data of the three foldcores*

## 3 Blast mitigation performance as sacrificial cladding

### 3.1 Numerical model and setup of cladding

The unit cell sizes of three types of the foldcores are scaled up twice of the tested specimen in order to achieve a reasonable height as cladding core. The foldcore height increases from 20 mm in the experiment to 40 mm, which is one of the common height for cladding core [Wu et al. 2011]. The top edge length of TTP, TSP and TPP is increased from 20 mm to 40 mm, and the bottom edge length increases from 40 mm to 80 mm accordingly. Miura-type foldcore is also constructed with the same unit cell dimension as TSP. The volumetric densities of these four types of foldcores are kept the same as 2.7%, and the wall thickness of each foldcore is calculated accordingly, as listed in Table 4. Due to the different shapes of each unit cell of the foldcores, the cladding base areas are not exactly the same. Number of unit cells for each type of cladding, as well the cladding mass are listed in Table 4.

**Table 4:** *Parameters of four types of foldcores*

| Foldcore $\rho_v =2.7\%$ | Wall thickness (mm) | Cladding base area (mm$^2$) | Cladding mass (g) |
|---|---|---|---|
| Miura | 0.62 | 102400 (16 unit cells) | 300 |
| TTP | 0.30 | 66504 (24 unit cells) | 195 |
| TSP | 0.53 | 102400 (16 unit cells) | 300 |
| TPP | 0.86 | 132683 (10 unit cells) | 388 |

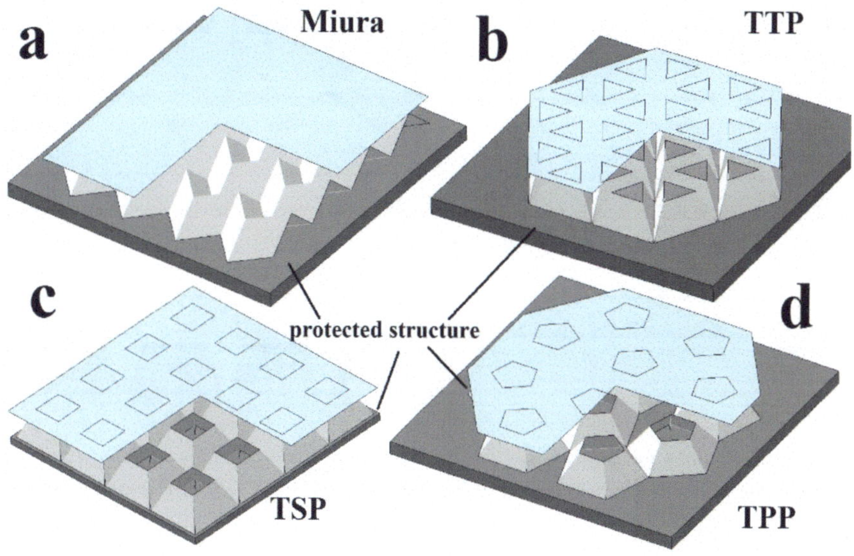


**Figure 6**: *Numerical model for sacrificial cladding with (a) Miura-type; (b) TTP; (c) TSP; (d) TPP; as core. Note: quarter of the top skin is cut for clear demonstration of the core*

The cladding setups with different cores are shown in Figure 6. The material model and material parameters are the same as in the previous section 2.2. Top layer is set to be 3 mm for all four claddings, whereas their size and shape are determined by the foldcore base area and the base shape. Keyword *LOAD BLAST ENHANCED is used for blast simulation where 2 kg of TNT is detonated at 1.5 m directly above the centre of top skin of cladding. A scenario of structure without any cladding is also modelled for comparison. Since the cladding core has a height of 40 mm, the stand-off distance for no-cladding scenario is 1.54m. The foldcore and cladding are simply placed on top of a rigid block with no glue or fixing, and the load-history curves of transmitted pressure to the protected structure are recorded for each cladding configurations. The outer edges of the foldcore are constrained for movement in both in-plane directions while the rotations are not restrained, which is to model the boundary of base plate in the experiment as shown in Figure 4. For the claddings with TTP and TSP as core, the outer bottom edges are fixed for movement along the in-plane directions. For the claddings with Miura-type and TPP as core, not all outer bottom edges shall be constrained due to the gap between adjacent unit cells. As shown in Figure 7, the bottom outer edges with fixed in-plane movement boundary condition are marked out in blue dash lines or blue lines. Under deformation, other sidewalls are allowed to slide on the base due to the gaps in

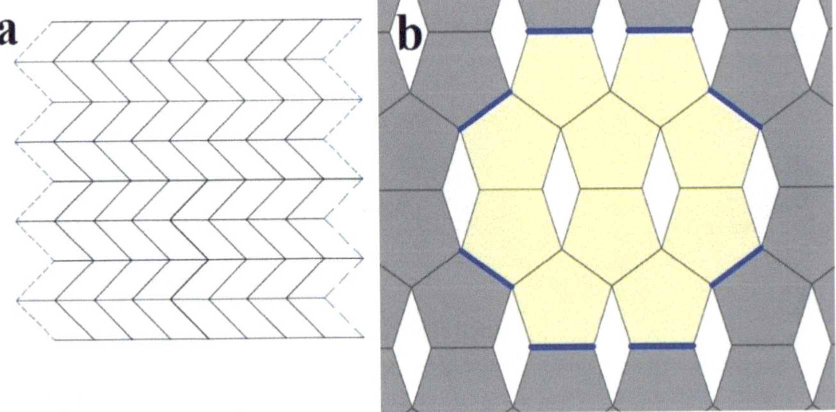

**Figure 7:** *Foldcore pattern and their fixed bottom edges in-plane movements of cladding with (a) Miura-type (marked out by blue dash line); (b) TPP (marked out by blue line)*

## 3.2 Structural response comparison

The time history curves of transmitted pressure to protected structure of different cladding configurations are shown in Figure 8. The parameters of blast mitigation capability of these cladding are listed in Table 5. The peak reflected blast pressure of the case without any cladding is 2.7 MPa. The peak pressure transmitted to protected structure is greatly reduced using four types of foldcores by 47.8% (Miura-type), 30% (TTP), 66.8% (TSP) and 67.7% (TPP), respectively. Clear prolonging of the loading to protected structure behind the cladding is also observed, where the positive phase of the blast loading for the scenario without any cladding is around 0.7 ms. The loading duration for other three cladding configurations has more than 1.2 ms.

Fluctuations and sharp rise of the transmitted pressure to protected structure at initial stage of blast loading are shown for the claddings with Miura-type and TTP as core. Similar results have been recorded for Miura-type foldcore and conventional square honeycomb structure under high strain rate loading [Pydah and Batra. 2017, Xue and Hutchinson. 2006]. The strain rate sensitivity of the structures are caused by the inertial effect and inertial stabilization effect of the connected sidewalls, which provide extra constraints under higher loading rate [Xue and Hutchinson. 2006]. As observed from Figure 2, from TTP to TSP to TPP foldcore, the inclining angle of the sidewall decreases and the size of the vertical interconnection connecting the adjacent sidewalls decreases as well. Under high speed out-of-plane crushing, higher inclining angle (closer to vertical) and larger vertical interconnections will lead to higher inertial effect and inertial stabilization effect. Therefore, the initial peak stress is on a decreasing trend from TTP to TSP to TPP foldcore subjected to high loading rate such as blast.

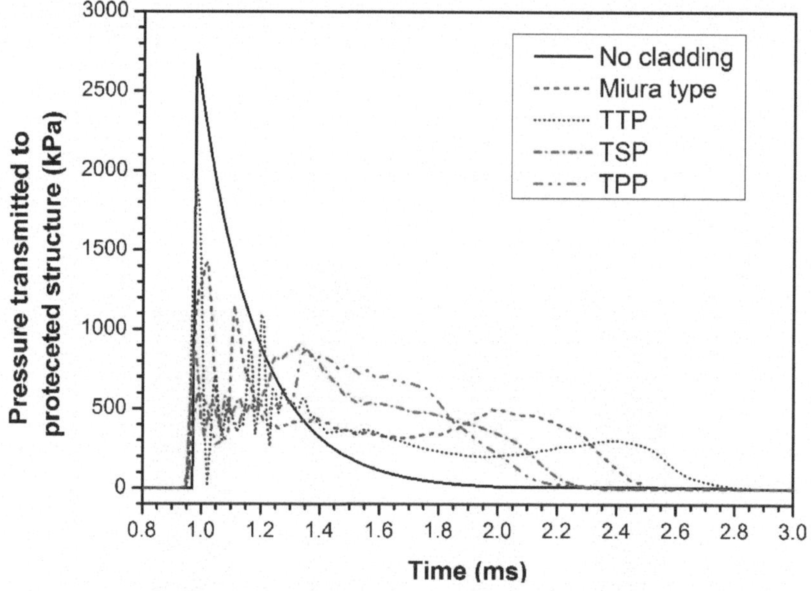


**Figure 8:** *Comparisons of transmitted pressure-time history curves of four cladding configurations under blast loading of 2 kg TNT at a stand-of-distance of 1.5 m*

The displacement of top skin centre point and the specific energy absorption by core are shown in Table 5 and Figure 9. The claddings with all three types of truncated pyramid foldcores have a larger specific energy absorption than Miura-type foldcore. The cladding configuration with TSP as core shows the highest specific energy absorption and the lowest displacement of top skin, which indicates the highest average stress during the deformation. Overall, cladding with TSP as core has the best performance out of these four types which have the same core density. The TSP foldcore greatly reduces the peak pressure transmitted to the protected structure behind the cladding. It also have the highest specific energy absorption and the smallest top layer centre displacement. This indicates its capability to mitigate slightly more intense blast than other three foldcores, as it has the smallest crushed distance at this blast intensity and can still undergo further deformation.

**Table 5:** *Parameters of structural responses of four cladding configurations under blast loading*

| Foldcore | No cladding | Miura | TTP | TSP | TPP |
|---|---|---|---|---|---|
| $\sigma_{peak}$ (kPa) | 2730 | 1426 | 1910 | 905 | 882 |
| Top skin centre point displacement (mm) | - | 20.1 | 26 | 17.9 | 21.2 |
| Specific energy absorption by core (J/g) | - | 3.37 | 3.41 | 4.14 | 3.61 |

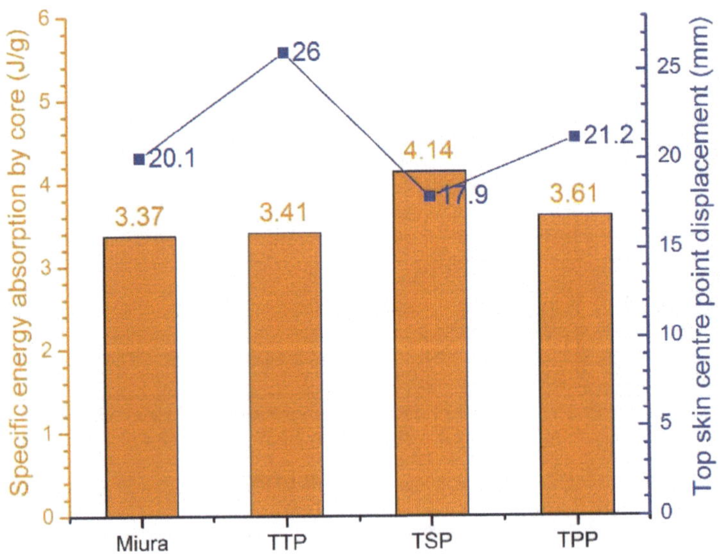


**Figure 9:** *Specific energy absorption by core and the displacement of top skin centre point of these four cladding configurations*

Deformation modes at the end of the blast event are shown in Figure 10. The damage modes of the three truncated pyramid foldcores are quite different. Multiple buckling of the sidewalls can be observed for TTP foldcore, which is similar to the deformation of tube structure under lateral loading. Because of its

high inclining angle, all sidewalls are in contact with adjacent unit cells. For TSP
foldcore, the top edges bends towards the centre of the unit cell and the sidewalls
buckle, but no multiple buckling is observed. As for TPP foldcore, due to the low
inclining angle, the top edges are bent towards centre and no obvious sidewall
buckling can be observed for the lower half of unit cell, which is different from the
deformation mode of TSP foldcore.

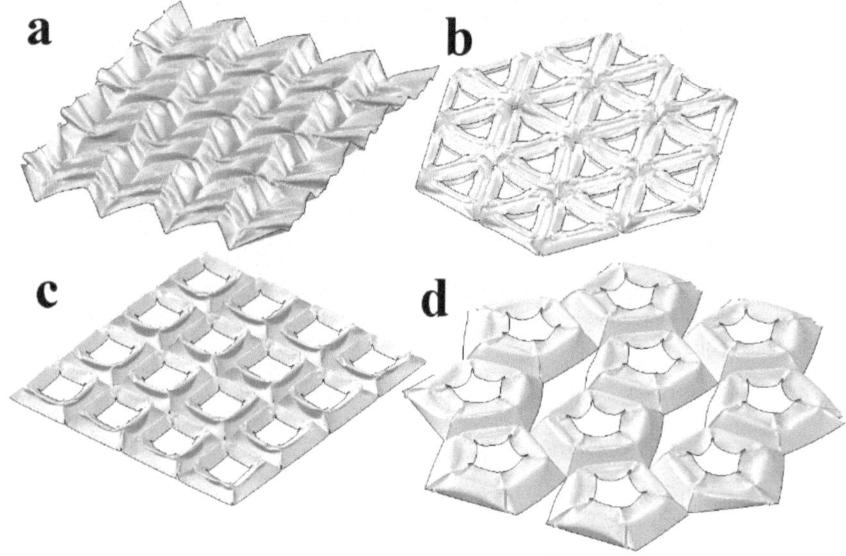

**Figure 10:** *Damage modes at the end of blast loading for (a) Miura-type; (b) TTP;
(c) TSP; (d) TPP*

# 4  Conclusions

The performance of sacrificial cladding with the proposed open-top
truncated pyramid structures as core is studied. Their structural responses
including peak transmitted pressure to protected structure, top layer centre
displacement and specific energy absorption of core are obtained numerically
and compared with Miura-type foldcore. Both TSP and TPP foldcores
demonstrate superior blast mitigation capability than Miura type foldcore
by further reducing the peak pressure transmitted to the protected
structure and higher specific energy absorption.

# 5  Acknowledgement

The authors acknowledge the support from Australian Research Council via
Discovery Early Career Researcher Award (DE160101116). This paper has been
awarded the 7OSME Gabriella & Paul Rosenbaum Foundation Travel Award.

# References


[Heimbs et al. 2010] S. Heimbs, J. Cichosz, M. Klaus, S. Kilchert and A. F. Johnson. "Sandwich structures with textile-reinforced composite foldcores under impact loads." Composite Structures 92 (2010). 1485-1497

[Mcshane et al. 2006] G. J. Mcshane, D. D. Radford, V. S. Deshpande and N. A. Fleck. "The response of clamped sandwich plates with lattice cores subjected to shock loading." European Journal of Mechanics - A/Solids 25 (2006). 215-229

[Mines. 2008] R. A. W. Mines. "On the characterisation of foam and micro-lattice materials used in sandwich construction." Strain 44 (2008). 71-83

[Ousji et al. 2017] H. Ousji, B. Belkassem, M. A. Louar, B. Reymen, J. Martino, D. Lecompte, L. Pyl and J. Vantomme. "Air-blast response of sacrificial cladding using low density foams: Experimental and analytical approach." International Journal of Mechanical Sciences 128-129 (2017). 459-474

[Côté et al. 2004] F. Côté, V. S. Deshpande, N. A. Fleck and A. G. Evans. "The out-of-plane compressive behavior of metallic honeycombs." Materials Science and Engineering: A 380 (2004). 272-280

[Radford et al. 2007] D. D. Radford, G. J. Mcshane, V. S. Deshpande and N. A. Fleck. "Dynamic Compressive Response of Stainless-Steel Square Honeycombs." Journal of Applied Mechanics 74 (2007). 658

[Radford et al. 2006] D. D. Radford, G. J. Mcshane, V. S. Deshpande and N. A. Fleck. "The response of clamped sandwich plates with metallic foam cores to simulated blast loading." International Journal of Solids and Structures 43 (2006). 2243-2259

[Hou et al. 2014] Y. Hou, R. Neville, F. Scarpa, C. Remillat, B. Gu and M. Ruzzene. "Graded conventional-auxetic Kirigami sandwich structures: Flatwise compression and edgewise loading." Composites Part B: Engineering 59 (2014). 33-42

[Imbalzano et al. 2016] Gabriele Imbalzano, Phuong Tran, Tuan D. Ngo and Peter V. S. Lee. "A numerical study of auxetic composite panels under blast loadings." Composite Structures 135 (2016). 339-352

[Chen and Hao. 2012] Wensu Chen and Hong Hao. "Numerical study of a new multi-arch double-layered blast-resistance door panel." International Journal of Impact Engineering 43 (2012). 16-28

[Chen and Hao. 2013] Wensu Chen and Hong Hao. "Numerical simulations of stiffened multi-arch double-layered panels subjected to blast loading." International Journal of Protective structures 4 (2013). 163-188

[Li et al. 2018] Zhejian Li, Wensu Chen and Hong Hao. "Numerical study of sandwich panel with a new bi-directional Load-Self-Cancelling (LSC) core under blast loading." Thin-Walled Structures 127 (2018). 90-101

[Langdon et al. 2010] G. S. Langdon, D. Karagiozova, M. D. Theobald, G. N. Nurick, G. Lu and R. P. Merrett. "Fracture of aluminium foam core sacrificial cladding subjected to air-blast loading." International Journal of Impact Engineering 37 (2010). 638-651

[Miura. 1985] Koryo Miura. "Method of packaging and deployment of large membranes in space." title The Institute of Space and Astronautical Science report 618 (1985). 1

[Herrmann et al. 2005] Axel S Herrmann, Pierre C Zahlen and Ichwan Zuardy. "Sandwich structures technology in commercial aviation." Sandwich structures 7: Advancing with sandwich structures and materials (2005). 13-26

[Pydah and Batra. 2017] Anup Pydah and R. C. Batra. "Crush dynamics and transient deformations of elastic-plastic Miura-ori core sandwich plates." Thin-Walled Structures 115 (2017). 311-322

[Gattas and You. 2014] J. M Gattas and Z You. "Quasi-static impact of indented foldcores." International Journal of Impact Engineering 73 (2014). 15-29

[Gattas and You. 2015] J. M Gattas and Z You. "The behaviour of curved-crease foldcores under low-velocity impact loads." International Journal of Solids and Structures 53 (2015). 80-91

[Fathers et al. 2015] R. K Fathers, J. M Gattas and Z You. "Quasi-static crushing of eggbox, cube, and modified cube foldcore sandwich structures." International Journal of Mechanical Sciences 101-102 (2015). 421-428

[Van Oorschot. 2017] N Van Oorschot. "Deformable and energy absorbing protection against ship collision." (2017).

[Hanssen et al. 2002] AG Hanssen, L Enstock and M Langseth. "Close-range blast loading of aluminium foam panels." International Journal of Impact Engineering 27 (2002). 593-618

[Zhou et al. 2016] Caihua Zhou, Bo Wang, Jiayao Ma and Zhong You. "Dynamic axial crushing of origami crash boxes." International Journal of Mechanical Sciences 118 (2016). 1-12

[Li et al. 2017] Zhejian Li, Wensu Chen and Hong Hao. "Numerical study of folded dome shape aluminium structure against flatwise crushing." 12th International Conference on Shock & Impact Loads on Structures (2017)

[Li et al. 2018] Zhejian Li, Wensu Chen and Hong Hao. "Crushing behaviours of folded kirigami structure with square dome shape." International Journal of Impact Engineering 115 (2018). 94-105

[Li et al. 2017] Zhejian Li, Wensu Chen and Hong Hao. "Blast resistant performance of multi-layer square dome shape kirigami folded structure,." 6th International Conference on Design and Analysis of Protective Structures (2017)

[Hao et al. 2018] Hong Hao, Zhejian Li and Wensu Chen. "Performance of sandwich panel with square dome shape folded kirigami core under blast loading." 13th International Conference on Steel, Space and Composite Structures (2018) Keynote

[ASTM. 2004] ASTM. "E8M-04 Standard Test Methods for Tension Testing of Metallic Materials (Metric) 1." (2004).

[Wu et al. 2011] Chengqing Wu, Liang Huang and Deric John Oehlers. "Blast Testing of Aluminum Foam–Protected Reinforced Concrete Slabs." Journal of Performance of Constructed Facilities 25 (2011). 464-474

[Xue and Hutchinson. 2006] Zhenyu Xue and John W. Hutchinson. "Crush dynamics of square honeycomb sandwich cores." International Journal for Numerical Methods in Engineering 65 (2006). 2221-2245

---

Zhejian Li
Centre for Infrastructural Monitoring and Protection, Curtin University, Kent Street, Bentley, WA 6102 Australia, e-mail: zhejian.li@postgrad.curtin.edu.au

Wensu Chen
Centre for Infrastructural Monitoring and Protection, Curtin University, Kent Street, Bentley, WA 6102 Australia, e-mail: wensu.chen@curtin.edu.au

Hong Hao
Centre for Infrastructural Monitoring and Protection, Curtin University, Kent Street, Bentley, WA 6102 Australia, e-mail: hong.hao@curtin.edu.au

# The Potential of Stress Oriented Foldings

*J. Musto, M. Trautz*

***Abstract:*** *The paper aims to increase the efficiency of rigid folding structures by adapting based on the stresses. In addition to a purely geometric alignment of the folds along the principal stress directions, approaches are presented to adapt the density of the folding structure to the stress intensity. The approaches presented are structurally analyzed and validated on the basis of three example systems.*

## 1  Introduction

Rigid foldings are an efficient design principle in which relatively high stiffness can be achieved with minimum material input. In nature the principle can be observed in various shapes within the flora and fauna, the leaves of deciduous trees and wings of insects to mention a few.

In lightweight constructions, this principle is implemented primarily with thin-walled, flat semi-finished products. The individual components are flat surfaces with low rigidity. The shear-resistant connection of the individual components into a folding structure results in a highly load-bearing construction. By folding up a flat surface, the effective static height is multiplied by a minimum of unwinding area. Due to the stiffening properties of the folding edges, normal stresses are transferred in a concentrated form via the folding edges. This leads to the question of whether the efficiency can be further increased if this load bearing property is used and the alignment of the foldings is adjusted based on the load. With the use of so-called Fold-Core Plates (Faltleichtbauplatten), it was confirmed (Musto & Trautz, 2017) that a geometrical alignment of the folds to the priciple moment directions produces a more load-bearing capacity. Fold-Core Plates consist usually of one or two layers of structural elements, where the arrangement of the folded elements on the folded layer is generated based on regular tessellation patterns.

In contrast to a purely geometric uniform folding structure, rigid folds are to be investigated whose folding edges are adapted based on the intensity and direction of the stresses in the system. The purpose of this paper is to examine the question of how, in addition to the principal moment direction, the stress intensity can be implicated in pattern generation and what effects result in terms of load-bearing capacity. After an initial compilation of the basics, three variants for pattern generation are presented and validated on the basis of three different systems with different loads and supporting conditions.

## 2 Alignment based on the stresses

By aligning the folds based on the stresses, the idea pursued is of transferring the load via the folding edges along "load transfer paths" in the system to the supports. The folding structure is based on a corresponding tessellation of the base area, which represents the planned geometry. This tessellation is implemented with the help of stress trajectories. Finally, the pyramidal folds can be constructed over the tessellated partial areas, whose base edges follow the stress trajectories (Figure 1).

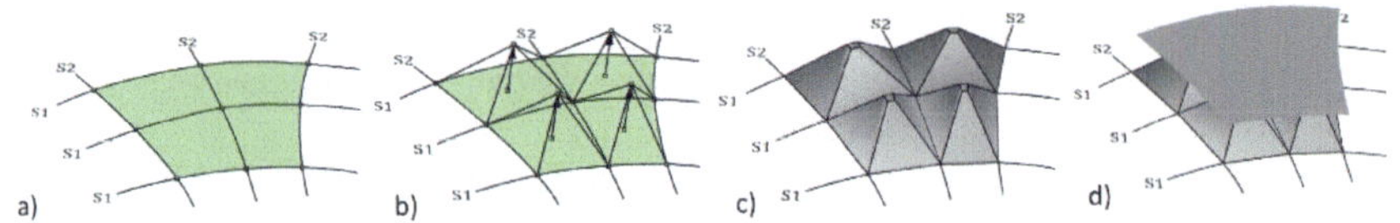

*Figure 1: Process of stress-oriented folding a) tessellation with stress trajectories S1/S2 b) unfolding through extrusion of area centres c) create truncated pyramid by cutting the tips d) completion with top plate*

The following subchapters cover the basics of stress states and stress trajectories required for the subsequent pattern generation approaches.

## 2.1    Stress state

Looking at an infinitesimal large and arbitrarily oriented free section cut of a two-dimensional body, the forces can be described by two components at each edge, a normal stress $\sigma$ and a shear stress $\tau$ (Figure 2). Due to the equilibrium of forces the stresses at each location can be reduced to the components of the symmetric matrix in (1).

$$\sigma = \begin{bmatrix} \sigma_x & \tau_{xy} \\ \tau_{xy} & \sigma_y \end{bmatrix} \tag{1}$$

The ratio of both stress components, shear stress and normal stress, is determined by the cutting direction or the angle of the element. The principal stress direction is that angle at which one of the two components reaches its extreme value. At this angle, either the principal normal stress or the principal shear stress results. The corresponding angle is referred to as the principal normal stress direction or principal shear stress direction. In the case of the principal normal stress, the shear stress becomes zero by definition, but this is not necessarily the case vice versa.

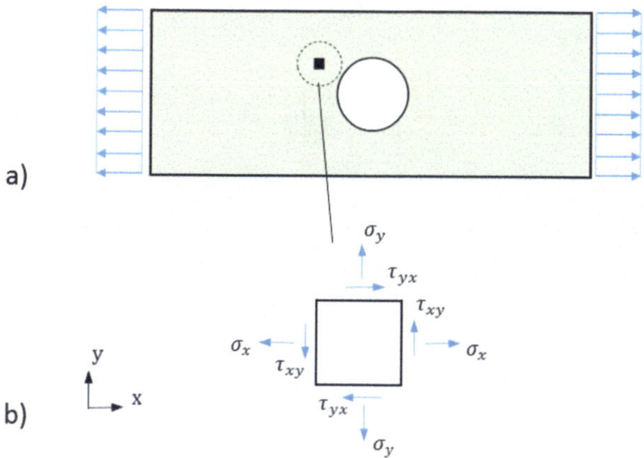

*Figure 2: Stress state a) Plane tension element with circular hole b) Planar stress state of an infinitesimal free section; The index normal stresses $\sigma$ indicates the direction of the stress. The first index of shear stresses specifies the coordinate axis that is perpendicular to the edge, and the second index is the coordinate direction in which the stress acts.*

Figure 3 schematically illustrates this for the case of the principal normal stresses. In this case, the section is oriented according to the principal normal stress direction, which is shown here as lines. The resulting principal normal stresses are called $\sigma_1$ and $\sigma_2$. The Corresponding angle $\varphi$ of the local axis $\xi$-$\eta$ to the global coordinate system indicates the direction of these principal stresses.

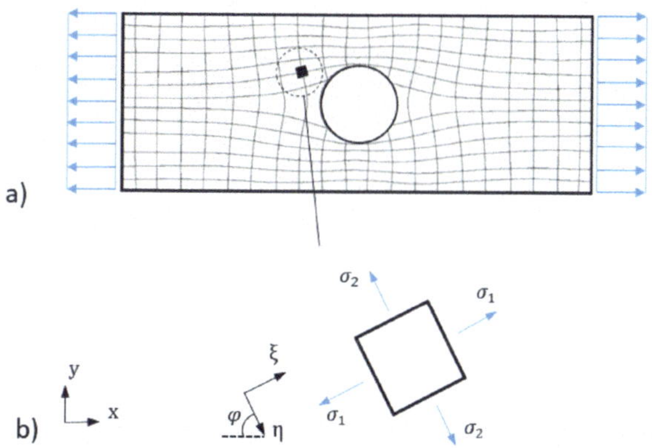

*Figure 3: Stress state a) Plane tension element with circular hole b) Infinitesimal free section with principal stresses $\sigma_1$ and $\sigma_2$ in corresponding direction $\xi$ and $\eta$.*

From equation (1) the tensor for the principal stress state of the normal stresses is calculated as follows

$$\sigma = \begin{bmatrix} \sigma_1 & 0 \\ 0 & \sigma_2 \end{bmatrix}.$$

(2)

The principal normal stresses $\sigma_1$ and $\sigma_2$ can be determined by the components of equation (1) with the following equation.

$$\sigma_{1,2} = \frac{\sigma_x + \sigma_y}{2} \pm \sqrt{\left(\frac{\sigma_x - \sigma_y}{2}\right)^2 + \tau_{xy}^2}$$

(3)

The associated principal normal stress direction $\varphi$ can be calculated using the following tangent function.

$$\tan 2\varphi^* = \frac{2\tau_{xy}}{\sigma_x - \sigma_y}$$

(4)

For the sake of completeness, it should be mentioned that the principal shear stresses and their directions can be determined analogously to the normal stresses, but these are not relevant for the objective of this work.

The stresses resulting from the transfer of bending moments in the components of the folding structure, in particular the folding edge, are mainly normal stresses (Figure 4). Hence the principal normal stresses and the corresponding normal stress directions are of importance for further consideration. Therefore, the shear stress components (in the plane direction) are not discussed in more detail and the terms are reduced to principal stresses and principal stress direction for better readability.

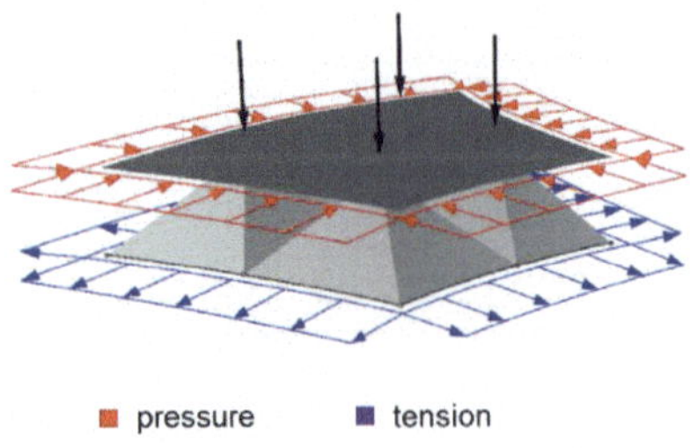



Figure 4: Stresses in a result of vertical load

## 2.2 Mesh generation with stress trajectories

Stress trajectories are a popular tool, especially in construction and mechanical engineering, to visualize the stresses in structural elements. Stress trajectories can be described as lines that illustrate the stress directions of a component. The trajectories are based on a tensor field, which consists of comprehensively determined principal stresses and their directions. The trajectories are constructed in the form of a polygonal line by tangentially connecting vectors in the same direction (Figure 5). Neither the intensities nor the signs of the stress components are taken into account.

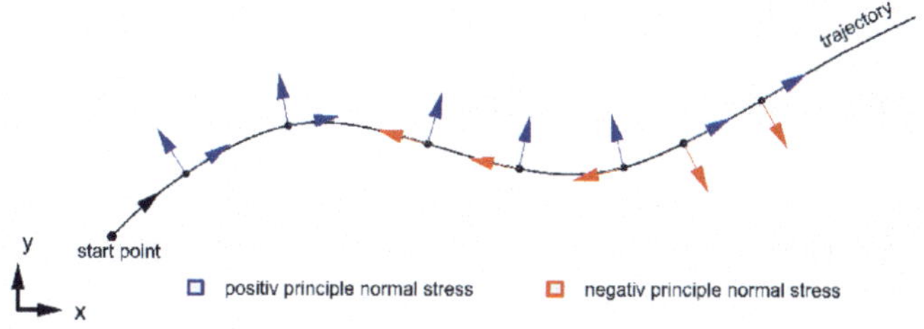


*Figure 5: Construction of a trajectory*

With the exception of singularities and isotropic points, which can lead to the closure of the trajectory, each trajectory ends at the edge of the part. The polygonal line is constructed until the direction given by the vector field results in intersection with one of the component's edges. The cutting angle of the trajectories and the edges depends principally on the external forces at this point: If this edge area is not subjected to an external load, the angle is 90° - in the case of an edge load, the cutting angle corresponds to the load angle.

### 2.2.1    Calculation of the trajectories

The trajectories were calculated for further application using a Python script which generates a polygonal line following the Runge-Kutta 4. Method. The Runge-Kutta-Method is a numerical solution method from mathematics for the approximation of solutions to differential equations, specifically the initial value problems. Initial value problems are equations in which the initial equation is satisfied by the initial values. With reference to the stress trajectories (Figure 6), the starting points $(x_0, y_0)$ determine which curves emerge.

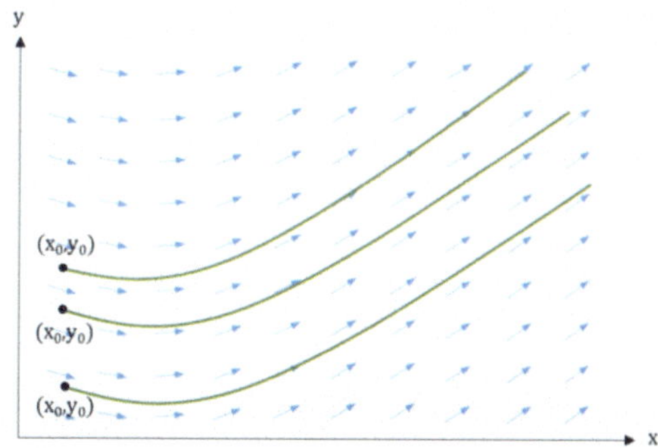


*Figure 6: Tensor field with different solution curves depending on the starting point*

## 2.2.2 Determination of trajectory density

As already explained, the trajectories do not allow conclusions to be drawn about stress intensity. Since the geometry of the folding structure is derived directly from the trajectory network according to the procedure in Figure 1, the distance or density of the trajectories can be used to directly determine the distance of the folding edges. The density, in turn, can be controlled by a corresponding number of starting points during the trajectory calculation.

The stress-orientation is influenced with two parameters. While the trajectories determine the geometrical direction, the stress intensities of the folds can be influenced by the density (Figure 7).

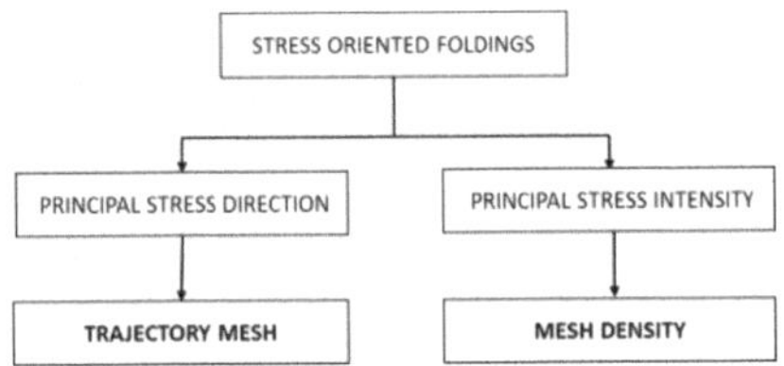


*Figure 7: Elements of stress alignment*

The aim in terms of intensity is to produce a folding density which results in the most homogeneous stress or utilization of the folding structure. This means (assuming a linear system behaviour) that the distances between the trajectories must be inversely proportional to the stress intensity. However, since the number of trajectories remains constant and is determined by the number of starting points over their entire length, a local change of the distance usually has an effect on the entire curve shape. The interaction between trajectories or trajectory density and the stresses will be discussed in more detail below.

Considering the course of the trajectories in Figure 3, it can be seen that the trajectories compress at the side of the hole as the load concentration increases. This phenomenon can often be observed, which is why in the literature a connection between trajectory density and stress intensity is often conveyed. In this context, there is often stress flow mentioned. Based on the continuity analogy of fluid mechanics, parallels are drawn which assume a constant "force flow" between two adjacent trajectories.

In general, this theory does not apply (Wegner, 1934; Deutler, 1941). The correlation can be explained by the Lamé-Maxwell equations as follows:

Consider the dashed element (Figure 8), which is limited from all sides by the principal stress lines. The stress change of $\sigma_1$ along the principal stress line 1 can be described as the ratio of the principal stress difference $(\sigma_1 - \sigma_2)$ to the curvature $\varphi_2$ of line 2 (Wolf, 1961). The same applies to the stress $\sigma_2$

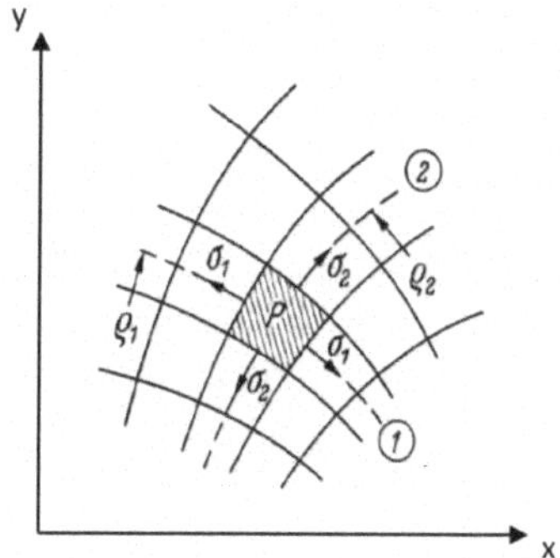

Figure 8: A section P limited by principal stress lines (Wolf, 1961)

$$\frac{\partial \sigma_1}{\partial s_1} + \frac{\sigma_1 - \sigma_2}{\varphi_2} = 0 \tag{6}$$

$$\frac{\partial \sigma_1}{\partial s_1} + \frac{\sigma_1 - \sigma_2}{\varphi_2} = 0 \tag{7}$$

However, this does not allow any conclusions about the resulting force between two adjacent principal stress lines. By introducing angular relations, the change in force between two adjacent principal stress lines can be calculated approximately with formula (8)/(9) according to (Rückert, 1992).

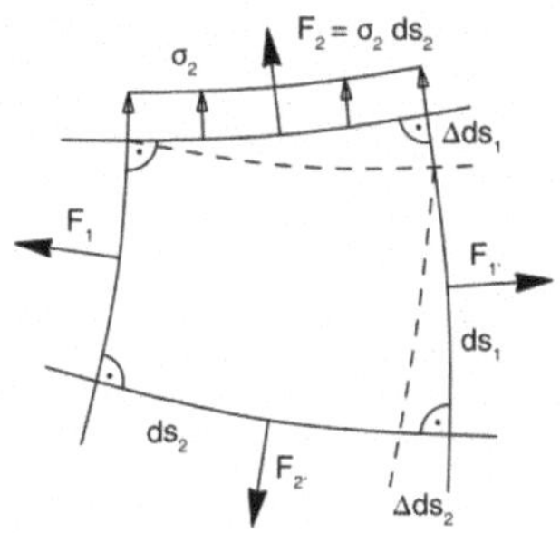

Figure 9: Resulting forces of an element limited by principal stress lines (Rückert, 1992)

$$\Delta F_1 = F_1' - F_1 = \sigma_2 \times \Delta d_{s1} \tag{8}$$

$$\Delta F_2 = F_2' - F_2 = \sigma_1 \times \Delta d_{s2} \tag{9}$$

The equations show that the principle of continuity between two adjacent principal stress lines is only guaranteed if they do not show any curvature or are parallel to each other ($\Delta d_{s1} = \Delta d_{s2} = 0$). Apart from a few special cases, this is only fulfilled in trivial cases, with a constant biaxial stress state (straight trajectories).

For the application of this paper, this means that a predefined force can only be controlled locally along two principal stress lines by the distance. It is not possible to ensure a constant "force flow" over the entire length of the principal stress lines without breaching the mechanical equilibrium conditions of the considered systems.

For this reason, the following approaches are limited to determining the distances along a line which is to be defined. Here, this line must be perpendicular to the observed stresses at each point (analogous to the transverse principal stress line). For this purpose, the characteristic load-bearing behaviour of symmetrical systems is used. This is because one of the two principal stresses inevitably runs along one of the symmetry axes (assuming symmetry of the geometry, supports and loads) and the other perpendicular to it. After assigning a symmetry axis to the principal stress perpendicular to it, the stress intensity of the folds can be defined by the distances of the stress trajectories along this symmetry axis. Using the integral of the stresses, the distances can be calculated in such a way that the same force results for each fold (at least along the symmetry axes).

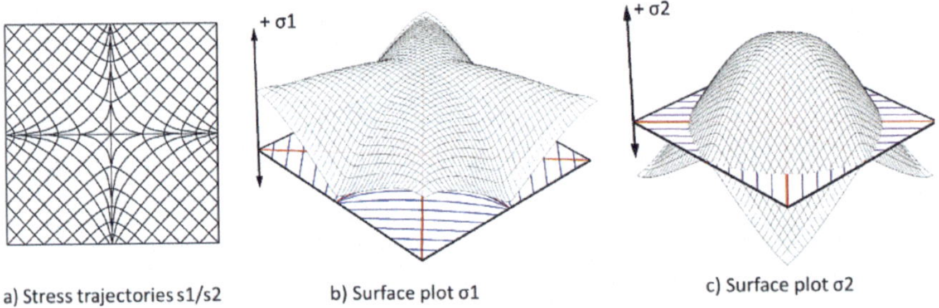



*Figure 10: a) stress trajectories of a simply supported plate with distributed load b) + c) surface plot of σ1/σ2 with trajectories S1/S2 (blue) und symmetry axes (red)*

Figure 10 illustrates the principal stress trajectories of both directions in (a) and separately in b) and c) with the corresponding stress intensities as a surface plot, using the example of a square plate with linear support conditions and distributed load. It can be seen that under the usual loads and given symmetry, according to appropriate allocations of principal stresses and symmetry axes, the maximum stresses extends along these symmetry axes. With this knowledge, a mesh can be created by using an appropriate distance rule along the symmetry axes, which achieves a homogeneous stress at least in the deciding areas of the systems with the highest stress. Therefore, three approaches for determining the distances were investigated, which will be explained further using the example above.

1. Approach 1 (Figure11) incorporates a distance rule, which creates a grid that is as uniform as possible without taking the stress intensity into account. Depending on the system and the result, either a simple division of both symmetry axes into n equal distances or a linked method to determine the starting points was used. In the linked procedure, the trajectories $S_1$ are generated in the first run with n equal distances along the symmetry axis (blue colour) until the system edges are reached. This endpoint is then considered to be the starting point (red colour) for the calculation of trajectories $S_2$.

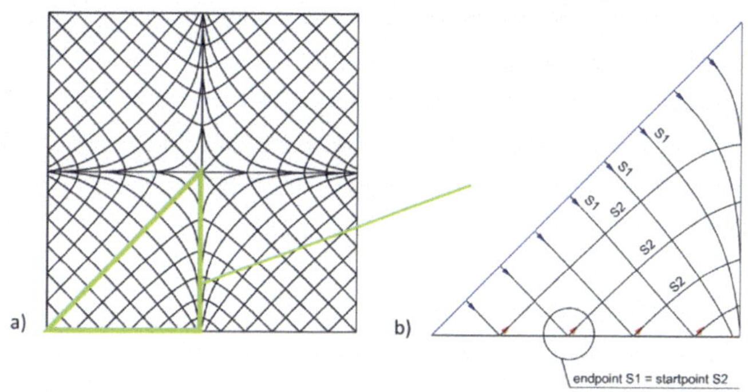


*Figure11: a) Principal stress trajectories of the entire plate*
*b) Two-step procedure for calculating the trajectories on the eighth-model*

2. The second approach (Figure 12) implies the intensities of the principal stresses in such a way that the distance of adjacent trajectories of a principal stress along a symmetry axis results in the same stress integral. The distances are selected in such a way that the resulting stresses along a symmetry axis are distributed evenly over the number of fields between two trajectories.

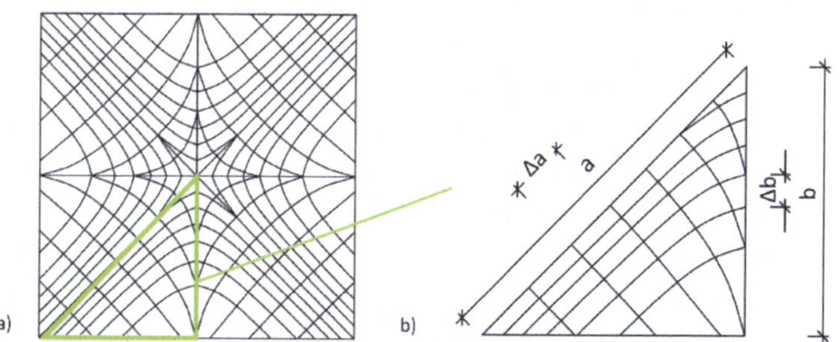


*Figure 12: Principal stress trajectories of the entire plate for approach No. 2*
*b) Spacing at the eighth-model*

$$\int_{\Delta a_n} \sigma_1 = \frac{1}{n} \int_a \sigma_1 \tag{10}$$

$$\int_{\Delta b_n} \sigma_2 = \frac{1}{n} \int_b \sigma_2 \tag{11}$$

3. In the third approach, compared to approach No. 2, the principal stresses $\sigma_1$ and $\sigma_2$ are not included separately. While in the case of approach No. 2, the stress integral can differ depending on the stresses $a_n$ and $b_n$, in this approach they are the same for the distance calculation. The resulting forces between all adjacent trajectories of both directions will be equal in size. This means that, depending on the prevailing stresses, a different number of starting points or trajectories in $S_1$ or $S_2$ direction can result.

$$\int_{\Delta a_n} \sigma_1 = \int_{\Delta b_n} \sigma_2 = \frac{1}{\frac{a}{\Delta a}+\frac{b}{\Delta b}} \left( \int_a \sigma_1 + \int_b \sigma_2 \right) \tag{12}$$

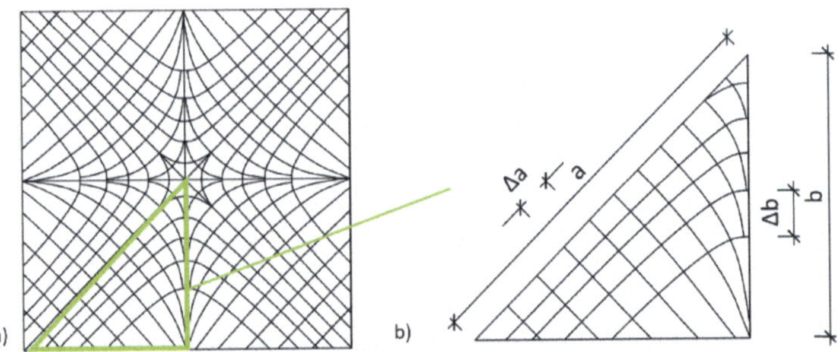


*Figure 13: Principal stress trajectories of the entire plate for approach No. 3*
*b) Spacing at the eighth model*

## 3  Calculation and validation

The theory that efficiency can be increased with the geometrically alignment of the fold edges according to the principal stress directions has already been confirmed in the studies mentioned above. In the last chapter, three approaches were presented with which a stress-oriented trajectory pattern was generated, which takes into account the geometric orientation as well as the intensities of the principal stresses. The results of the approaches will be examined using three systems with different support conditions and loads (Figure 14).

### 3.1  Basis of calculation

In order to be able to clearly quantify the effects (abbreviations explained in section 2) just stress analyses of the folding layer were considered .The top layer of the construction is not considered. The following figure shows the three systems to be examined with the trajectory meshes (example only for approach no. 1 as section 3.2.2).

The initial system for all three approaches was the regular folding structure with a frequency of L/20 (20 folds/span of a system) and a height of L/20 (1/20 of the span of a system). The value L/20 for folding density and construction height is selected as the mean value of the relevant ranges which, according to findings from preliminary studies and the parameter study in (Della Puppa & Trautz, 2016), range between L/10 and L/30.

In order to be able to guarantee a fair comparison of both folding structures, all variants of the stress-oriented folding structure are generated with a density, which leads to the same material usage (equal total areas an sheet thickness) of the uniform structure (± 4%).

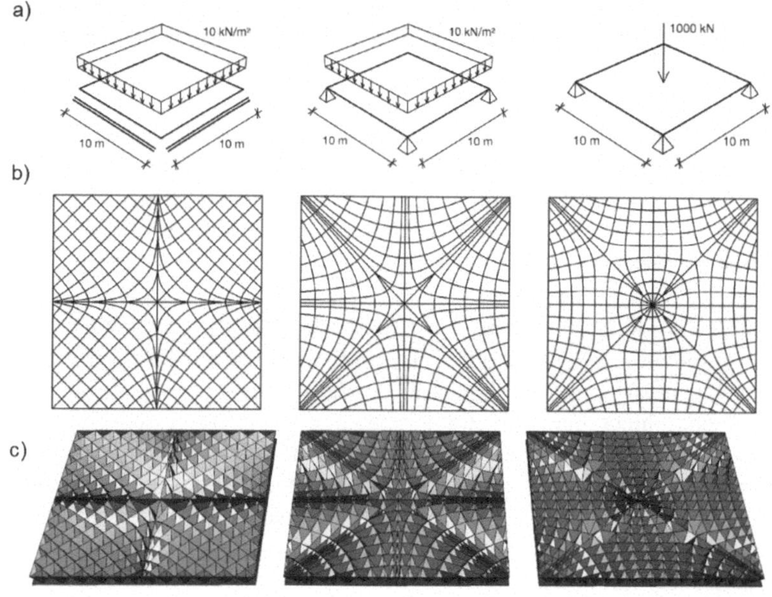


*Figure 14: a) Investigated systems b) stress trajectories c) bottom view of the structure*

## 3.2 Results

The results of the structural analyses using the ANSYS program are summarized in the following table. For each of the three systems, all four configurations are compared with each other. In addition to the maximum stress, the arithmetic mean value of the stresses is also calculated according to formula (13) in order to not focus the evaluation only on individual stress peaks, but also on stress behaviour of the entire structure.

$$\sigma_{avarage} = \frac{1}{n}\sum_{i=1}^{i=n} \sigma_{vmises} \; ; \; n = number \; of \; FEM - Nodes \qquad (13)$$

Based on the idea that if the stress summation of all FEM nodes of two structures with the same system (load, support conditions, dimensions) are different, these differences can only be traced back the structures behaviour. A lower stress summation means that in order to support the same loads, the resulting stresses in the system are lower. The same applies to the arithmetic mean, which is in a linear relationship to stress sum, but is more appropriate due to the number size.

Comparing only the maximum stresses of the systems with each other, it is noted that they are less pronounced in all stress-orientated variants, with the exception of system 1.2. The fact that the maximum stress in system 1.2 is higher than in system 1.0 can be justified by a local stress concentration, because the arithmetic average of all variants (also system 1.2) are lower in all three systems than those of the regular folding. From this it can be concluded that all three stress-orientated variants are bearing the same load with. Looking at the deformation, which in some cases is significantly lower for stress-oriented systems, it also shows that the rigidity of the structure is adapted based on the stresses. If one looks at the results with to the loads and supports of the plates in perspective, it can be seen that in concentrated load transfer, the effects of stress-orientation are greater.

Using the three variants, an approach was successively pursued which incorporates the stress intensity in a more detailed or differentiated manner. A comparison of the results of the three variants of the stress oriented configurations shows that only for system 2 a considerable reduction of the stresses was observed. Why this does not happen with the other systems was determined by an analysis of the graphical stress results. The established approaches aim at adapting the folds to the principal normal stresses. This can result in the folding density becoming so low due to the normal stresses in areas with a high shear forces, that the stresses caused by those forces become decisive. As with a spatial tetrahedron framework, in which the shear forces are transferred via pressure/tension (normal jumps) in the infill, the shear forces in this structure generate normal stresses in the pyramid edges.

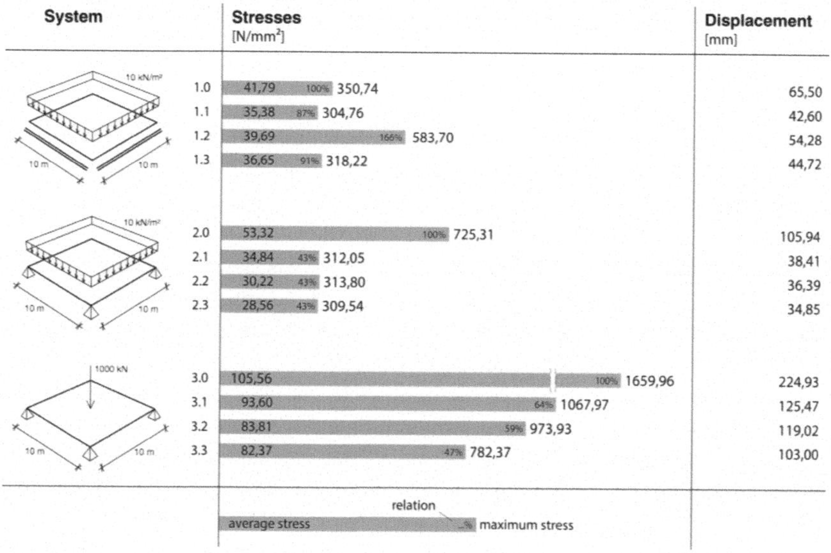

| System | Stresses [N/mm²] | | Displacement [mm] |
|---|---|---|---|
| 1.0 | 41,79 | 100% 350,74 | 65,50 |
| 1.1 | 35,38 | 87% 304,76 | 42,60 |
| 1.2 | 39,69 | 166% 583,70 | 54,28 |
| 1.3 | 36,65 | 91% 318,22 | 44,72 |
| 2.0 | 53,32 | 100% 725,31 | 105,94 |
| 2.1 | 34,84 | 43% 312,05 | 38,41 |
| 2.2 | 30,22 | 43% 313,80 | 36,39 |
| 2.3 | 28,56 | 43% 309,54 | 34,85 |
| 3.0 | 105,56 | 100% 1659,96 | 224,93 |
| 3.1 | 93,60 | 64% 1067,97 | 125,47 |
| 3.2 | 83,81 | 59% 973,93 | 119,02 |
| 3.3 | 82,37 | 47% 782,37 | 103,00 |

relation
average stress __% maximum stress

*Figure 15: Results of structural calculations; System X.0: regular folding structure System X.1 - X.3: stress orientated folding pattern with configuration 1-3 as section 2.1.2*

In summary, it can be concluded that the stress-orientation leads to, in comparison to the regular folding structure, stress reductions of 13% to 57% and considerably stiffer systems. The more concentrated a load-bearing behaviour is, the stronger this effect becomes. The investigations showed that the efficiency of load-oriented folds can be further increased if, in addition to the stresses from the bending moments, the shear forces are considered in the form finding process.

# 4  Summary and outlook

In addition to the entirely geometric orientation, this paper pursued a further optimization of the structure by adapting the folding density based on the stress intensity.

After an initial compilation of the basics of the stress states and trajectories that were necessary for the subsequent considerations, three approaches for generating a trajectory network were presented. While one of the three variants implies the entirely geometric alignment according to the principal stress directions, the other two variants involve adjustment of the foldings to the intensity of the stresses. All three configuration variants were structurally analyzed on the basis of three systems with different load and supporting conditions and were compared to an equivalent regular folding structure.

The results showed that due to the stress-orientation of the folds, the stresses are reduced and the structure becomes significantly stiffer. The positive effects are greater when the load transfer in the system is more concentrated. The results of the pattern variants, which were also generated based on the principal stress intensities, showed that the efficiency of a stress-oriented folding structure can be further increased, if the shear forces are also taken into account during pattern generation.

# References


Della Puppa, G., & Trautz, M. (2016). *Effizienz von Faltleichtbauplatten.* Aachen: Shaker Verlag.

Herkrath, R., & Trautz, M. (2011). Starre Faltungen als Leichtbauprinzip im Bauwesen. *Bautechnik, 88*(2), S. 80-85.

Musto, J., & Trautz, M. (2017). Stress Oriented Foldings as an Optimized Lightweight System. *International Association for Shell and Spatial Structures (IASS).* Hamburg.

Rückert, K. J. (1992). Entwicklung eines CAD-Programmsystems zur Bemessung von Stahlbetontragwerken mit Stabwerksmodellen. Stuttgart.

Trautz, M. (2009). Das Prinzip des Faltens. *Detail – Zeitschrift für Architektur und Baudetail, 12*, S. 1368 ff.

Wegner, U. (1934). Über den Zusammenhang zwischen Strömungs- und Spannungsproblemen. In R. A. Grammel, *Ingenieur-Archiv* (Bd. V., S. 449-469). Springer Verlag.

Wolf, H. (1961). *Spannungsoptik.* Berlin/Göttingen/Heidelberg: Springer Verlag.


---

System X.1 - X.3: stress orientated folding pattern with configuration 1-3 as section 3.1.2


Univ. Prof. Dr.-Ing. Martin Trautz

Chair of Structures and Structural Design, Schinkelstr. 1 – 52062 Aachen, e-mail: trautz@trako.arch.rwth-aachen.de

# Local Actuation of Tubular Origami

*S. W. Grey, F. L. Scarpa, M. Schenk*

**Abstract**: *This paper is an experimental and numerical investigation of the elastic decay of localised actuation in a tubular engineering origami system. A numerical model of a Miura-ori tube subject to localised actuation is developed, using experimentally determined material properties, and verified by comparing against a locally actuated experimental model. Results indicate that the ratio between the facet and folds stiffness influences the distance over which localised actuation decays in Miura-ori tubes. Furthermore, a 'spring-back' effect is observed where an actuation causes an opposite sense deformation further along the tube.*

## 1   Introduction

Origami promises the ability to achieve three-dimensional geometries from flat sheets of material. In order to achieve this, origami-inspired systems could be designed with distributed embedded actuators, allowing the shape to be controlled through a combination of local actuations. A crucial constraint to the development of such systems is the current understanding of the propagation of local actuation in origami systems. The current work focuses on Miura-ori tubes [Filipov et al. 15] with the aim of simplifying the propagation of an actuation to a single direction, along the length of the tube.

Engineering origami structures are not restricted to rigid-foldable mechanisms with a single degree of freedom, such as the Miura-ori pattern. For instance, patterns that are de-velopable, but not rigid-foldable, have been used to generate complex 3D shapes or introduce multistability [Dudte et al. 16, Silverberg et al. 15]. In this paper we focus on rigid-foldable patterns which also experience deformations between the fold lines, due to the nature of the loading.

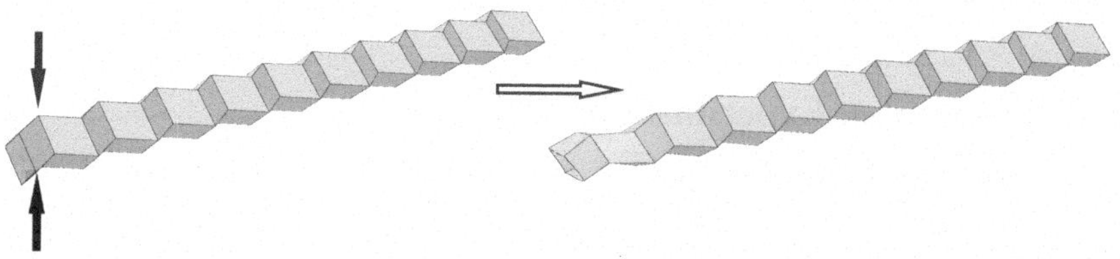

**Figure 1:** *A localised actuation, here compressing a unit cell, at one end of the Miura-ori tube propagates along its length. The effect decays away from the source, eventually returning to the unperturbed shape. This elastic decay is a result of facet deformations, which alter the unit cell kinematics.*

Using the simplest model to capture the mechanics of origami, rigid origami [Tachi 09], the Miura-ori pattern has a single degree of freedom. This drives the kinematic models that capture the global behaviours of Miura-ori sheets [Schenk and Guest 13, Wei et al. 13]. However, folds are not perfect hinges, and introducing a folding stiffness transforms the Miura-ori from a mechanism into a structure. Furthermore, in physical structures the facets have a finite stiffness and can, therefore, deform. This effectively increases the number of degrees of freedom at the vertices, and introduces an elastic decay of any actuation. Figure 1 shows a representative actuation of a Miura-ori tube, achieved by compressing the left most unit cell; the actuation has a reduced effect in each unit cell until eventually returning to the undeformed shape. This elastic decay is due to deformations between the fold lines, which rigid origami cannot capture. The aim of this paper is to better understand this elastic decay, and the influence of the elastic properties of the folds and facets on the decay.

Finite Element Analysis (FEA) may be used to investigate the physical behaviour of origami structures. Using shell elements to model the facets, FEA has been applied to linear-elastic static problems [Liu et al. 15, Gattas and You 15], as well as analyses involving plastic deformations [Ma and You 13, Schenk et al. 14]. The folds are modelled as smooth, bent sections, or by shell elements connected at a fixed angle. The use of FEA can provide insight into the behaviour, but reduced-order models are often preferable for providing a physical understanding. Rigid origami, which represents the folds as hinges, can be augmented with a pseudo-fold across the short diagonal of the facets to represent their bending [Schenk and Guest 11]. The fold lines, representing both facet deformation and folds, can then be given a torsional stiffness. The in-plane deformations are modelled by using bars on the fold lines [Filipov et al. 17]; further fold lines and nodes can be introduced to capture in-plane properties more accurately [Filipov et al. 15]. This bar and hinge model has been extended to non-linear analyses, and a specific implementation, called MERLIN [Liu and Paulino 17], is used in this work for qualitative comparison with FEA.

Evans *et al.* investigate the decay of local perturbations in Miura-ori sheets, using a lattice mechanics approach [Evans et al. 15]. The effect of a local perturbation is related to the unit cell geometry and the magnitude of the actuation, but the decay length is implied to be independent of the fold and facet stiffness properties. Chen *et al.* also describe a strictly geometric dependence of propagation of a local perturbation in a quasi-1D origami strip [Chen et al. 16]. In contrast, the effect of material properties, specifically fold and facet stiffness, on the elastic decay of local actuation is the focus of this paper.

This paper is organised as follows. Section 2 describes the numerical and experimental models used. Section 3 investigates the elastic decay in Miura-ori tubes using a finite element model, and explores the effect of the ratio of the fold and facet stiffness and the unit cell geometry. The paper concludes with a discussion and conclusions in Sections 4 and 5.

## 2 Numerical & Experimental Model

Three methods are used in the analysis of the Miura-ori tube: (1) an experimental model to provide data for and verify (2) a numerical FEA model for quantitative analysis and (3) MERLIN for a reduced order qualitative analysis.

### 2.1 Experimental Model

Three separate experiments are detailed: (1) tensile testing of card to determine its in-plane properties, (2) a uniaxial compression of a Miura-ori tube to determine the fold stiffness, and (3) a local squeezing of a Miura-ori tube, as shown in Figure 1, to provide validation for the numerical models.

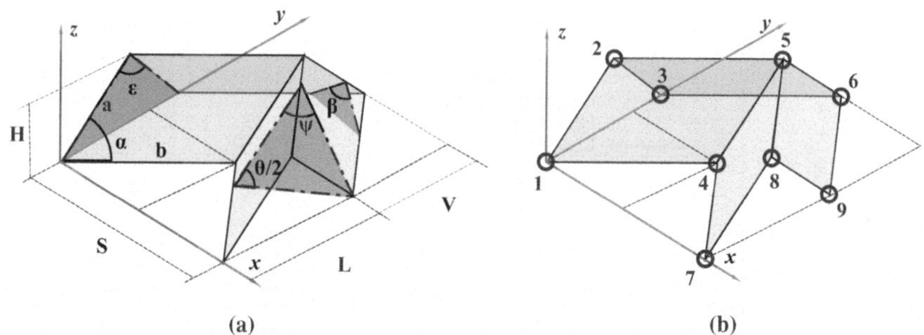

(a)           (b)

**Figure 2:** *(a) The geometric parameters of a Miura-ori unit cell; (b) the node numbering system within a unit cell; nodes 2, 5, and 8 on the mirror (negative z) side of the tube are labelled 2M, 5M, and 8M.*

### 2.1.1   Material and Manufacture

The samples are manufactured using Canford 300 gsm $\pm 5\%$ card, with a thickness of 380 $\mu$m $\pm 7.5\%$. The thickness and mass per unit area of the card are measured using multiple calliper measurements (following ASTM D646-13) as 370 $\mu$m and 297 gsm respectively, which are both within the manufacturing tolerance. The tensile modulus is measured (following ASTM D828-16), and is orthotropic and bi-linear. The moduli are $E_1 = 2300$ Nmm$^{-2}$ and $E_2 = 6100$ Nmm$^{-2}$ along and perpendicular to the length of the flattened tube respectively. The moduli are constant up to approximately 0.5% strain before transitioning to a lower stiffness up to the failure of the coupon. The initial stiffness is used for the numerical models due to the relatively low strains in the facets. The Poisson's ratios are measured during the same tests and are found to be $v_{12} = 0.17$ and $v_{21} = 0.4$. The material shear stiffness is estimated following [Jones 67].

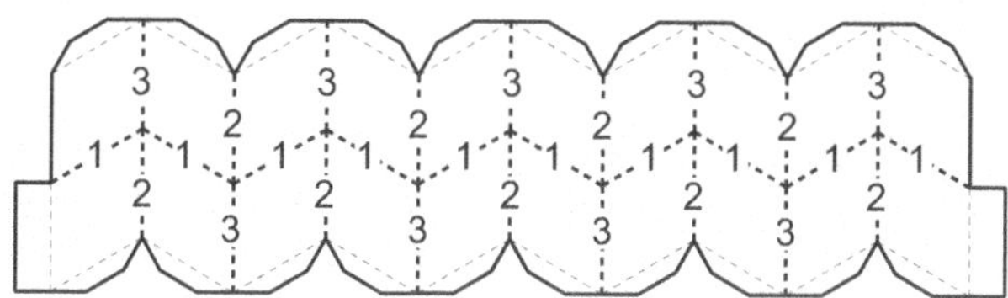

**Figure 3:** *One of the halves of the Miura tubes that are bonded together. Numbers represent the order of folding.*

The mechanical properties of folds in paper have been investigated on a fold in isolation. Methods include pulling the fold apart [Lechenault et al. 14, Pradier et al. 16, Yasuda et al. 13], pushing the fold together [Abbott et al. 14], or actuating the fold with a known force perpendicular to the material [Francis et al. 13]. Investigations into the effect of folding on the micro-structure of paper show that it delaminates during folding, causing changes in mechanical properties [Yasuda et al. 13, Beex and Peerlings 09]. This damage has been observed to accumulate over numerous cycles of folding and unfolding [Francis et al. 13, Reid et al. 17]. Therefore, a consistent manufacturing and testing procedure for any experimental samples is important.

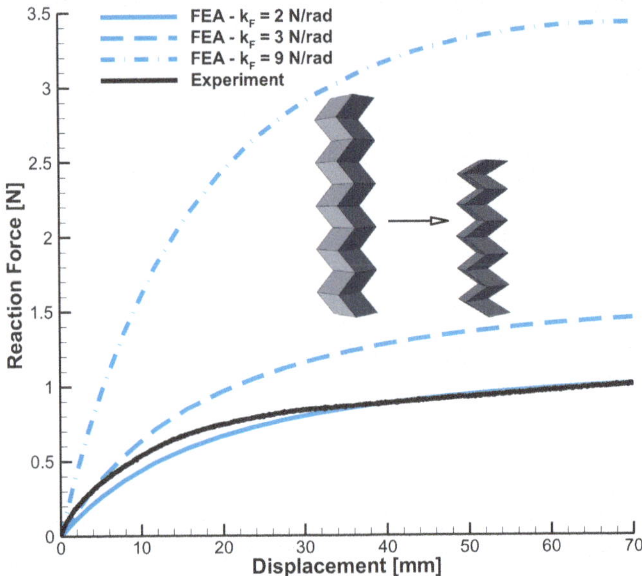

**Figure 4:** *The experimental results compared against finite element results for a range of fold stiffness values.* $k_F = 2$ N/rad *and* 3 N/rad *are selected as the most appropriate fold stiffness values for comparison with experimental results. However, a single linear fold stiffness cannot capture the entire range of motion.*

The Miura-ori tubes are constructed using the repeating cell shown in Figure 2a. Two mirrored strips, as shown in Figure 3, are cut using a laser cutter with a unit cell characterised by lengths $a = b = 30$ mm and angle $\alpha = 60°$. The fold lines are perforated during the laser cutting; all folds have 2 mm long perforations between 2 mm gaps. Due to the orthotropic nature of the card, the tubes are consistently orientated when they are cut out. Once the halves have been cut, the halves are first pre-creased. The folding follows a predefined sequence, indicated by the numbers in Figure 3. After pre-creasing, the halves are fully compressed and extended in the $x$-direction before being allowed to rest in their minimum-energy state. The two halves are then bonded together using double sided adhesive tape on the tabs and, finally, fully compressed and fully extended again in the $x$-direction, before letting the tube assume its rest configuration. The sample is then ready for testing.

### 2.1.2   Fold Stiffness Measurement

Rather than measuring an individual fold line, the stiffness of the folds is determined from a uniaxial compression test of the Miura-ori tube, where rigid origami kinematics is assumed. The Miura-ori tube is placed between two glass plates, and attached using double sided adhesive tape on tabs at each end. The glass plates reduce the friction induced as, during uniaxial loading, the cross-sections at the ends deform and rotate. The tubes are globally compressed in the $x$ direction, defined in Figure 2a, at a constant rate of 1 mm/s for 70 mm and the resulting reaction force is recorded, using a 1 kN load cell. The deformation is shown in the schematic inset in Figure 4. The resulting force and displacement data is used to calculate an approximate, averaged linear stiffness of the folds.

The calculation of the fold stiffness is based on the principle of virtual work. To simplify these calculations some assumptions have been used. Firstly, rigid origami assumptions mean that the deformation consists purely of changes in fold angles and these are the same across every unit cell. Further, it is assumed that all folds have the same stiffness and that the folds of the tabs connecting the tube to the glass plates have a negligible contribution to the reaction force. Finally, it is assumed that the tube starts in a zero-energy state and that there are no energy losses in the system. This has been evaluated in literature [Yasuda et al. 13] and is mainly due to high stresses at the vertices. The current experimental and finite element models are not connected at the vertices, thereby reducing the local stress concentrations.

Using these assumptions, the external work and internal strain energy can be equated to find the stiffness of the folds. The external work, $U$, is a function of the applied force $F$ and displacement $\Delta S$ from the initial rest state.

$$U = F\Delta S \tag{1}$$

Taking a constant fold stiffness per unit length, $k_F$, for all fold lines, the internal strain energy of a single unit cell of the tube is calculated as:

$$U = 2ak_F\left((\psi - \psi_0)^2 + (\beta - \beta_0)^2 + (\theta - \theta_0)^2\right) \tag{2}$$

where $\psi_0$, $\beta_0$, and $\theta_0$ are the angles at the initial rest conditions of the tube, and $a = b$. Differentiating Equations 1 and 2 with respect to $S$ and equating, enables the calculation of the fold stiffness:

$$k_F = \frac{F}{4a\left((\psi - \psi_0)\frac{d\psi}{dS} + (\beta - \beta_0)\frac{d\beta}{dS} + (\theta - \theta_0)\frac{d\theta}{dS}\right)} \tag{3}$$

where $\psi$, $\beta$, $\theta$, $d\psi/dS$, $d\beta/dS$, and $d\theta/dS$ are found in terms of $S$ using unit cell kinematics [Schenk and Guest 13, Wei et al. 13].

Using the experimental force and displacement data, a distribution for $k_F$ can be found ranging from 2 to 9 N/rad. An FE analysis of the uniaxial deformation is compared to the experimentally derived force-displacement relationship for different values of linear fold stiffness in Figure 4. A fold stiffness, $k_F$, between $2-3$ N/rad more closely matches the experimental results over the range of motion; therefore, these are selected for comparison between FEA and experimental local actuations. As all folds are assumed to have the same linear stiffness, it is not possible to derive a relationship between the moment and angle on each fold. Furthermore, Figure 4 shows that a constant fold stiffness cannot accurately describe the entire motion of the compression test. Nonetheless, for small deformations, the constant fold stiffness provides sufficient insight into the structural response of the Miura-ori tube. In future, the stiffness of individual folds will be measured to provide the nonlinear moment-angle relationship for large rotations, as well as study the effect of material fibre direction on the fold stiffness.

**Figure 5:** *The local actuation is achieved by using beads attached to nylon wire to compress opposing vertices of the tube. The deformed geometry is captured using a laser scanner, which generates a point cloud that is processed to find the fold lines and vertices.*

### 2.1.3 Local Actuation

The experimental local actuation tests consist of compressing the unit cell at one end of a tube. This is achieved by pulling together the two central vertices using a pair of beads on a nylon wire. First, a knot is tied in the nylon wire and a bead threaded onto it, before threading the wire through the vertices; this is repeated with another nylon wire and bead from the other side of the tube. The tube is then flattened to its maximum extension in the $x$-direction, allowed to return to a rest state, and the length measured.

Next, the tube is placed upright, where the $x$-direction is upwards. The nylon wires are pulled tight to achieve the actuation, in the bottom unit cell, and pinned in place. The outer geometry is captured as a point cloud using a 3D laser scanner, as shown in Figure 5. From this point cloud the folds can be identified by finding the points with the highest and lowest $y$ and $z$ values for a slice along the tube. The average $y$-position of each fold can then be found. The vertices are found as the points of a fold furthest from the average $y$-position. The calculated vertices are superimposed onto the point cloud and checked visually. The angle $\varepsilon$ is calculated by finding the angle between relevant vectors defined by the vertices. A value for this angle is found for both sides of the tube, which are averaged to give the value for the unit cell. A total of 3 tubes of 10 unit cells were used for the experiments, and each tube was actuated once.

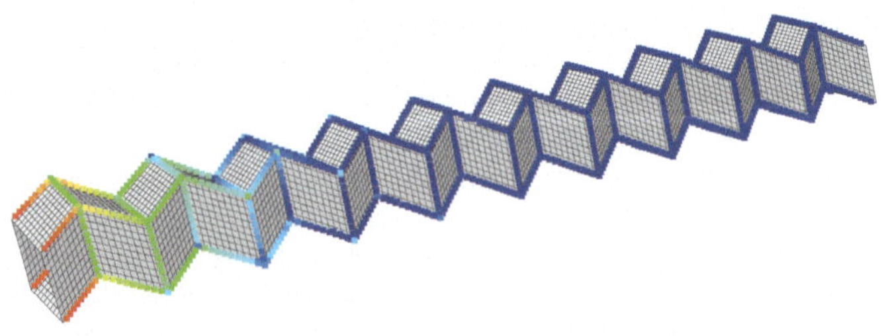

**Figure 6:** *FEA model of the Miura-ori tube; colours along the fold lines indicate the change in fold angle during actuation.*

## 2.2  Numerical Models

The FEA model, implemented in ABAQUS/Standard, consists of facets joined together by torsional springs along the fold lines. The facets are meshed using shell elements (S4R). A convergence study found that a mesh density of $25 \times 25$ elements on each facet is sufficient to capture the angle $\varepsilon$ to within 1% of a mesh with double the elements on each edge, at every point along the tube. A linear-elastic, orthotropic material stiffness is assumed for the paper facets. The torsional springs along the fold lines use hinge elements (CONN3D2) with a linear torsional stiffness; the resulting model is shown in Figure 6. Section 2.1 details how the stiffness properties are determined. Furthermore, a gravity load is included for comparison with experimental results. The tube is constrained such that it is pinned on node 4 and is fixed in the $x$-direction at nodes 5 and 5M; node numbers are defined in Figure 2b. A displacement in $z$ is prescribed, to move nodes 5 and 5M closer. The initial state of the tube in both the numerical and experimental tests is defined by angle $\theta = 90°$.

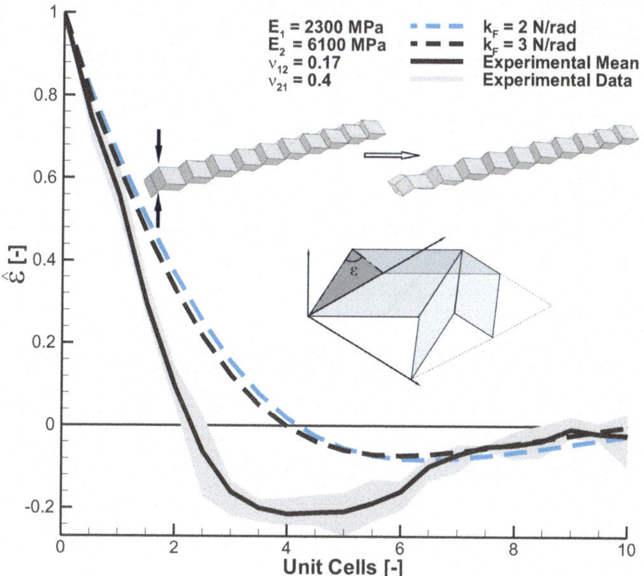



**Figure 7:** *The angle between the folds reduces along the length of the tube in both the FE analysis and experimental tests. FEA results are shown for different values of fold stiffness. The shaded area indicates the maximum/minimum experimentally determined values.*

All results are normalised using the initial conditions and the maximum value, where $\varepsilon_0$ represents the undeformed configuration:

$$\hat{\varepsilon} = \frac{\varepsilon - \varepsilon_0}{\max(\varepsilon) - \varepsilon_0} \qquad (4)$$

Figure 7 shows that the non-linear FEA and experimental tests match closely at the source of actuation and towards the end of the tube, returning to the undeformed shape. Both also have a negative region which physically represents a deformation in the opposite direction to that of the actuation. This effect, termed a 'spring-back', is discussed further in Section 4. Although gravity accentuates the effect, as shown in Figure 7, it is also present when the effect of gravity is removed; see Figure 8. The global mechanical response is independent of material anisotropy, and is also found with a linear analysis. The linear approximation is valid up to an actuation compressing the tube by 3/4 of its initial height $H$, with analyses at compressions of $H/4$ and $H/2$ showing the same normalised response. Therefore, an actuation of $H/4$ is used for further analyses. The reduced-order MERLIN model also captures the behaviour; a qualitative comparison is shown in Figure 8 as representative stiffness and area values have not been established for MERLIN.

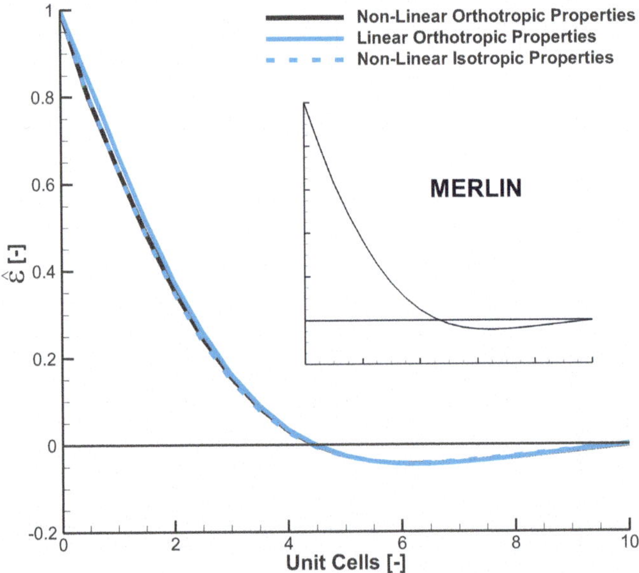



**Figure 8:** *The behaviour of the tube in response to local actuation can be effectively captured with isotropic material properties and a linear FE analysis. Actuating the unit cell with a $H/2$ displacement shows a nominally identical response using linear and non-linear analyses. For a $3H/4$ actuation the linear and non-linear responses start to diverge. The reduced-order non-linear MERLIN model (inset with an axes range equal to the main plot) also captures the behaviour.*

Despite these similarities, the 'spring-back' effect is greater in the experimental results, which could be due to differing stiffness in each fold. As highlighted in Section 2.1, the card material is orthotropic, and the fold stiffness will depend on the orientation of the fold with respect to that of the card. Future work will test single folds in isolation to characterise this effect. Nonetheless, despite the difference in magnitude of the 'spring-back' in the experimental and numerical tests, the important features of the behaviour are captured in the FEA model. Consequently, it is used to further investigate the behaviour of the Miura-ori tubes.

## 3 Elastic Decay in Miura tubes

The FEA model is used to investigate the effect of (1) the elastic stiffness properties and (2) unit cell geometry (thickness, size, facet angle) on the local actuation of an origami tube. In the following analyses, the orthotropic nature of the paper is neglected, and gravity is removed. The effective, isotropic Young's modulus is taken as $E = \sqrt{E_1 E_2} = 3750$ MPa and effective isotropic Poisson's ratio is taken as $v = \sqrt{v_{12} v_{21}} = 0.261$.

### 3.1 Elastic Properties

Two key material parameters in the mechanical response of origami structures are the fold stiffness and facet bending stiffness [Schenk and Guest 11]. Rather than specifying the fold stiffness directly, we make use of a length scale $L^*$ introduced by [Lechenault et al. 14],

$$L^* = \frac{B}{k_F} \quad \text{where} \quad B = \frac{Et^3}{12(1-v^2)} \tag{5}$$

which relates the fold stiffness $k_F$, per unit length of fold, to the material flexural stiffness $B$. $L^*$ provides a compact measure of the ratio between the mechanical properties of the facets and folds, to characterise the overall behaviour of the origami structure. Larger values of $L^*$ correspond to an origami system that is closer to rigid origami, with perfect hinges and rigid facets.

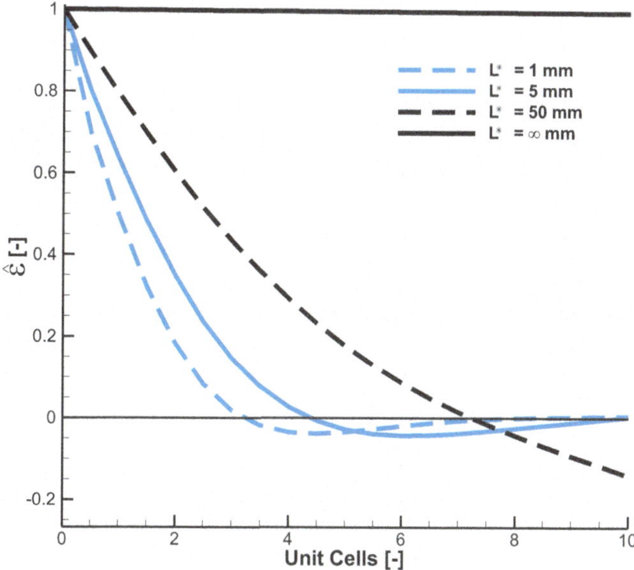


**Figure 9:** *Reducing $L^*$ causes a reduction in the decay length, as the folds become stiffer in comparison to the facets. In all cases $a = b = 30$ mm, $\alpha = 60°$, $t = 0.37$ mm, and $\theta = 90°$.*

Figure 9 shows FEA models with varying length scales; $L^*$ is varied by changing the fold stiffness and keeping the material properties and geometry of the facets constant. Included is a model with perfect hinges, leading to an infinite $L^*$, which behaves as rigid origami and has a negligible elastic decay. Conversely, smaller length scales show an increased localisation of the actuation, as the bending stiffness of the facets reduces. However, for analyses where the actuation completes within the length of the tube, the magnitude of any 'spring-back' remains approximately constant. For the test with $L^* = 50$ mm the decay does not complete within the length of the tube, which results in a significant change in the response of the structure. The role of edge effects on decay length will be explored in future work. When the local actuation of a Miura-ori strip (one half of the tube) is investigated, the effect is no longer observed. This suggests that the 'spring-back' phenomenon is a result of the implied boundary condition of symmetry about a central $x - y$ plane in the tube, which is not present in the strip. The implications of this are discussed further in Section 4.

## 3.2   Unit Cell Geometry

A limitation of the length scale is that it does not account for the facet geometry, and therefore does not effectively capture the facet bending stiffness. The facet stiffness is here taken as the equivalent stiffness of a pseudo-fold across the short diagonal, $k_B$. [Filipov et al. 17] provide an empirical formulation for the bending stiffness per unit length, $k_B$, of a facet

$$k_B = \left(0.55 - 0.42\frac{\Sigma\gamma}{\pi}\right) B \left(\frac{1}{tD_s^2}\right)^{1/3} \tag{6}$$

where $D_s$ is the shortest diagonal length, $t$ the material thickness, $B$ the material flexural stiffness, and $\Sigma\gamma$ is the sum of the exterior angles of the facet. For the parallelogram facets considered here, this equals $2\alpha$, where $\alpha$ is defined in Figure 2a.

The ratio of the fold and facet bending stiffness $k_B/k_F$ has previously been used to characterise the mechanical response of origami systems [Schenk and Guest 11].

$$\eta = \frac{k_B}{k_F} = \frac{E}{12k_F(1 - v^2)}\left(0.55 - 0.42\frac{2\alpha}{\pi}\right)\left(\frac{t^4}{D_s}\right)^{2/3} \tag{7}$$

Equation 7 captures the fold stiffness, elastic material properties, and facet geometry as a product of the effect of material and geometric properties. If the in-plane ($E$) and fold stiffness ($k_F$) are changed proportionally to maintain a constant $\eta$ identical behaviour is observed for significantly different stiffness values, for a fixed geometry.

However, varying the facet geometry will affect the elastic decay. Figure 10 shows the effect of changing facet dimensions and parallelogram angle, while maintaining a constant ratio $D_s/t = 81$, a constant $\eta = 0.15$, and a constant initial configuration $\theta$. In contrast to the elastic properties, the geometry of the unit cell has a more complicated effect, which is not fully captured by the formulation in Equation 7. Nonetheless, the salient features, localisation of actuation in origami with finite facet stiffness and a 'spring-back' effect, are shown not to be specific to a particular geometry.

The ratio between the material flexural stiffness of the facets and the elastic stiffness of the folds can be used to describe the decay of localised actuation in Miura-ori tubes of fixed geometry. However, the unit cell geometry also affects the elastic decay and Equation 7 does not adequately capture this relationship. Future work will aim to better capture this dependence on the unit cell geometry in order to develop a formulation to describe the elastic decay of localised actuation in Miura-ori tubes.

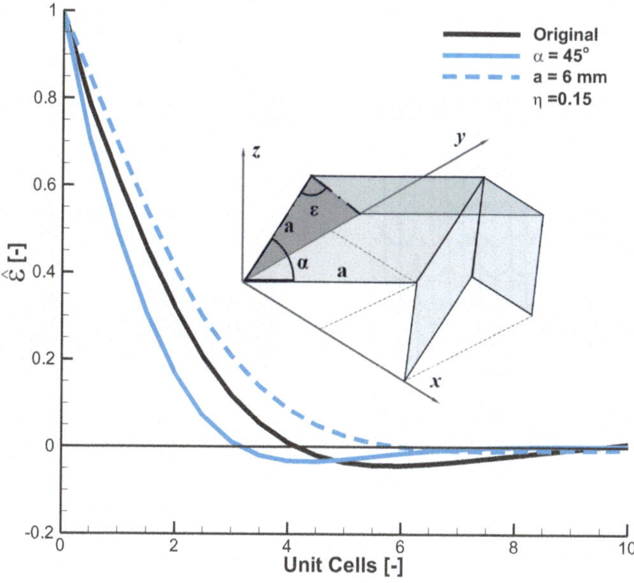

**Figure 10:** *The behaviour of the decay is sensitive to both the unit cell shape and size. The original comparison FEA test has $a = 30$ mm and $\alpha = 60°$ and all of the tests have $D_S = 81t$, which is identical to the experimental tests. Changes to $\alpha$ and $a$ are with the other parameter kept constant at the original value; $k_F$ is also changed to maintain a constant $\eta = 0.15$.*

# 4  Discussion

Numerical and experimental results show a good correlation, as shown in Figure 7; however, the experimental measurements exhibit a greater magnitude of 'spring-back'. The reason for this discrepancy is not currently understood. The magnitude of the 'spring-back' is shown to be insensitive to small changes in fold stiffness. Therefore, the effects of variations in fold stiffness due to orientation with respect to the grain of the card, or due to the non-linear stiffness of the fold lines, are not expected to be significant. Investigating the causes of this difference will be the subject of future work.

This unexpected 'spring-back' behaviour in the Miura-ori tube invites an analogy with the response of a long thin-walled cylinder to an applied ring load, where the deflection decays in an oscillatory fashion away from the point of application [Calladine 83]. That analysis is itself analogous to that of a beam on an elastic foundation [Hartog 52]. For the thin-walled cylinder the tulastic foundation is provided by the generated hoop stress. In the Miura-ori tubes the facets also experience in-plane strains, which provide the elastic foundation for the 'spring-back' in the elastic decay. Initial work suggests that the decay curves in the Miura-ori tube can be closely approximated by the equations for a beam on elastic foundation, and future work will aim to characterise this behaviour analytically.

The effect of in-plane and fold/facet stiffness on the localised actuation response is not limited to Miura-ori tubes. Figure 11 shows the effect of local actuation of a single unit cell

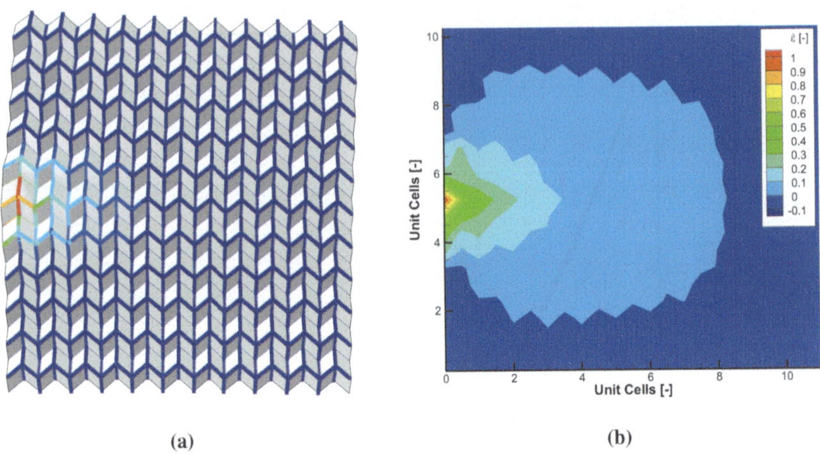

(a)          (b)

**Figure 11:** *(a) a single unit cell of a Miura-ori sheet is actuated; colours indicate the change in fold angle throughout the sheet; (b) the change in angle $\hat{\varepsilon}$ again shows a spring-back effect, where unit cells have an oppositely signed deformation to the original actuation.*

in a large ($11 \times 11$) Miura-ori sheet; all folds in the unit cell are opened simultaneously. The finite facet stiffness results in a localisation of the actuation, and a 'spring-back' effect is again observed. Here, the sheet around the actuated unit cell provides the elastic foundation. Similar to the Miura-ori tubes, changing the elastic properties of the folds and facets in a Miura-ori sheet causes a different profile of decay. This is in contrast to previous work on inhomogeneous deformations [Evans et al. 15] which suggests that the decay of a local perturbation depends only on the geometry of a unit cell, the initial conditions of the sheet, and the size of the perturbation.

# 5   Conclusions

This paper investigates the effect of the elastic properties of the facets and folds of Miura-ori tubes on the elastic decay of a localised actuation. This is done numerically using a Finite Element model, which is verified experimentally against paper samples.

Elastic properties representing a larger departure from idealised rigid origami lead to increasing localisation of an actuation. For a given geometry this effect can be described by the ratio of the in-plane stiffness of the facets and the torsional stiffness of the folds. This behaviour is linear for small actuations and is not influenced by the orthotropic material properties of paper. Furthermore, a new phenomenon in origami has been observed: a 'spring-back' effect where a local actuation causes a deformation of the opposite sense at a point elsewhere in the system. This effect is not specific to Miura-ori tubes and has also been observed in a Miura-ori sheet.

## Data Statement

All supporting data are provided within this paper.

1254

## Acknowledgments

This work was supported by the Engineering and Physical Sciences Research Council through the EPSRC Centre for Doctoral Training in Advanced Composites for Innovation and Science (grant number EP/L016028/1).

This paper has been awarded the 7OSME Gabriella & Paul Rosenbaum Foundation Travel Award.

## References

[Abbott et al. 14]  A. C. Abbott, P. R. Buskohl, J. J. Joo, G. W. Reich, and R. A. Vaia. "Characterization of creases in polymers for adaptive origami structures." In *ASME 2014 Conference on Smart Materials, Adaptive Structures and Intelligent Systems*, 2014.

[Beex and Peerlings 09]  L. A. A. Beex and R. H. J. Peerlings. "An experimental and computational study of laminated paperboard creasing and folding." *International Journal of Solids and Structures* 46:24 (2009), 4192–4207.

[Calladine 83]  C. R. Calladine. *Theory of Shell Structures*. New York: Cambridge University Press, 1983.

[Chen et al. 16]  Bryan Gin-ge Chen, Bin Liu, Arthur A. Evans, Jayson Paulose, Itai Cohen, Vincenzo Vitelli, and C. D. Santangelo. "Topological Mechanics of Origami and Kirigami." *Physical Review Letters* 116:13 (2016), 135501.

[Dudte et al. 16]  L. H. Dudte, E. Vouga, T. Tachi, and L. Mahadevan. "Programming curvature using origami tessellations." *Nature Materials* 15:5 (2016), 583–588.

[Evans et al. 15]  A. A. Evans, J. L. Silverberg, and C. D. Santangelo. "Lattice mechanics of origami tessellations." *Physical Review E* 92:1 (2015), 013205.

[Filipov et al. 15]  E.T. Filipov, T. Tachi, and G. H. Paulino. "Origami tubes assembled into stiff, yet reconfigurable structures and metamaterials." *Proceedings of the National Academy of Sciences* 112:40 (2015), 12321–12326.

[Filipov et al. 17]  E. T. Filipov, K. Liu, T. Tachi, M. Schenk, and G. H. Paulino. "Bar and hinge models for scalable analysis of origami." *International Journal of Solids and Structures* 124 (2017), 26–45.

[Francis et al. 13]  K. C. Francis, J. E. Blanch, S. P. Magleby, and L. L. Howell. "Origami-like creases in sheet materials for compliant mechanism design." *Mechanical Sciences* 4:2 (2013), 371–380.

[Gattas and You 15]  J.M. Gattas and Z. You. "The behaviour of curved-crease foldcores under low-velocity impact loads." *International Journal of Solids and Structures* 53 (2015), 80–91.

[Hartog 52]  J. P. Den Hartog. *Advanced Strength of Materials*. New York: McGraw-Hill Book Company Inc., 1952.

[Jones 67]  A. R. Jones. "An experimental investigation of in-plane elastic moduli of paper." Ph.D. thesis, Lawrence University, 1967.

[Lechenault et al. 14]  F. Lechenault, B. Thiria, and M. Adda-Bedia. "Mechanical response of a creased sheet." *Physical Review Letters* 112:24 (2014), 1–5.

[Liu and Paulino 17]  K. Liu and G. H. Paulino. "Nonlinear mechanics of non-rigid origami: an efficient computational approach." *Proceedings of the Royal Society A: Mathematical, Physical and Engineering Science* 473:2206 (2017), 20170348.

[Liu et al. 15] S. Liu, G. Lu, Y. Chen, and Y. W. Leong. "Deformation of the Miura-ori patterned sheet." *International Journal of Mechanical Sciences* 99 (2015), 130–142.

[Ma and You 13] Ji. Ma and Z. You. "Energy Absorption of Thin-Walled Square Tubes With a Prefolded Origami Pattern–Part I: Geometry and Numerical Simulation." *Journal of Applied Mechanics* 81:1 (2013), 011003.

[Pradier et al. 16] C. Pradier, J. Cavoret, D. Dureisseix, C. Jean-Mistral, and F. Ville. "An Experimental Study and Model Determination of the Mechanical Stiffness of Paper Folds." *Journal of Mechanical Design* 138:4 (2016), 041401.

[Reid et al. 17] A. Reid, F. Lechenault, S. Rica, and M. Adda-Bedia. "Geometry and design of origami bellows with tunable response." *Physical Review E* 95:1 (2017), 013002.

[Schenk and Guest 11] M. Schenk and S. D. Guest. "Origami Folding: A Structural Engineering Approach." In *Origami 5: Fifth International Meeting of Origami Science, Mathematics, and Education*, pp. 291–304, 2011.

[Schenk and Guest 13] M. Schenk and S. D. Guest. "Geometry of Miura-folded metamaterials." *Proceedings of the National Academy of Sciences* 110:9 (2013), 3276–3281.

[Schenk et al. 14] M. Schenk, S. D. Guest, and G. J. McShane. "Novel stacked folded cores for blast-resistant sandwich beams." *International Journal of Solids and Structures* 51:25-26 (2014), 4196–4214.

[Silverberg et al. 15] J. L. Silverberg, J. Na, A. A. Evans, B. Liu, T. C. Hull, C. D. Santangelo, R. J. Lang, R. C. Hayward, and I. Cohen. "Origami structures with a critical transition to bistability arising from hidden degrees of freedom." *Nature materials* 14:4 (2015), 389–93.

[Tachi 09] T. Tachi. "Simulation of Rigid Origami." In *Origami 4*, pp. 175–187. A K Peters/CRC Press, 2009.

[Wei et al. 13] Z. Y. Wei, Z. V. Guo, L. Dudte, H. Y. Liang, and L. Mahadevan. "Geometric mechanics of periodic pleated origami." *Physical Review Letters* 110:21 (2013), 1–5.

[Yasuda et al. 13] H. Yasuda, T. Yein, T. Tachi, K. Miura, and M. Taya. "Folding behaviour of Tachi-Miura polyhedron bellows." *Proceedings of the Royal Society A: Mathematical, Physical and Engineering Sciences* 469:2159.

---

Steven Grey
Bristol Composites Institute (ACCIS), Department of Aerospace Engineering, University of Bristol, Bristol, BS8 1TR, United Kingdom, e-mail: steven.grey@bristol.ac.uk

Fabrizio Scarpa
Bristol Composites Institute (ACCIS), Department of Aerospace Engineering, University of Bristol, Bristol, BS8 1TR, United Kingdom, e-mail: f.scarpa@bristol.ac.uk

Mark Schenk
Bristol Composites Institute (ACCIS), Department of Aerospace Engineering, University of Bristol, Bristol, BS8 1TR, United Kingdom, e-mail: m.schenk@bristol.ac.uk

# Elastic Buckling of Thin-walled Cylinders with Pre-embedded Diamond Patterns

*X. Yang, S. Zang, J. Ma, Y. Chen*

***Abstract:*** *Elastic buckling of thin-walled structures are sensitive to imperfection and therefore difficult to be precisely predicted. In this paper, unfolded diamond patterns were pre-embedded on the surfaces of thin-walled cylinders with the aim of controlling their buckling behaviours when axially loaded. Three buckling modes were observed from quasi-static experiments, and the correlation between pattern geometry and buckling mode were obtained. By properly selecting geometric parameters, the buckling mode could be precisely controlled. A local snap-through behaviour was also identified and explained by combination of experimental and analytical approaches. This proposed method shows great promise to achieve stable and controllable deformation modes for thin-walled structures.*

## 1  Introduction

Thin-walled tubes are frequently used in various branches of structural engineering, e.g., load-bearing components in civil structures, rocket body shells, and energy absorption devices, due to their low cost, high manufacturability and excellent load-carrying efficiency [Calladine 83]. Under axial compression, long tubes generally undergo global Euler buckling [Calladine 83]. Tubes with medium length, on the other hand, initially buckle locally and then progressively collapse following a certain mode depending on tube geometry and material properties [Lu and Yu 03].

When axially compressed, a circular cylinder could buckle into two modes, i.e., the axisymmetric mode, or the ring mode [Singace et al. 95], and the non-axisymmetric mode, or the diamond mode [Yoshimura 55] with a number of circumferential lobes. These deformation modes are mainly determined by two geometric parameters, which are tube diameter $D$ to wall thickness $t$ ratio, $D/t$,

and tube diameter $D$ to tube height $H$ ratio, $D/H$ [Guillow and Lu et al. 01].

Around 1960s, a theoretical model of the axisymmetric mode was first built by Alexander [Alexander 60] to obtain a physical understanding of the buckling and post-buckling behaviours, which was further improved to better match experimental results [Arbramowicz and Jones 84, Wierzbicki et al. 92]. Study on the diamond mode was first conducted by Pugsley and Macaulay [Pugsley and

Macaulay 60], whose theory was proved less successful and largely empirical. Later, the diamond mode was also investigated by Johnson et al. [Johnson and Soden 77] based on experiments of PVC tubes with a given number of pre-folded diamond lobes. Singace [Singace 99] also conducted theoretical studies of the diamond mode based on Alexander's theoretical model of the ring mode. Despite numerous studies for decades, the exact buckled shape and mechanical properties are still difficult to predict and control due to the strong sensitivity to geometric and load imperfections.

To achieve stable and controllable buckling, numerous approaches have been attempted. Chen and Ozaki, Mozafari et al. and Wu et al. introduced different corrugations on the surface of thin-walled cylinders to trigger a stable ring mode [Chen and Ozaki 09, Mozafari et al. 18, Wu et al. 16]. Another method is to use non-uniform wall thickness to form graded structures proposed by Zhang et al. and Sun et al. [Zhang et al. 15 and Sun et al. 14]. Studies on employing holes, grooves and dents to trigger the collapse of tubular structures can be traced back to 1990s [Han et al. 07, Lee et al. 99]. In addition, applying buckling initiators can also help to achieve predictable buckling modes, for example, a device used in the landing gears of aircrafts [Airoldi and Janszen 05].

Origami patterns, traditionally a Japanese art of paper folding, have been widely applied in the design of thin-walled tubes. They are pre-folded on tube surfaces as a mode inducer to generate desirable deformation modes and mechanical properties. For example, Yoshimura patterns was utilized to produce beverage cans due to its foldability [Han, et al. 04]. In Yang's research [Yang et al. 16], specimens of thin-walled cylinders with and without diamond origami patterns (Yoshimura Pattern) were tested, which validated the numerical results. Ma and Hou also introduced a new kite-shape pattern by pre-folding the surface of a square tube to alter the collapse mode [Ma et al. 16]. A novel device, named as the origami crash box [Ma and You 13] was proposed to induce a special collapse mode. In Song's study [Song et al. 12], a new origami patterned tube was proposed to trigger novel and predictable deformation modes to obtain ideal physical properties.

Recently, we noticed an interesting tubular structure named origami vase. As can be seen from Fig. 1(a), the origami vase starts from a flat paper board with diamond patterns indented on it. Subsequently it is folded along the crease lines (Fig. 1(b)) and flattened (Fig. 1(c)), ending up with a roughly flat sheet again. At last, it is rolled and the two edges are joined (Fig. 1(d)) to form a cylinder. Remarkably, the tube surface buckles locally during the rolling process, and each diamond unit naturally deforms into a curved shape resembling those in the diamond mode of thin-walled cylinders. Different from those origami tubes mentioned above, the patterns on the origami vase are initially pre-embedded but not folded, and are later triggered by external loading, which in this case, are bending of tube surface during rolling. Inspired by the origami vase, in this paper we experimentally investigate the axial compression of thin-walled cylinders with those pre-embedded diamond patterns, with the aim of precisely control the buckling mode of the cylinders.

The layout of the paper is as follows. A geometric analysis of the diamond pattern and the fabrication procedure of the cylinders are shown in Section 2. In Section 3, the experimental setup for quasi-static axial compression of the cylinders is described, and the experimental results are presented and analysed. Subsequently the local snap-through behaviour of the cylinder is investigated experimentally and analytically. Finally, the conclusion is given in Section 5, which ends the paper.

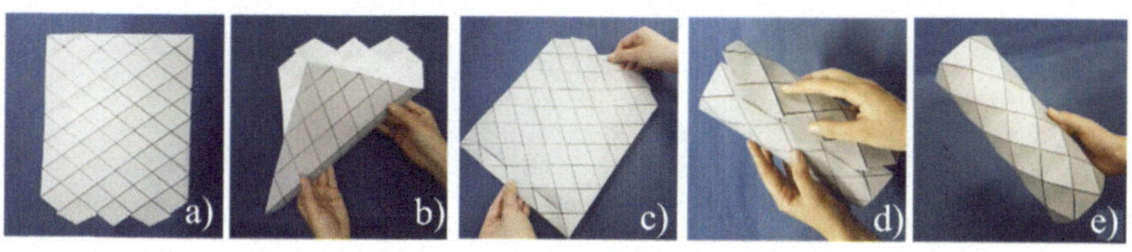


**Figure 1:** *Steps of forming an origami vase. (a) A thin paper board with pre-pressed diamond pattern. (b) Folding the paperboard along the creases in one direction. (c) The shape of the paper board after folding along creases. (d) Rolling the paper board into a cylinder. (e) A finished origami vase.*

# 2  Design and Fabrication

## 2.1  Geometry

The diamond patterns employed in this paper are similar to the well-known Yoshimura pattern [Yoshimura 55] omitting the horizontal diagonal. It can be determined by two geometric parameters, which are the number of circumferential lobes $n$ and the slant angle $\alpha$ to the horizontal line. Therefore, we generally name our models with these two geometric parameters in the form of "$n-\alpha$". For example, a model with $n=7$ and $\alpha=60°$ is named as "$7-60°$". The developed surface of one specimen $6-55°$ is shown in Fig. 2(a). The length of the vertical diagonal $h$, the length of the horizontal diagonal $b$, and the apex angle $\beta$ of each diamond lobe could also be derived as

$$\beta = \pi - 2\alpha \tag{1}$$

$$b = \frac{\pi D}{n} \tag{2}$$

$$h = \frac{\pi D \tan \alpha}{n} \tag{3}$$

To retain as many complete lobes or half of the lobes as possible in the finite size, we select the proper number of axial lobes $m$ to achieve slightly varied height $H$ of each cylinders due to the difference of diamond lobes.

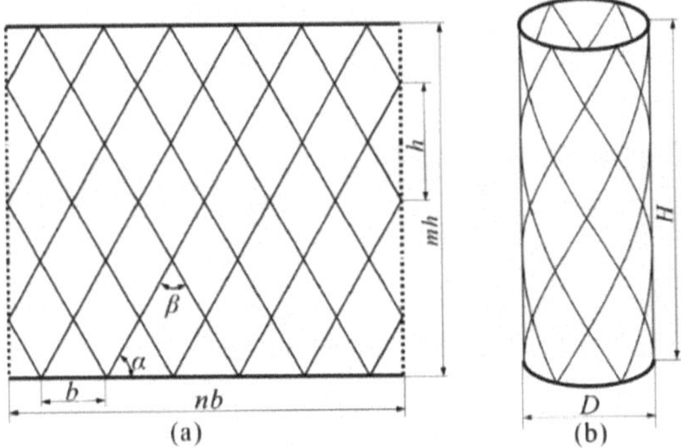

**Figure 2:** *(a) Diamond planar pattern, (b) A cylinder with diamond patterns (m=3, n=6).*

## 2.2   Fabrication

A total of 37 various diamond patterned ones with systematically varied geometric parameters were fabricated for axial compression. For variable patterned tubes on the tests, their number of circumferential lobes $n$ ranged from 4 to 9, while their slant angle $\alpha$ varies from 30° to 70° in this paper. In addition, a conventional cylinder with $D = 88.49$mm and the modified height $H_0 = 210$mm were also investigated for comparison. The geometries of all the specimens and their buckling modes under quasi-static compression are listed in Table 1. The symbols "◇", "○" and "△" in Table 1 represented three buckling modes of various tubes, i.e., the "unfollow" mode, the "completely follow" mode and the "partly follow" mode respectively, which will be elaborated in Section 3.3. Paper boards with a thickness $t = 0.27$mm were adopted for cylinder fabrication. The pattern creases were precisely indented on both sides of the paper by a spherical nib with a radius 0.3mm (Fig. 3(a)). Subsequently each crease were folded back and forth several times, see Fig. 3(b). Finally the paper was rolled and glued into a cylindrical shape (Fig. 3(c)).

Furthermore, an additional specimen $5 - 50°$ was made of polypropylene (PP) sheets with identical thickness in order for digital image correlation experiments. In this case, the diamond patterns were engraved by a laser cutter in the depth of 0.15mm on one side of the sheet. Then the same procedure with paper cylinders was followed to form a cylinder.

***Table 1:*** *Geometries of cylinders with diamond patterns and buckling modes.*

| Model | $\alpha$ [deg] | $n$ | $m$ | $H$ (mm) | Mode |
|---|---|---|---|---|---|
| CIR | 35 | 6 | —— | 210 | △ |
| 4-30° | 30 | 4 | 5.5 | 220.7 | △ |
| 7-30° | 30 | 7 | 9 | 206.4 | △ |
| 4-45° | 45 | 4 | 3 | 208.5 | ○ |
| 5-45° | 45 | 5 | 4 | 222.4 | ○ |
| 6-45° | 45 | 6 | 4.5 | 208.5 | ○ |
| 7-45° | 45 | 7 | 5.5 | 218.4 | ○ |
| 8-45° | 45 | 8 | 6 | 208.5 | ○ |
| 9-45° | 45 | 9 | 7 | 216.2 | ○ |
| 4-50° | 50 | 4 | 2.5 | 207.1 | ○ |
| 5-50° | 50 | 5 | 3 | 198.8 | ◇ |
| 5−50° * | 50 | 5 | 3 | 198.8 | ◇ |
| 6-50° | 50 | 6 | 4 | 220.9 | ◇ |
| 7-50° | 50 | 7 | 4.5 | 213.0 | ○ |
| 8-50° | 50 | 8 | 5 | 207.1 | ○ |
| 9-50° | 50 | 9 | 5.5 | 202.5 | ○ |
| 4-55° | 55 | 4 | 2 | 198.5 | ◇ |
| 5-55° | 55 | 5 | 2.5 | 198.5 | ◇ |
| 6-55° | 55 | 6 | 3 | 198.5 | ◇ |
| 7-55° | 55 | 7 | 3.5 | 198.5 | ◇ |
| 8-55° | 55 | 8 | 4 | 198.5 | ◇ |
| 9-55° | 55 | 9 | 5 | 220.6 | ○ |
| 4-60° | 60 | 4 | 2 | 240.8 | ◇ |
| 5-60° | 60 | 5 | 2 | 192.6 | ◇ |
| 6-60° | 60 | 6 | 2.5 | 200.6 | ◇ |
| 7-60° | 60 | 7 | 3 | 206.4 | ◇ |
| 8-60° | 60 | 8 | 3.5 | 210.7 | ◇ |
| 9-60° | 60 | 9 | 4 | 214.0 | ◇ |
| 4-65° | 65 | 4 | 1.5 | 223.6 | ○ |
| 5-65° | 65 | 5 | 2 | 238.5 | ◇ |
| 6-65° | 65 | 6 | 2 | 198.7 | ◇ |
| 7-65° | 65 | 7 | 2.5 | 212.9 | ◇ |
| 8-65° | 65 | 8 | 3 | 223.6 | ◇ |
| 4-70° | 65 | 4 | 1.5 | 223.6 | ○ |
| 5-70° | 70 | 5 | 1.5 | 229.1 | ○ |
| 6-70° | 70 | 6 | 1.5 | 191.0 | ◇ |
| 7-70° | 70 | 7 | 2 | 218.2 | ○ |
| 8-70° | 70 | 8 | 2 | 191.0 | ○ |

*indicates the specimen made of thin PP board for DIC test.

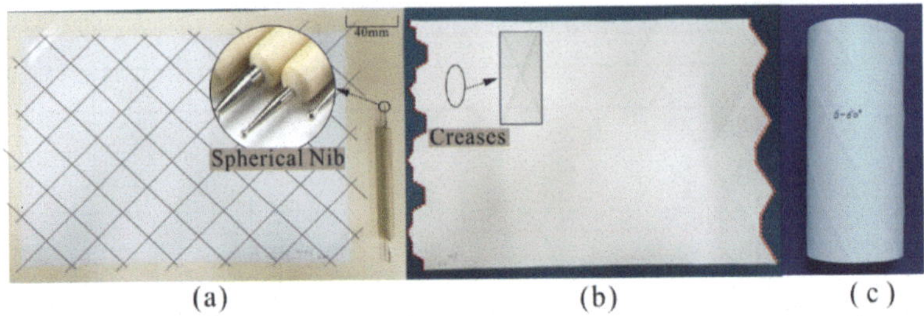

(a)           (b)           ( c )

**Figure 3:** *Fabrication process of a paper cylinder: (a) Indenting creases with a spherical nib. (b) Folding each crease back and forth. (c) Rolling the paper to form a cylinder (n=6, α =60°).*

## 3 Experimental Results and Discussion

### 3.1 Experimental Setup

To study the relationship between pattern geometry and buckling mode, quasi-static axial compression tests were conducted on specimens by an Instron Universal Testing Machine (type 5982) at a loading speed of 2mm/min. In the tests, a specimen was sandwiched between two plastic caps with circular grooves of 3.6mm in depth (Fig. 4(a)), which formed pinned boundary conditions. The deformation process was recorded by a standard digital camera (Canon 70D).

**Figure 4:** *(a) Intron experimental setup and (b) VIC-3D experimental setup.*

In addtion, a digital image correlation system CSI Vic-3D9M was adopted to capture the exact deformed configurations of specimen $5-50°*$ during the test

(Fig. 4(b)) at an interval of 200 milliseconds. Speckles with a density of about $3.6/mm^2$ were sprinkled on the surface of the specimen prior to the test. The displacements of 3606 feature points which were uniformly distributed were calculated to obtain an accurate geometric reconstruction of the area of interest.

## 3.2 Compression of the conventional cylinder

The conventional cylinder CIR is first studied as a benchmark. As can be seen in Fig. 5(a), it buckles into the non-axisymmetric mode with 6 circumferential lobes and the slant angle $\alpha_0 = 35°$. As to force versus displacement (Fig. 5(b)), the force rises approximately in a linear manner, reaching an initial peak force $P_{I0}$ of 194.06N when the diamond lobes are well formed. Further compression leads to a drop in force as the lobes are crushed.

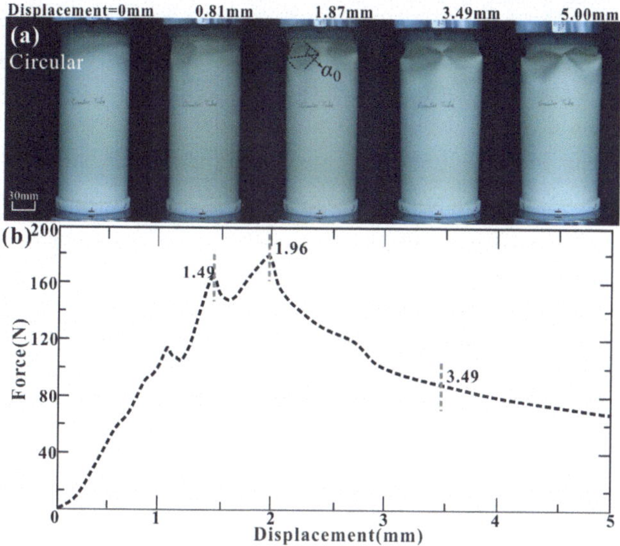


**Figure 5:** (a) Deformation process and (b) force vs. displacement curve of the CIR.

## 3.3 Compression of patterned cylinders

Patterned tubes with various geometric parameters were tested under the axial compression. All specimens intentionally kept an unbuckled shape as shown in Fig. 3(c) prior to tests. Under the quasi-static compression, patterned cylinders could naturally generate curved unit cells around their surfaces forming various buckled surface textures. In general, their buckling modes can be divided into three different buckling types, as shown Fig. 6(a), (b), (c), which sensitively depends on geometric parameters $n$ and $\alpha$ of the diamond patterns. For example, for cylinders with a fixed number of 7 circumferential lobes, they could buckle into all the three different modes when the slant angle $\alpha$ values from $30°$ to $60°$. The corresponding force versus displacement curves are shown in Fig. 6(d).

In all three experiments, our interest are focused on the $7-60°$ specimen (Fig. 6(c)) with the "completely follow" type represented by "$\diamond$" in Table 1, which collapses completely along the pre-embedded creases. It indicated that the induced geometric imperfections can precisely control their buckling modes. Its progressive buckling process can be explained as follows.

Initially, progressive buckling starts from a few random diamond lobes in the upper portion of the model and follows their surrounding straight creases into a curved unit cell. At this stage, the load increases with fluctuations, which occurs every time a unit cell has been buckled inwards from its original cylindrical shape. Its force response can be explained that the formation of a curved unit cell is finished and becomes absorbed into the general restoration of shape.

Then their buckling tendency extends both axially and circumferentially until all the lobes buckle into curved unit cells forming a stable texture. Every diamond unit on its surface is able to deform into a curved unit cell aligned orderly in both the axial and circumferential directions. Meanwhile, the compression at this moment reaches the peak force $P_1$ (64.96N) (Fig. 6(d)) during the entire deformation. Compared with $P_{10}$ (179.06N) of the conventional cylinder, the reduction could be attributed to the weakening of the initial shell stiffness caused by the introduced geometric imperfections.

Under further compression, some neighbouring lobes of the model are triggered to be buckled together forming a bigger quadrilateral curved surface (the fourth picture in Fig. 6 (c)). Meanwhile the force response reduces remarkably with strong fluctuations, which occurs every time neighbouring curved unit cells merge together into a bigger curved shape.

Finally, the model is crushed entirely in chaos with irreversible deformation (the fifth picture in Fig. 6(c)), which means that some parts of the model has entered into plastic response, and their force reach a plateau correspondingly.
The deformation processes of the other two types are also shown in Fig. 6(a), (b), different from the "completely follow" one.

The patterned tube with $n = 7$ and $\alpha = 30°$, as shown in Fig. 6(a), deforms in a similar manner to the conventional thin-walled cylinder, which can be identified from their deformation process in Fig. 5(a) and Fig. 6(b). This model buckles without following the pre-fabricated creases, which indicates the negligible influence of this pattern on the buckling mode. Therefore, we call it the "unfollow" type, represented by "$\triangle$" in Table 1.

The third type is called "partly follow" type and represented by "$\bigcirc$"in Table 1, which means the patterned tube can only collapse along some of the pre-embedded creases. Some units of the $7-45°$ specimen, as shown in Fig. 6(b), naturally deform into the curved cells, as marked in black dashed lines. However, the other area, as highlighted by the shadow in Fig. 7, keeps unbuckled in their initial cylindrical shape during the whole elastic deformation process.

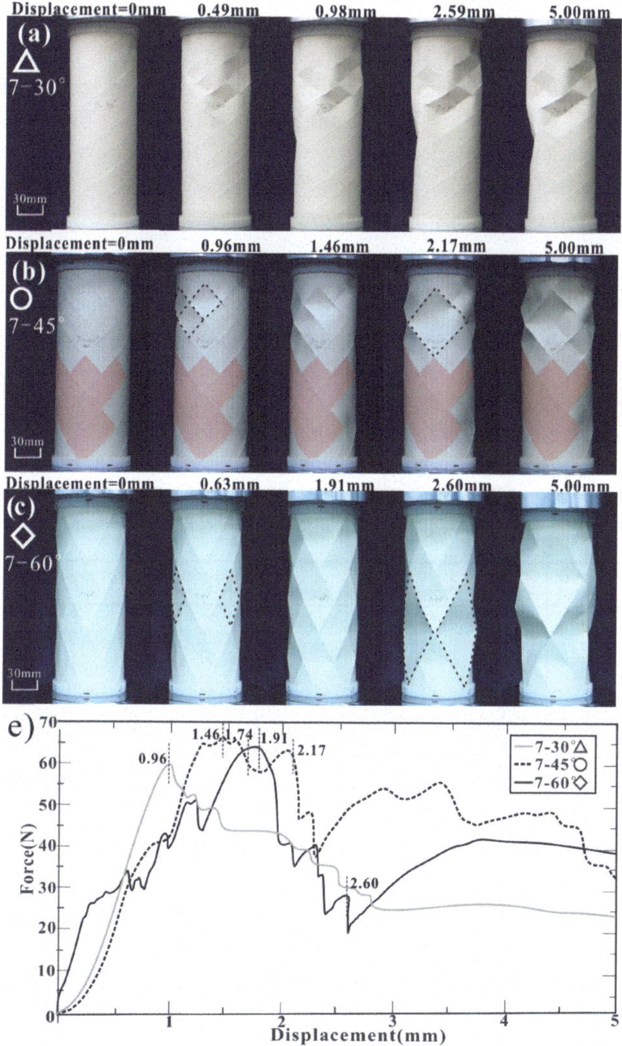



**Figure 6 :** *Deformation of patterned specimens where buckled areas are highlighted by black dashed lines and unbuckled areas are indicated by the shadow. (a) An "unfollow" model with n=7 and α =30°. (b) A "completely buckle" model with n=7 and α =45°. (c) A "partly follow" model with n=7 and α =60°. (d) Force vs. displacement curves of the specimens.*

In the early stage, a few random single units of the "partly follow" type buckled inwards into the curved cells, similar to the process of the "completely follow" one (the second picture in Fig. 6(b) and (c)). The axial compression (Fig. 6(e)) increases to its peak value in a relatively steady state, compared with the "completely follow" specimen, due to fewer buckled units. The third pictures in Fig. 6(a), (b), (c) show the deformed textures when the load reaches the peak. Then, the buckling extended to the diamond lobes surrounding the buckled area. At the same time, the force decreases with strong fluctuations. At last, when irreversible deformation occurs, the compressive loads stabilise at a low lever, which is around half of their initial peak force $P_1$ .

## 3.4  Effect of geometric parameters

Previous experiments above have proved that the geometric imperfections induced by pre-embedded creases played a crucial role in changing their buckling modes, which are sensitively affected by the planar design parameters $n$ and $\alpha$, as shown in Table 2. Some general rules can be found as follows.

***Table 2:*** *The relationship between geometric parameters and buckling types*

| $n$ \ $\alpha$ | 30° | 45° | 50° | 55° | 60° | 65° | 70° |
|---|---|---|---|---|---|---|---|
| 4 | △ | ○ | ○ | ◇ | ◇ | ◇ | ○ |
| 5 | △ | ○ | ◇ | ◇ | ◇ | ◇ | ○ |
| 6 | △ | ○ | ◇ | ◇ | ◇ | ◇ | ◇ |
| 7 | △ | ○ | ○ | ◇ | ◇ | ◇ | ○ |
| 8 | △ | ○ | ○ | ◇ | ◇ | ○ | ○ |
| 9 | △ | ○ | ○ | ○ | ○ | | |

When the slant angle $\alpha$ is less than $45°$ , cylinders with short and wide diamond patterns deform similarly to the non-axisymmetric mode of a conventional cylinder, completely "unfollowing" their pre-pressed creases (Fig. 6(a)), which demonstrates the insignificant influence of these patterns in changing the deformation modes. However, when their slant angle $\alpha$ exceeds $45°$ , diamond patterns become longer and narrower. These tubes can deform partly or completely following the pre-fabricated creases under quasi-static axial compression. Therefore, the slant angle $\alpha$ of $45°$ is a critical value to determine whether these cylinders may follow the pre-pressed diamond creases.

When selecting proper parameters $n$ and $\alpha$, as highlighted in the shadow area in Table 2, the stable and precisely controllable buckling modes can be obtained. The planar geometric parameters of these "completely follow" models are centrally distributed around the slant angle of $\alpha = 60°$ and the number of circumferential lobes $n = 6$, which is equal to the number of naturally formed lobes $n_0$. The selective area ranges from $n = 4$ to 8 and $\alpha = 55°$ to $65°$.

With $n$ surpasses 8, the diamond units are too small and dense to deform into a single curved surface and can only keep their initial state or form a large curved surface with neighbouring units buckling into the "partly follow" type.

In general, when the slant angle $\alpha$ is less than $45°$, the pre-pressed crease cannot change their buckling mode. When the slant angle is larger than $70°$, a patterned cylinder may not deform into a stable state either. Therefore, we can conclude that to obtain the controllable buckling mode, the number of the circumferential lobes can change from 4 to 8, while the slant angle should vary from $55°$ to $65°$.

# 4    Snap through of the diamond unit

## 4.1 Observation and DIC experiments

It has been shown in Fig. 6(c) that when proper geometric parameters are selected, all the diamond units will buckle inwards. When the cylinder is unloaded, the diamond units still remain the buckled shape. Furthermore, when these buckled diamond units are carefully pressed back, they will return to the initial cylindrical shape. This observation indicates that a diamond unit has two stable configurations, i.e., the initial cylindrical shape and the buckled shape, which can be switched to each other by a snap-through behaviour. To investigate this bistable behaviour in details, model 5-55°* made from PP sheet was axially compressed, and a DIC system was used to capture the configurations of a representative diamond unit (ABCD ,as highlighted in the shadow in Fig. 7(a)) during the whole loading process

To compare the curved configuration in axial and circumferential directions, two fixed rectangular coordinate systems were established on the highest apex of the unit A and the midpoint O of the straight line between the two middle apexes B,C respectively, as shown in Fig. 7(a). Since all the units on the cylindrical shell do not deform simultaneously, the distance between points A and C was used to characterize the compression of the unit. In addition, the shapes of the axial diagonal AC and the circumferential one BD were also extracted from the experiment to demonstrate the change in curvature of the unit, as shown in Fig. 7(c) and (d) respectively. The deformation process of the single unit is shown in Fig. 7(b).It can be noticed that the bending area gradually moved down from the top portion to the central area until the unit reached its stable buckled shape. At the same time, the initial curvature of the cylinder gradually turned into a straight line in the circumferential direction. It is noticed that under the axial compression, a diamond unit has a tendency to bend along the horizontal direction from its initial cylindrical state, which means its main curvature alters from the circumferential direction to the axial direction.

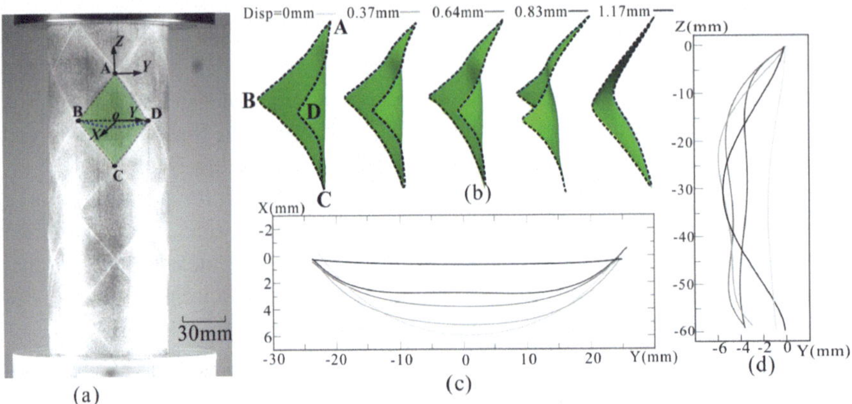



**Figure 7:** *DIC experimental results. (a) The initial specimen with pre-pressed creases under axial compression and detailed pictures in two directions with corresponding fixed coordinate systems. (b) The progressive formation of the representative unit diamond of the diagonal AC on the YZ plane. (c) Shape of the axial diagonal line AC projected on the YZ plane. (d) Shape of the circumferential diagonal line BD projected on the XY plane.*

## 4.2 Theoretical analysis

Having obtained the shape transformation of the unit during the buckling process, the next step is to calculate the variation in strain energy. The repeatable switch between the two stable configurations indicates that the diamond shell is subjected to very small strain and remains in the elastic range during this recoverable deformation. Therefore, according to Timoshenko and Woinowsky-Krieger's theory [Timoshenko and Woinowsky-Krieger 59], the theoretical value of the pure bending energy in the analysed area $U$ can be obtained as follows:

$$U = \int_{-t/2}^{t/2} \frac{1}{2}(\sigma_x \varepsilon_x + \sigma_y \varepsilon_y) A dz \tag{4}$$

in which, $t$ is the thickness of the uniform shell, $\varepsilon_x$ and $\varepsilon_y$ are the elastic strain in the horizontal and vertical directions of the diamond units, respectively, (Fig. 7(a)) depending on the magnitude of the curvatures of the bent plate $\kappa_x$ and $\kappa_y$. These strains also proportional to the distance $z$ from the neutral surface, which can be derived as

$$\varepsilon_x = z\kappa_x \qquad \varepsilon_y = z\kappa_y \tag{5}$$

$\sigma_x$ and $\sigma_y$ are the elastic stress based on the Young's Modulus $E$ and the Poisson's ratio $v$ of the PP board. These variables are determined by:

1268

$$\sigma_x = -\frac{E}{1-v^2}z\kappa_x - \frac{vE}{1-v^2}z\kappa_y \tag{6}$$

$$\sigma_y = -\frac{E}{1-v^2}z\kappa_y - \frac{vE}{1-v^2}z\kappa_x \tag{7}$$

Eqs. (6-7) are proper for thin shells, where its thickness $t$ is rather small compared to its curvature. Substituting Eqs. (5-7) to Eq. (4), the bending strain energy of the whole unit $U$ becomes

$$U = \frac{Et^3}{24(1-v^2)}\int_{\text{unit}}(\kappa_x^2 + \kappa_y^2 + 2v\kappa_x\kappa_y)\ \mathrm{d}A \tag{8}$$

To obtain its curvatures $\kappa_x$, $\kappa_y$, we reconstructed the continuous curved surface based on discrete speckle points measured in DIC experiments. Curvefitting Software in MATLAB 2014b is applied to build a binary five times polynomials equation to fit the curved surface. Its accuracy can be valued by the parameter R-square. When R-square is approaching to 1, the fitted surface are more dependable. In our fittings, the average R-square is 0.9882, which is highly convincing. Since the curvatures on the unit are not identical, we evenly divide the selected unit into 2500 tiny meshes $a$ with the equivalent curvatures $\kappa_x$, $\kappa_y$. The bending energy of the unit $U$ can be modified from Eq. (8) as follows:

$$U = \frac{Et^3}{24(1-v^2)}\sum_{i=1}^{2500}((\kappa_y^{i\,2} + \kappa_x^{i\,2} + 2v\kappa_x^i\kappa_y^i)a_i \tag{9}$$

In this experiments, the material properties of polypropylene were obtained according to the Hartmann's tensile tests of dog-bone samples at different temperatures [Hartmann et al. 87]. The mechanical properties in calculation are: the Young's Modulus $E = 1.260\text{GPa}$, the Poisson ratio $v = 0.30$. The measured average thickness of the sheet is $t = 0.27\,\text{mm}$.

Substituting all experimental parameters into Eq. (9), their relationship between elastic strain energy and the corresponding axial displacement marked by the dots , as shown in Fig. 8. A fitted curve is obtained by the Polyfit function in MATLAB (2014b) in order to illustrate the variation trend more clearly.

During the manufacturing process, pre-stresses have been introduced to the analysed area. Therefore, the initial curvatures of the analysed area are $\kappa_x = 0.023/\text{mm}$ in the horizontal direction and $\kappa_y = 0/\text{mm}$ in the vertical direction. In the early stage, the strain energy of the analysed area increases progressively when the buckling starts to occur in the axial direction, until it reaches the maximum energy $E_{\max} = 6.92\text{N}\cdot\text{mm}$ with the corresponding deformed configurations, as shown in Fig. 8. Simultaneously, the curvature in the circumferential direction gradually decreases from its initial cylindrical shape.

Then, the unit suddenly snaps through into a buckled shape with almost zero circumferential curvature $\kappa_x$ and largely decreased axial curvature $\kappa_y$. This releases the stored strain energy, leading to the lowest point $E_{\min} = 3.14 \text{N} \cdot \text{mm}$. However, under further axial loading, the buckled unit is continuously bent in the axial direction and begin to store these energy with a progressive increase. The path clearly illustrates its snap-through phenomenon. It proves that when the load is released at the lowest point in the process, the unit could never follow the path and climb back to the original stable cylindrical shape without inputting external energy. Therefore, the diamond unit can stably keep its two stable configurations.

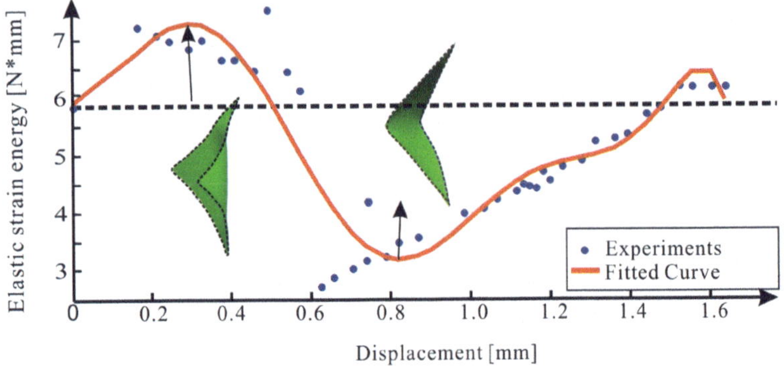


**Figure 8:** *Bending energy-displacement curves for one curved unit.*

# 5  Conclusions

In conclusion, we have designed a thin-walled cylinder with pre-pressed diamond patterns to achieve a predictable and controllable buckling mode under quasi-static axial compression. Models with various geometric parameters have been investigated experimentally to study their deformation processes and corresponding mechanical properties. The results show that by properly selecting pattern geometric parameters, a stable buckling mode, in which all the diamond units bend inward to form a curved texture on the surface, can be triggered. In addition, the deformation process and strain energy variation of a representative diamond unit have been studied in details, which confirms the snap-through behaviour of the diamond unit. By using the proposed method, it is possible to pre-design the buckled configuration of a cylinder upon specific requirements. Further investigation can be extended to circular tubes with various thicknesses and different boundary conditions to study the feasibility of our proposed method.

## Acknowledgements

This work was also supported by National Natural Science Foundation of China Projects No. 51721003 and 51575377. This paper has been awarded the 7OSME Gabriella & Paul Rosenbaum Foundation Travel Award.

# References


[Calladine 83] C. R. Calladine. Theory of Shell Structures, Cambridge University Press, 1983.

[Singace et al. 95] A. A. Singace, H. Elsobky, T. Y. Reddy, "On the eccentricity factor in the progressive crushing of tubes." International Journal of Solids and Structures 32(1995) 3589-3602.

[Guillow and Lu et al. 01] S. R. Guillow, G. Lu, R. H. Grzebieta. "Quasi-static axial compression of thin-walled circular aluminium tubes." International Journal of Mechanical Sciences 43(2001)2103-2123.

[Alexander 60] J. M. Alexander. "An approximate analysis of the collapse of thin cylindrical shells under axial loading." The Quarterly Journal of Mechanics and Applied Mathematics 13(1960)10-15.

[Arbramowicz and Jones 84] W. Arbramowicz and N. Jones, "Dynamic axial crushing of cylinders." Journal of impact engineering 2(1984) 263-281.

[Wierzbicki et al. 92] T. Wierzbizi, S.U. Bhat, W. Abramowicz, D. Brodkin, "Alexander revisited-a two folding elements model of progressive crushing of tubes." International Journal of Solids and Structures, 29(1992)3269-3288.

[Pugsley and Macaulay 60] S. A. Pugsley and M. Macaulay, "The large-scale crumpling of thin cylindrical columns." The Quarterly Journal of Mechanics and Applied Mathematics 13(1960) 1-9.

[Johnson et al. 77] W. Johnson, P. D. Soden and S. T. S. Al-Hassani. "Inextensional collapse of thin-walled tubes under axial compression." The Journal of Strain Analysis for Enigineering Design 12(1977) 317-330.

[Singace 99] A. A. Singace, "Axial crushing analysis of tubes deforming in the multi-lobe mode." The Journal of Mechanical Sciences 41(1999) 865-890.

[Chen and Ozaki 09] D. Chen and S. Ozaki, "Numerical study of axially crushed cylindrical tubes with corrugated surface." Thin-walled Structures 47(2009) 1387-1396.

[Wu et al. 16] S. Wu，G. Li，G. Li，X. Wu，Q. Li, "Crashworthiness analysis and optimization of sinusoidal corrugation tube." Thin-walled Structures 105(2016) 121-134.

[Mozafari et al. 18] H. Mozafari, S. Lin, G. Tsui, L. Gu, "Controllable energy absorption of double sided corrugated tubes under axial." Composites Part B: Engineering 34( 2018) 9-17.

[Zhang et al. 15] X. Zhang, H. Zhang, Z. Wen, "Axial crushing of tapered circular tubes with graded thickness." International Journal of Mechanical Science 92(2015) 12-23.

[Sun et al. 14] G. Sun, F. Li, Q. Li, "Crashing analysis and multi objective optimization for thin-walled structures with functionally graded thickness." International Journal of Impact Engineering 64(2014) 62-74.

[Han et al. 07] H. Han, F. Taheri and N. Pegg, "Quasi-static and dynamic crushing behaviors of aluminium and steel tubes with a cutout." Thin-walled Structures 45(2007) 283-300.

[Lee et al. 99] S. Lee, C. Hahn, M. Rhee and J. E. Oh, "Effect of triggering on the energy absorption capacity of axially compressed aluminium tubes." Material&Design 20(1999) 31-40.

[Airoldi and Janszen 05] A. Airoldi and G. Janszen, "Adesign solution for a crashworthy landing gear with a new triggering mechanism for the plastic collapse of metallic tubes." Aerospace Science and Technology  56(1989) 113-120.

[Han et al. 04] J. Han, K. Yamazaki and S. Nishiyama, "Optimization of the crushing characteristics of triangulated aluminium beverage cans." Structural and Multidisciplinary optimization 28(2004) 47-56.

[Yang et al. 16]K. Yang, S. Shen, J. Zhou, Y. Xie, "Energy absorption of thin-walled tubes with pre-folded origami patterns: Numerical simulation and experimental verification." Thin-walled Structures 103(2016) 33-44.

[Ma et al. 16] J. Ma, D. Hou, Y. Chen and Z. You, "Quasi-static axial crushing of thin-walled tubes with a kite-shape rigid origami pattern: Numerical simulation and experimental verification." Thin-walled Structures 100(2016) 38-47.

[Song et al. 12] J. Song, Y. Chen and G. Lu, "Axial crushing of thin-walled structures with origami patterns." Thin-walled Structures 54(2012) 65-71.

[Ma and You 13] J. Ma and Z. You, "Energy absorption of thin-walled square tubes with a pre-folded origami pattern-part I: geometry and numerical simulation." Journal of Applied Mechanics." Journal of Applied Mechanics 2013; 81(1): 11003.

[Timoshenko and Woinowsky-Krieger 59] S. Timoshenko and S. Woinowsky-Krieger, "Theory of plates and shells." McGraw-Hill Book Company, Inc., New York.1959.

[Hartmann et al. 87] B. Hartmann, G. Lee, W. Wong, "Tensile Yield in Polypropylene." Polymer Engineering and Science 27(1987) 823-828.

---

X. Yang,
School of Mechanical Engineering, Tianjin University, Tianjin 300072, China,
e-mail: xiaochen_yang@tju.edu.cn.

S. Zang,
 School of Mechanical Engineering, Tianjin University, Tianjin 300072, China,
e-mail: sx_zang@tju.edu.cn

J. Ma,
School of Mechanical Engineering, Tianjin University, Tianjin 300072, China,
e-mail: jiayao.ma@tju.edu.cn.

Y. Chen,
 School of Mechanical Engineering, Tianjin University, Tianjin 300072, China,
e-mail: yan_chen@tju.edu.cn

# Reality Check - Mechanical Potential of Tessellation-based Foldcore Materials

*M. Grzeschik, Y. Klett, P. Middendorf*

**Abstract**: *This study presents the results of mechanical testing of folded structures used as sandwich core materials, also called foldcores. Foldcores in a Miura-ori configuration were manufactured from aerospace-grade composite materials, and their compressive and shear strengths and stiffnesses were measured. A comparison to state-of-the-art foams and honeycombs was carried out. Results show interesting characteristics of the tested foldcore material especially with regard to shear performance.*

## 1   Introduction

Tessellation-based sandwich core materials have been under investigation for use in lightweight sandwich structures for several decades by now. Early studies [Miura 75] mention a high potential of the use of tessellations in sandwich constructions, but also some factors that may hinder application, including a lack of experience with such structures. Especially in demanding application areas like aerospace, the material choice is guided by a very conservative approach, and introduction of new materials can easily take decades of development, testing and certification.

After a considerable hiatus, the research into tessellation-based materials was invigorated by new demands on core materials that were (and are) not met by currently used state-of-the-art core materials like honeycombs and foams [Chaliulin 99, Klett and Drechsler 07, Klett and Drechsler 11]. Considerable effort has since then gone into characterizing this young material familiy.

New methods to simulate, produce and test folded structures have been developed in the last decade, and a lot of experience with regard to the properties of e.g. Miura-ori has been gathered (see e.g. [Heimbs et al. 07, Grzeschik and Drechsler 11, Fischer 12, Schenk and Guest 11, Gattas and You 14]). However, the large variety of examined tessellations, materials, and applied test methods results in numerous data sets that are difficult to compile into a coherent picture.

This study compares the performance of one foldcore type and two common aerospace core materials of equal density using standardized testing methods, and provides a full set on compressive and shear properties. This enables a 1:1 comparison of the performance spectra of the competing materials, and helps design engineers not intimately familiar with foldcores to judge their potential quickly, objectively and comprehensively.

## 2  Sandwich Core Properties and Testing

Sandwich structures can be used to build lightweight structures with high specific strength and stiffness. The sandwich gets its properties from the combination of the facesheets and the enclosed core. Primary requirements on the core side include strength, stiffness and density, but also secondary aspects touching on cell size, base material, fire/smoke/toxicity (FST), handling, environmental impact, longevity, and – in a category all by itself – cost.

Foldcores have been identified as a material that can satisfy several of these requirements [Klett 13, Kilchert 13, Johnson et al. 15, Sturm 16], but are not yet state-of-the-art. One reason for this seemingly slow adoption by the industry lies in the conservative nature of many high-end sandwich applications in aerospace, another one in the considerable challenges to manufacture foldcores in industrially relevant quantities. In addition, the large amount of possible combinations of geometry and materials have by no means been fully screened.

Independent of the material and configuration used however, the most basic data points to characterize the weight-specific mechanical performance of a sandwich core are easily accounted for, and are universal for any core material:

1. Compressive strength: How much flatwise load can a core take before collapsing? This is e.g. relevant for aircraft floor panels, which are subjected to impact loads caused by high heels or trolleys.

2. Compressive stiffness: How much does the core deform under flatwise load?

3. Shear strength: How much transversal shear can the core take? This determines maximum bending loads and possible deflections of a given sandwich panel.

4. Shear stiffness: Again relevant for bending, how much will a panel deflect under bending loads?

5. Density: What is the mass of a given piece of core?

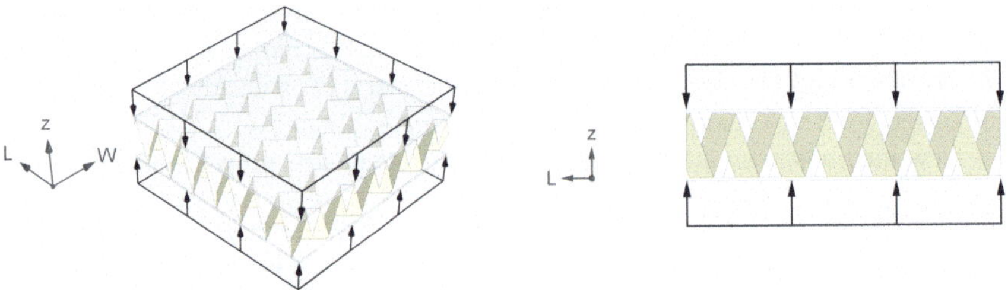

Figure 1:  Schematic setup for sandwich flatwise compression testing. The core sample is fixed to the (stiff) facesheets, and out-of-plane compression load is introduced.

The most common testing thus is for flatwise compression and transversal shear as shown schematically in Figures 1 and 2. Notably, for shear testing, the properties of the material may depend on the in-plane orientation of the load vector. At least

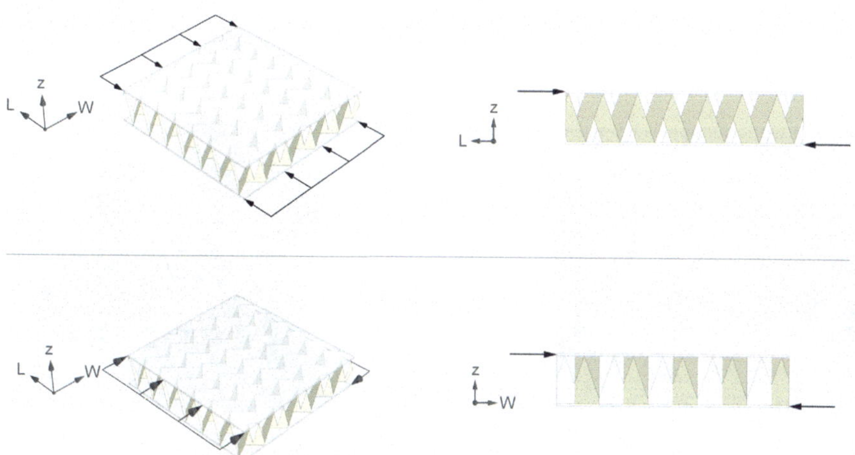


Figure 2: Schematic setup for transversal sandwich shear testing. The core sample
is fixed to the (stiff) facesheets, and in-plane shear load is introduced. Top: Shear
loading in L-direction. Bottom: Shear loading in W-direction.

two perpendicular directions need to be tested to map anisotropic behavior, which
– depending on the core structure – can be quite pronounced. Shear testing is much
less frequently performed due to the additional overhead for sample preparation
[Klett et al. 16].

Further common testing methods include 3- or 4-point bending [ASTM C393 16,
ASTM D7250 16, DIN 53293 82]. These are not part of this study, because they examine specific combinations of core and facesheets, and do not deliver information
on the core itself.

## 3   Setup

For lightweight construction, different core materials can only be usefully compared if they feature similar densities. Mechanical performance does not scale
linearly with density, especially for lightweight core materials with densities well
below $72\,\mathrm{kg/m^3}$ [Hexcel Composites 18], and there are no reliable extrapolations
methods available to compare different materials at different densities.

To get a good picture of foldcore properties in comparison to standard materials, a density of $32\,\mathrm{kg/m^3}$ was chosen. Core materials with this density are commonly used for aircraft linings, and provide a good reference point. The candidates
for the comparison are a polymethacrylimide (PMI) foam (designation PMI32)[1]
and a honeycomb made from aramid paper and phenolic resin coating (designation HC32)[2]. Both materials feature a density of $32\,\mathrm{kg/m^3}$, mechanical data was

---

[1] ROHACELL® 31A [Evonik 18]
[2] HexWeb® A1-32-6 [Hexcel Composites 18]

taken from manufacturer data sheets. Examples of these materials together with a comparable foldcore are shown in Figure 3.

Figure 3: Samples of the three core material contenders in this study: Foldcore, PMI foam and honeycomb.

A Miura-ori cell geometry with a density of $32\,\text{kg/m}^3$ and a height of $12.7\,\text{mm}$ was designed (designation FC32), using an aramid paper with phenolic resin impregnation and a grammage of $116\,\text{g/m}^2$. The resulting geometry parameters for the unit cell are shown in Figure 4. The unit cell was designed to provide a fairly isotropic shear behavior (from empirical evidence by setting $S = V$), and the cell size was chosen to represent a compromise between decent manufacturing suitability and good compressive performance.

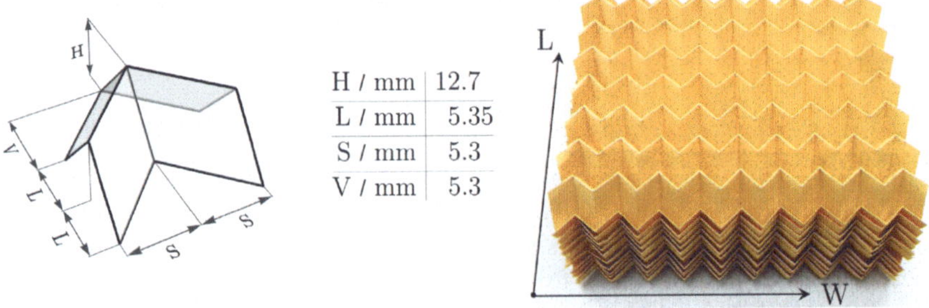

| | |
|---|---|
| H / mm | 12.7 |
| L / mm | 5.35 |
| S / mm | 5.3 |
| V / mm | 5.3 |

Figure 4: Left: Characteristic geometry parameters for the FC32 Miura unit cell (geometry not to scale). Right: A stack of folded samples before bonding to the facesheets, with W- and L-directions indicated.

The foldcore sample design was performed according to [Klett et al. 16]. Samples were prepared in a prototyping process using mechanical scoring [Klett 13], and the folded samples were adhesively bonded[3] to stiff facesheets. Figure 4 shows a stack of ten compression samples before bonding.

Compression testing was carried out according to [DIN 53291 82], and shear

---

[3] Adhesive: A20/B20 [Lange und Ritter 18]

testing was carried out according to [DIN 53294 82], using a universal testing machine with a crosshead speed of 0.5 mm/min. For every one of the three tests, six samples were tested. Exemplary test setups for both compression and shear test are shown in Figure 5.

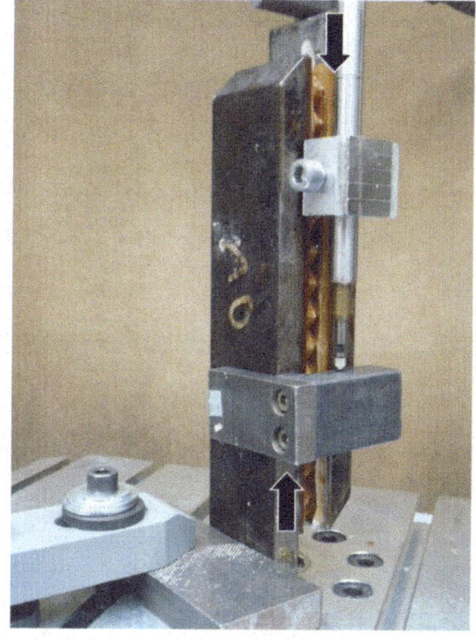

Figure 5: Test setups. Left: Flatwise compression of a stabilized foldcore sample. Right: Transversal shear test of a foldcore sample in W-direction. Loading directions are shown by black arrows.

# 4   Results

The results for the compression and shear tests are displayed in Table 1 together with the according data for the PMI foam and the honeycomb as provided by the manufacturer. Standard deviations are only supplied for the foldcore tests, because material data sheets do not provide those. Stiffnesses were evaluated by determining the slope of a secant between 30 % and 70 % of the evaluated strength $\sigma_{max}$ (or $\tau_{max}$ in case of shear) as illustrated in Figure 6, providing robust and conservative results.

## 4.1   Compression

The compression test results are in line with previous assessments. Figure 6 shows the strain-stress curves for all tested FC32 samples. The spread in strength and stiffness is considerable. Samples C-6 and especially C-4 display higher than average strength and stiffness. After an initial settling phase, the stress-strain curves exhibit a linear section. Compression testing is particularly sensitive to manufacturing imperfections and especially to nonparallel facesheets. The fairly large spread here hints at a considerable optimization potential with regard to core quality and uniformity, and sandwich assembly.

| Material | FC32 | PMI32 | HC32 |
|---|---|---|---|
| Compressive Strength / MPa | 0.74 | 0.4 | 1.2 |
| Standard Deviation / MPa | 0.09 | – | – |
| Compressive Stiffness / MPa | 31.5 | 36.0 | 75.0 |
| Standard Deviation / MPa | 7.3 | – | – |
| Shear Strength W / MPa | 0.67 | 0.4 | 0.4 |
| Standard Deviation / MPa | 0.02 | – | – |
| Shear Stiffness W / MPa | 63.2 | 13.0 | 19.0 |
| Standard Deviation / MPa | 11.7 | – | – |
| Shear Strength L / MPa | 0.53 | 0.4 | 0.7 |
| Standard Deviation / MPa | 0.01 | – | – |
| Shear Stiffness L / MPa | 60.8 | 13.0 | 29.0 |
| Standard Deviation / MPa | 6.1 | – | – |

Table 1: Results of the foldcore testing (means) together with datasheet properties for PMI foam and honeycomb from [Evonik 18, Hexcel Composites 18]. Standard deviations for PMI and honeycomb materials are not available from datasheets.

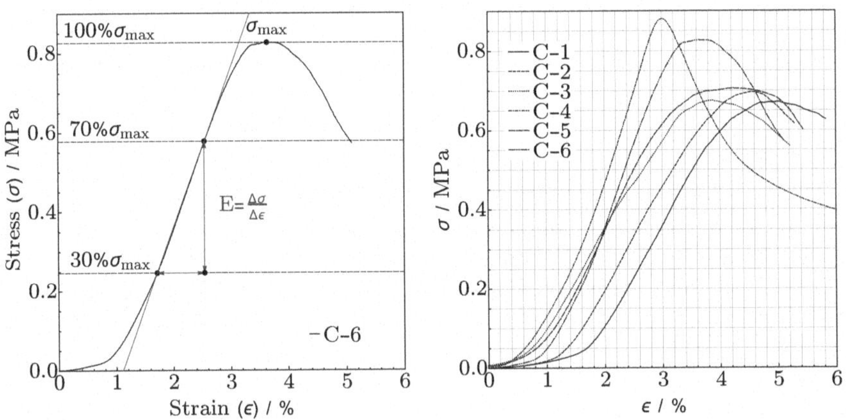


Figure 6: Left: Test data for compression sample C-6 with illustration of secant stiffness evalution between 30 % and 70 % of $\sigma_{max}$. Right: Data of all compressive samples for the FC32 material.

A comparative graph is shown in Figure 7. The foldcore samples reach about 62 % of the honeycomb performance, and beat the PMI foam by about 85 %. The excellent honeycomb performance is inherent in the honeycomb geometry, which provides optimal face alignment with regard to compression. Stiffness-wise, this

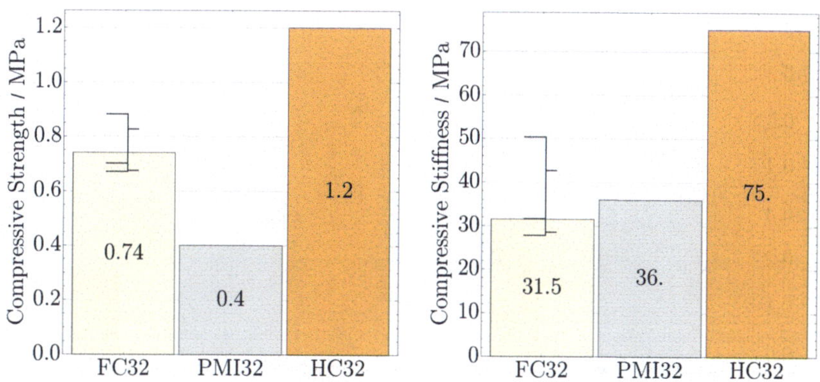

Figure 7: Compressive strength and stiffness for foldcore (FC), PMI foam (PMI) and honeycomb (HC) with a density of $32\,kg/m^3$. Brackets indicate minima and maxima together with first, second and third quartiles.

effect is even more pronounced, as the foldcore performs to about 42 % of the honeycomb, and to about 88 % of the foam.

## 4.2 Shear

The picture from the compression analysis is somewhat reversed when looking at the shear properties. Here, the previously suboptimal face alignment of the Miura unit cell works in favor of the foldcore performance.

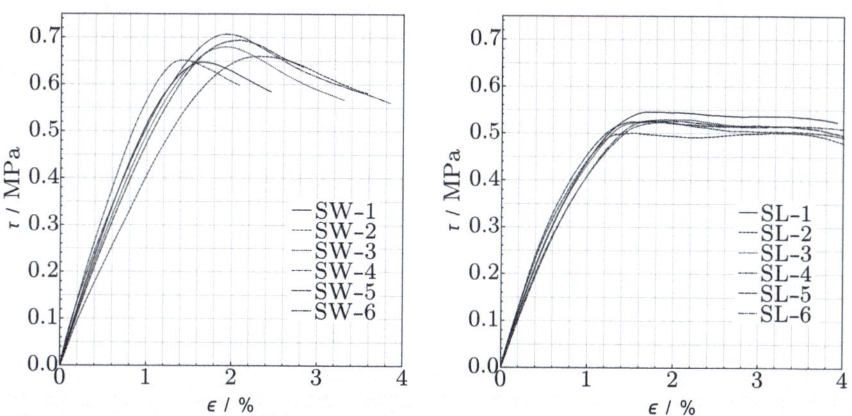

Figure 8: Test data of the shear testing for the FC32 material. Left: Shear in W-direction (SW). Right: Shear in L-direction (SL).

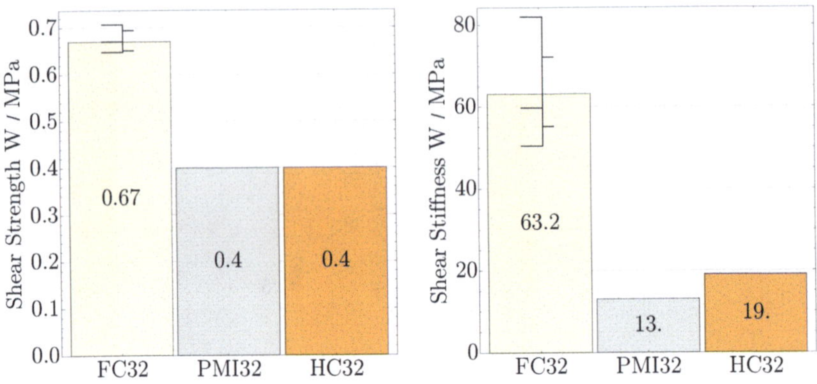

Figure 9: Shear strength and stiffness in W-direction for foldcore, foam and honeycomb.

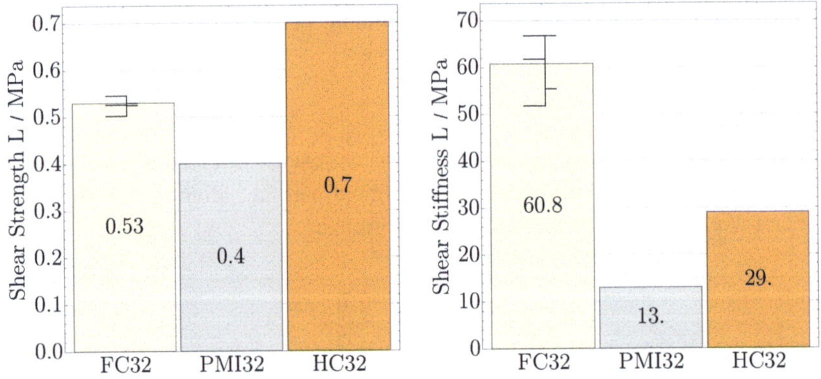

Figure 10: Shear strength and stiffness in L-direction for foldcore, foam and honeycomb.

The strain-stress curves from shear testing are presented in Figure 8. In general, there is a distinct linear slope for both directions. The W-direction features a larger spread in modulus, but overall the results for shear are noticeably more uniform than those for compression. Compared to the W-direction, the L-direction shows a pronounced stress plateau after peak stress is reached. This is an interesting effect, which merits closer inspection in future research.

As shown in Figure 9, for W-strength, the foldcore outperforms both foam and honeycomb by about 68 %. For W-stiffness the effect is even larger, with foldcore

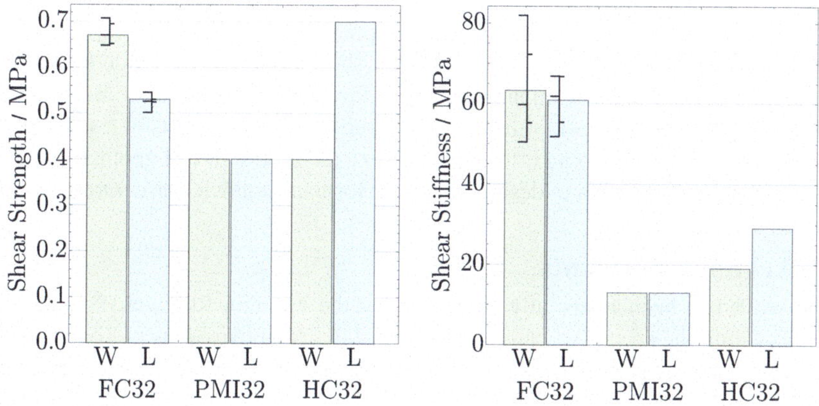

Figure 11: Shear W and L strengths and stiffnesses for foldcore, foam and honeycomb.

stiffness resulting in about 486 % of the foam, and about 332 % of the honeycomb properties.

Figure 10 shows that in the L-direction, the honeycombs provide the highest strength, with the foldcore reaching 133 % of the foam, and about 76 % of the honeycomb strengths. The difference between W- and L-direction is due to an intrinsic pronounced anisotropic shear behavior of the honeycombs[4]. With regard to stiffness, the foldcore again seriously outperforms foam and honeycombs by about 468 % and 210 %, respectively.

All shear results have been consolidated in Figure 11, and allow a number of interesting observations. First of all, strength-wise, foldcores are on average on par with the honeycombs and above the foam, with the honeycomb displaying a stronger anisotropic behavior. Second, foldcore stiffness is quite similar in W- and L-direction, prompting a near-isotropic behavior, again offset against a strong anisotropic streak in the honeycombs. Third and most strikingly, the tested foldcore excels in shear stiffness. While good shear performance has been hinted at very early on [Miura 75], the presented results seriously underline this aspect, and not very subtly prompt for application in shear-intensive scenarios.

## 5   Conclusion

We have presented test results on the compressive and shear performance of foldcores that were manufactured to allow a fair weight-specific comparison with two other high-performance aerospace core materials. The comparison shows that fold-

---

[4]This anisotropy leads to the fact that in many applications, the weaker direction is used for dimensioning, because a correct alignment of honeycomb inlays cannot be assumed safely.

cores provide an interesting property profile. Compression-wise, decent performance of the tested foldcore can be shown, with a comparably large spread for strength and stiffness hinting at considerable potential for further quality improvements. With regard to shear, and especially shear stiffness, the foldcore results are promising, and indicate an unprecedented level of performance for materials in this density range. Together with the many other interesting aspects of tessellation-based foldcores and ongoing efforts to optimize the used unit cell geometries, this should open up new ways to design highly performant sandwich structures.

# 6 Acknowledgements

This work has been financially supported by the German Research Foundation (DFG) within the special research grant SFB1244 "Adaptive Skins of Tomorrow". Also, the authors gratefully acknowledge support by the yellow faction - you know who you are.

# References

[ASTM C393 16]  ASTM C393. *Standard Test Method for Core Shear Properties of Sandwich Constructions by Beam Flexure.* American Society for Testing and Materials, 2016. Available online (`http://www.astm.org/Standards/C393.htm`).

[ASTM D7250 16]  ASTM D7250. *Standard Practice for Determining Sandwich Beam Flexural and Shear Stiffness.* American Society for Testing and Materials, 2016. Avail-able online (`http://www.astm.org/Standards/D7250.htm`).

[Chaliulin 99]  I.V. Chaliulin. *Technological schemes for sandwich structures production.* ISBN 5-7579-0295-7, KSTU Kazan, 1999.

[DIN 53291 82]  DIN 53291. *Testing of sandwiches; Compression test perpendicular to the faces.* Deutsches Institut fuer Normung, 1982.

[DIN 53293 82]  DIN 53293. *Testing of sandwiches; Bending test.* Deutsches Institut fuer Normung, 1982. Available online (`https://www.beuth.de/en/standard/din-53293/939627`).

[DIN 53294 82]  DIN 53294. *Testing of sandwiches; Shear test.* Deutsches Institut fuer Normung, 1982. Available online (`https://www.beuth.de/en/standard/din-53294/939687`).

[Evonik 18]  Evonik. "Product Information ROHACELL A." Technical report, 2018. Available online (`http://www.rohacell.com/product/peek-industrial/downloads/rohacell%20a%20product%20information.pdf`).

[Fischer 12]  Sebastian Fischer. *Rechnerische Ermittlung der mechanischen Eigenschaften von Faltkernen.* ISBN 978-3-8440-1258-3, Universität Stuttgart, 2012.

[Gattas and You 14]  Joseph M Gattas and Zhong You. "Quasi-static impact response of single-curved foldcore sandwich shells." In *ASME 2014 International Design Engineering Technical Conferences and Computers and Information in Engineering Conference.* American Society of Mechanical Engineers, 2014.

[Grzeschik and Drechsler 11] Marc Grzeschik and Klaus Drechsler. "Isometrically folded high performance core materials." In *PFAM XIX, 978-0-473-18178-9*, 2011.

[Heimbs et al. 07] S. Heimbs, P. Middendorf, S. Kilchert, A.F. Johnson, and M. Maier. "Numerical Simulation of Advanced Folded Core Materials for Structural Sandwich Applications." In *First CEAS European Air and Space Conference, Deutscher Luft- und Raumfahrt Kongress, Berlin*, pp. 2889–2896, 2007.

[Hexcel Composites 18] Hexcel Composites. "A1 & A10 High Strength Nomex Aramid / Phenolic Resin." Technical report, 2018. Available online (http://www.hexcel.com/user_area/content_media/raw/A1A10_eu.pdf).

[Johnson et al. 15] A. Johnson, S. Kilchert, S. Fischer, and N. Toso-Pentecôte. *Structural Integrity and Durability of Advanced Composites: Innovative Modelling Methods and Intelligent Design*, Chapter 29 Design and performance of novel aircraft structures with folded composite cores, pp. 793–825. Woodhead Publishing, 2015.

[Kilchert 13] S. Kilchert. *Nonlinear finite element modelling of degradation and failure in folded core composite sandwich structures.* Universität Stuttgart, 2013. doi:dx.doi.org/10.18419/opus-3932. Available online (http://dx.doi.org/10.18419/opus-3932).

[Klett and Drechsler 07] Yves Klett and Klaus Drechsler. "Design of Multifunctional Folded Core Structures for Aerospace Sandwich Applications." In *1st CEAS European Air and Space Conference*, 2007.

[Klett and Drechsler 11] Y. Klett and K. Drechsler. "Designing Technical Tessellations." In *Origami$^5$, ISBN 1568817142*, edited by P. Wang-Iverson, R. J. Lang, and M. Yim, pp. 305–322. CRC Press, 2011.

[Klett et al. 16] Y. Klett, M. Grzeschik, and P. Middendorf. "Comparison of Compressive Properties of Periodic Non-flat Tessellations." In *Origami$^6$; Part 2, ISBN 978-1-4704-1876-2*, edited by K. Miura, T. Kawasaki, T. Tachi, R. Uehara, R. J. Lang, and P. Wang-Iverson, pp. 371–384, 2016.

[Klett 13] Yves Klett. *Auslegung multifunktionaler isometrischer Faltstrukturen für den technischen Einsatz.* ISBN 978-3-8439-1025-5, Institut für Flugzeugbau, Universität Stuttgart, 2013. Available online (http://www.dr.hut-verlag.de/9783843910255.html).

[Lange und Ritter 18] Lange und Ritter. "Produktübersicht Download Klebeharze." Technical report, 2018. Available online (http://www.lange-ritter.de/fileadmin/user_upload/Downloads/Produkte/Epoxidharze/Epoxidharze_Klebeharze_01.pdf).

[Miura 75] K. Miura. "New structural form of sandwich core." *Journal of Aircraft, 12(5):437.* doi:10.2514/3.44468.

[Schenk and Guest 11] M. Schenk and S. D. Guest. "Origami Folding - A structural engineering approach." In *Origami$^5$*, edited by P. Wang-Iverson, R. J. Lang, and M. Yim, pp. 291–303. CRC Press, 2011.

[Sturm 16] R. Sturm. *Untersuchung der Schadensinitiierung doppelschaliger Rumpfpaneele unter crashrelevanter Druck-Biege-Belastung.* Universität Stuttgart, 2016. Available online (http://elib.dlr.de/114484/1/DLR-Forschungsbericht_2016-18.pdf).

---

Marc Grzeschik
Institute of Aircraft Design, Pfaffenwaldring 31, 70569 Stuttgart,
e-mail: grzeschik@ifb.uni-stuttgart.de

Yves Klett
Institute of Aircraft Design, Pfaffenwaldring 31, 70569 Stuttgart,
e-mail: klett@ifb.uni-stuttgart.de

Peter Middendorf
Institute of Aircraft Design, Pfaffenwaldring 31, 70569 Stuttgart,
e-mail: middendorf@ifb.uni-stuttgart.de

# Design Methods and Analysis of the Mechanical Properties of Resch Pattern Foldcores

*X.Y. Lv, X. Zhou*

**Abstract:** *Foldcore sandwich structures are a promising alternative to conventional honeycomb sandwich structures in the aerospace industry. This paper presents a parametric study on the mechanical properties of a variety of Resch-based foldcore models virtually tested under quasi-static compression using the finite element method. The geometrical design methods based on constraint equations are discussed in this paper to establish the three types (hexagonal pattern, quadrangular pattern, and triangular pattern) of Resch pattern models. The hexagonal pattern exhibits the best performance under compression case. Parametric analysis established that increasing the length of the inner polygon in the initial crease pattern could significantly improve the compression property.*

## 1   Introduction

Sandwich structures offer a wide range of advantages and their use is continuously increasing in the aerospace and other industrial sectors. Specifically, they exhibit excellent performance in compression with low weight. Existing composite cores for sandwich structures include honeycomb cores and folded cores, also known as foldcores. Honeycomb cores have excellent specific strength, stiffness and energy absorption properties. However, they are also known to have the condensed water accumulation problem which hinders their widespread use on large aircraft structures such as the fuselage [Smith and Doe 17].

Foldcores, formed by folding a sheet material into a periodic three-dimensional (3D) pattern according to the principle of origami, are regarded as a promising alternative to honeycomb cores. Current research on foldcores mainly focused on the Miura pattern. This paper investigates a new type of foldcore design rooted from the generalized Resch patterns that are fundamentally different from the Miura pattern. The generalized Resch patterns have three main types: the hexagonal pattern, the quadrangular pattern, and the triangular pattern [Tachi et.al 13]. The foldcore's geometry is defined by a repetitive arrangement of unit cells that are determined by a set of geometric parameters [Klett et.al 07, Hachenberg et.al 03]. The mathematical models of the Resch pattern foldcores were the unit cells. The mechanical properties of the foldcores are found to depend on these parameters.

To fully understand how these parameters affect the mechanical properties of the Resch pattern foldcores, a parametric analysis of different geometric parameters generated by the established design method is performed with the finite element (FE) method. This virtual testing approach has successfully been applied to other foldcore structures [Heimbs et.al 07]. In order to evaluate the compression performance of Resch pattern foldcores, effective mechanical properties under out-of-plane compression is required.

## 2   Geometrical design method

Before the virtual testing of Resch Pattern foldcores, methods to establish the mathematical models to describe the folded geometries of the generalized Resch patterns must be developed. It was found that all Resch pattern models have the same structural unit. Thus, if the relationship between the structural unit and the parameters of the crease pattern can be obtained, the entire Resch pattern can then be solved using a rotation matrix.

Since the solution procedures for the three types of Resch pattern are similar, the hexagonal pattern is taken as an example. The structural unit of the hexagon pattern is shown in Fig. 1(a). It contains three facets, which can be determined by five points: $A$, $B'$, $C$, $O$ and $O_2$. The coordinates of points $O$ and $O_2$ can be obtained using the bottom polygon. The coordinate system is established in Fig. 3 (b). Hence, to fully define the three facets, the coordinates of points $A$, $B'$ and $C$ need to be determined, which yields 9 unknowns to be solved.

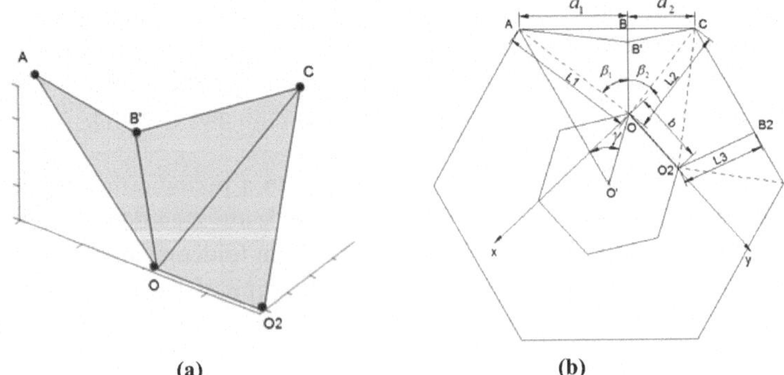

(a)           (b)

*Figure 1: (a) The structural unit of the hexagonal pattern; (b) The unit cell of the hexagonal pattern.*

derived for the folding process. In the first step, the lengths and angles of the crease pattern must be obtained. The crease pattern of a unit cell of the hexagon pattern is shown in Fig. 1(b), where $b$ is the length of the inner polygon, $\gamma$ is the relative angle between the two polygons, and $a_1$ and $a_2$ defines the division point location on the outer polygon. Let $\angle AOB = \beta_1$, $\angle BOC = \beta_2$, $OA = L_1$, $OC = L_2$ and $OB = L_3$. To make points $A$, $B$ and $C$ coplanar, $OB$ must be shortened to $OB'$. The ratio $v$ between $OB'$ and $OB$ is referred to as the shrinkage ratio, which is related to the height $h$ of the folded pattern. Here, the height of the foldcore is treated as a prescribed quantity. Therefore, the known parameters are $h$, $b$, $a_1$, $a_2$ and $\gamma$, and the five unknown parameters to be solved are $L_1$, $L_2$, $L_3$, $\beta_1$ and $\beta_2$. Five equations are required to solve the unknown parameters

According to the geometric relationship of plane the crease pattern, the following constraint equations can be obtained:

$$\cos\gamma = \frac{(a_1 + a_2)^2 + b^2 - L_1^2}{2(a_1 + a_2)b} \tag{1}$$

$$\cos\beta_1 = \frac{L_1^2 + L_3^2 - a_1^2}{2L_1 L_3} \tag{2}$$

$$\cos\beta_2 = \frac{L_2^2 + L_3^2 - a_2^2}{2L_2 L_3} \tag{3}$$

$$\cos(\beta_1 + \beta_2) = \frac{L_1^2 + L_2^2 - (a_1 + a_2)^2}{2L_1 L_2} \tag{4}$$

$$\cos^{-1}\left(\frac{L_1^2 + b^2 - (a_1 + a_2)^2}{2L_1 b}\right) = \cos^{-1}\left(\frac{L_1^2 + b^2 - L_2^2}{2L_1 b}\right) + \frac{\pi}{3} \tag{5}$$

Incidentally, the derivation of Eq. (5) depends on the shape of the bottom polygon. Hence, the quadrangular and triangular pattern will have slightly different versions of Eq. (5). The other equations remain the same.

To define the folded geometries, the coordinates of points $A$, $B'$ and $C$ need to be determined. Let the coordinates of $A$, $B'$ and $C$ be $(x_a, y_a, z_a)$, $(x_b, y_b, z_b)$ and $(x_c, y_c, z_c)$, respectively. Using geometrical relationships, the following constraint equations can be obtained

$$L_2^2 = x_c^2 + y_c^2 + z_c^2 \tag{7}$$

$$L_3^2 = x_b^2 + y_b^2 + z_b^2 \tag{8}$$

$$\overrightarrow{OA} \cdot \overrightarrow{OB} = L_1 L_2 \cos \beta_1 \tag{9}$$

$$\overrightarrow{OB} \cdot \overrightarrow{OC} = L_2 L_3 \cos \beta_2 \tag{10}$$

$$\left(x_a - \frac{\sqrt{3}}{2}b\right)^2 + \left(y_a - \frac{b}{2}\right)^2 + z_a^2 = L_2^2 \tag{11}$$

$$x_c^2 + (y_c - b)^2 + z_c^2 = L_1^2 \tag{12}$$

$$(y_c - y_a)(vz_b - z_a) = (z_c - z_a)(vy_b - y_a) \tag{13}$$

$$(x_c - x_a)(vz_b - z_a) = (z_c - z_a)(vx_b - x_a) \tag{14}$$

The coordinates of the five points $A$, $B'$, $C$, $O$ and $O_2$ can be obtained by solving Eqs. (1-14) in MATLAB. Subsequently, the geometrical model of the whole unit cell of the hexagon pattern can be determined according to the rotation matrix method.

## 3  FE Model

The FE models of the three types of foldcore designs based on the hexagonal, quadrangular and triangular patterns are shown in Fig. 2. The FE analysis was performed in the FE solver Abaqus/Explicit due to its ability to cope with large nonlinear deformations, post-buckling behaviors and complex contact conditions. The considered material for the foldcore models is polyethylene terephthalate (PET). The density, elastic modulus, Poisson's ratio and thickness of the material are taken as 1.35 g/cm$^3$, 2665.15 MPa, 0.39 and 0.3 mm, respectively. The plastic data of the material is listed in Table 1.

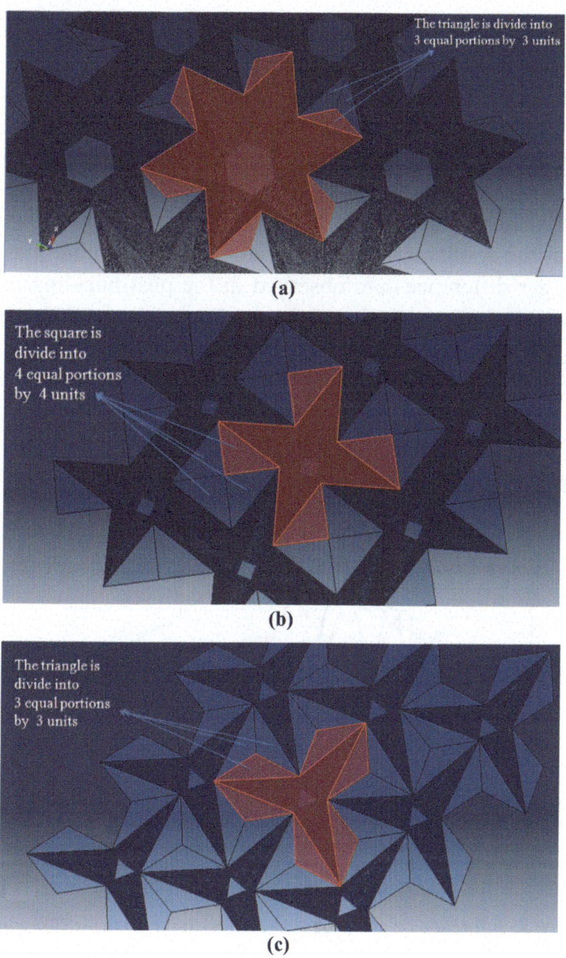


**Figure 2:** *(a) Hexagonal model; (b) Quadrangular model; (c) Triangular model*

**Table 1:** *The plastic data of PET*

| Plastic strain | Yield stress (MPa) | Plastic strain | Yield stress (MPa) |
|---|---|---|---|
| 0 | 28.58 | 0.007 | 66.88 |
| 0.001 | 43.82 | 0.008 | 67.93 |
| 0.002 | 51.68 | 0.009 | 68.85 |
| 0.003 | 56.79 | 0.010 | 69.46 |
| 0.004 | 60.64 | 0.011 | 69.86 |
| 0.005 | 63.31 | 0.012 | 70.17 |
| 0.006 | 65.26 | 0.013 | 70.34 |

whereas excessively dense meshes will significantly increase the computational cost. Hence, in the process of mesh generating, the mesh size must be carefully investigated. A convergence analysis of the mesh sizes ranging from 2 mm to 0.25 mm was performed for the hexagonal pattern under flatwise compression. The mesh sizes have an error margin of 10%. The results are shown in Fig. 3. It can be seen that the resultant stress-strain curves show similar elastic behaviors initially, but larger differences are observed in the post-buckling region. This is due to the inability of coarse meshes to correctly represent folding behavior. The results converged for mesh sizes below 0.5 mm. In the following analysis, all models are meshed with the uniform mesh size of 0.45-0.55 mm.

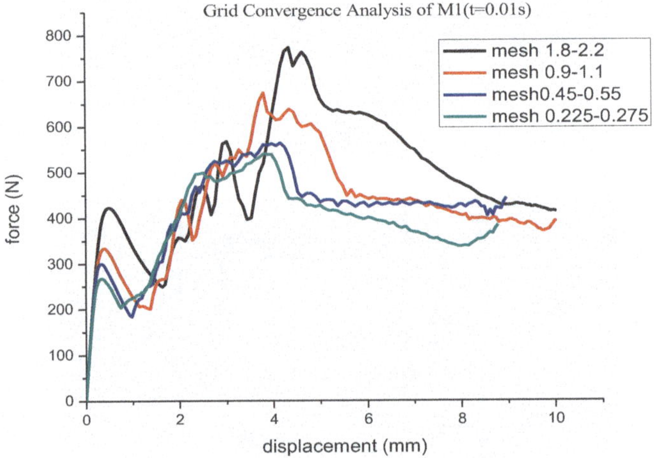


*Figure 3: Influence of mesh size on compressive stress-strain curve*

In addition to the mesh sizes, the simulation time also influences the results. Typical material tests are performed at a quasi-static loading rate. However, performing quasi-static virtual tests in the simulation would be costly, since the explicit integration of dynamic simulation codes usually leads to very small time steps, resulting in an immense computational time. In order to obtain the reasonable simulation time for quasi-static results, a time convergence study was performed at various loading speeds. The results of the time convergence study are shown in Fig. 4. According to the figure, the simulation time is chosen as 1 s for all subsequent analysis.

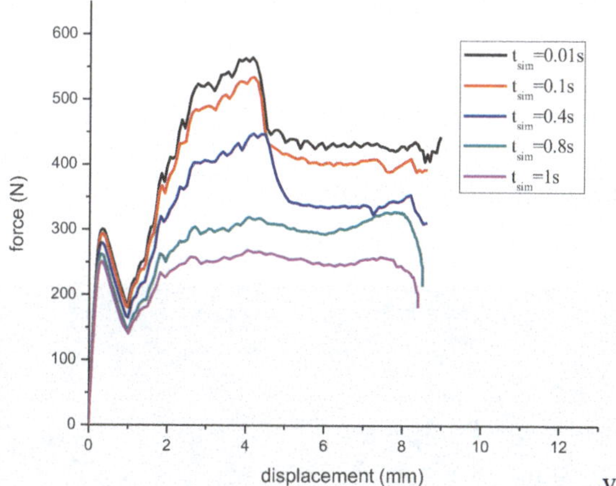



***Figure 4:*** *Influence of simulation time on compressive stress-strain curve*

## 4  Parametric Analysis

In this section, virtual tests were performed to investigate the influences of different geometric parameters on the mechanical properties of Resch Pattern foldcores subjected to quasi-static compression loading conditions. In the virtual test, two rigid plates RP1 and RP2, parallel to the *x-y* plane, were attached to two ends of the model in the thickness direction using the tie constraint, as shown in Fig. 5. Rigid plate RP1 was fixed both translationally and rotationally, and rigid plate RP2 was displaced by half of the thickness towards RP2, resulting in a maximum loaded compressive strain of 50%. In the present study, all models have a fixed height $h$ of 20 mm. According to the time convergence study, the loading rate was chosen as 10 mm/s to ensure quasi-static results. The analysed parameters include the length of the inner polygon $b$, the relative angle between the two polygons $\gamma$, and the ratio between $a_1$ and $a_2$. Specifically, for the hexagon pattern, fifteen models, denoted as M1 to M15, are considered. For the quadrangular pattern, twelve models, denoted as MF1 to MF12, are considered. For the triangular pattern, eleven models, denoted as MT1 to MT11, are considered. The detailed model parameters of these models are listed in Table 2.

In order to compare the mechanical properties between different structures, the specific stress $F_{spe}$ was used, which is defined as

$$F_{spe} = \frac{F}{S_u \rho_e} \tag{15}$$

is the effective density of the unit cell, defined by

$$\rho_e = \frac{m}{S_u h} \tag{16}$$

where $m$ is the total mass of the unit cell.

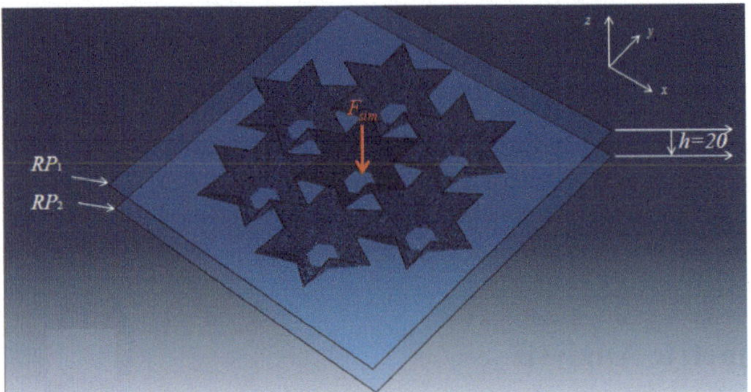

**Figure 5:** *Virtual compression loading.*

**Table 2:** *The model parameters for models M1-M15, models MF1-MF12 and models MT1-MT11*

| Model | $b$ (mm) | $a_1$ (mm) | $a_2$ (mm) | $\gamma$ (deg) |
|---|---|---|---|---|
| M1 | 9.29 | 21.55 | 21.55 | 30 |
| M2 | 9.29 | 21.55 | 21.55 | 35 |
| M3 | 9.29 | 21.55 | 21.55 | 40 |
| M4 | 9.29 | 21.55 | 21.55 | 45 |
| M5 | 9.29 | 21.55 | 21.55 | 50 |
| M6 | 9.29 | 21.55 | 21.55 | 55 |
| M7 | 9.29 | 21.55 | 21.55 | 60 |
| M8 | 11.34 | 21.55 | 21.55 | 30 |
| M9 | 12.47 | 21.55 | 21.55 | 30 |
| M10 | 13.61 | 21.55 | 21.55 | 30 |
| M11 | 14.74 | 21.55 | 21.55 | 30 |
| M12 | 15.88 | 21.55 | 21.55 | 30 |
| M13 | 9.29 | 18.148 | 24.95 | 30 |
| M14 | 9.29 | 19.28 | 23.815 | 30 |
| M15 | 9.29 | 20.418 | 22.68 | 30 |

| MF2 | 6.67 | 33.33 | 33.33 | 55 |
| MF3 | 6.67 | 33.33 | 33.33 | 65 |
| MF4 | 6.67 | 33.33 | 33.33 | 75 |
| MF5 | 6.67 | 33.33 | 33.33 | 85 |
| MF6 | 8.33 | 33.33 | 33.33 | 45 |
| MF7 | 10.00 | 33.33 | 33.33 | 45 |
| MF8 | 11.67 | 33.33 | 33.33 | 45 |
| MF9 | 13.33 | 33.33 | 33.33 | 45 |
| MF10 | 6.67 | 23.33 | 43.33 | 45 |
| MF11 | 6.67 | 26.67 | 40 | 45 |
| MF12 | 6.67 | 30 | 36.67 | 45 |
| MT1 | 15 | 60 | 60 | 60 |
| MT2 | 15 | 60 | 60 | 70 |
| MT3 | 15 | 60 | 60 | 80 |
| MT4 | 15 | 60 | 60 | 90 |
| MT5 | 17 | 60 | 60 | 60 |
| MT6 | 19 | 60 | 60 | 60 |
| MT7 | 21 | 60 | 60 | 60 |
| MT8 | 23 | 60 | 60 | 60 |
| MT9 | 15 | 45 | 75 | 60 |
| MT10 | 15 | 50 | 70 | 60 |
| MT11 | 15 | 55 | 65 | 60 |

The specific compressive stress versus strain curves of models M1-M15 are plotted in Fig. 6. In the stress-strain curves, a logarithmic scale is used for the strain axis to better illustrate the regions of small strains. The results of M1 are used as reference. According to Fig. 6, the behaviors of the foldcores under compression are characterized by four distinct stages, i.e. pre-buckling (stage I), yielding (stage II), reversing (stage III) and fluctuation (stage IV). In the first stage, the folded cores behave linear-elastically up to an average strain of 1.62% where the yield of the material begins to propagate. In the yielding stage, the reacting forces of RP1 decline sharply. In the reversing stage, the reacting force has a sharp rise which looks like another elastic stage. As for the fluctuation stage, the stress of most regions of the material has reached the yield point, so the reacting forces show a trend of fluctuating downward. To further reveal the phenomena observed in Fig. 6, we plot the stress contours of M4 during compression in Fig. 7. The vertices begin to yield at point A, where the reacting force reaches the peak. At point B, the high stress regions start to propagate towards the side. Then the yield regions extend into the facets. This is the reason

F_spe-compressive strain(relative angle γ)
M1(30)
M2(35)
M3(40)
M4(45)
M5(50)
M6(55)
M7(60)
F_spe[Mpa/ρ]
I
II
III
IV
compressive strain [%]

(a)

F_spe-compressive strain(the length b)
M1(9.2986)
M8(11.3398)
M9(12.47378)
M10(13.60776)
M11(14.74174)
M12(15.87572)
F_spe[Mpa/ρ]
I
II
III
IV
compressive strain [%]

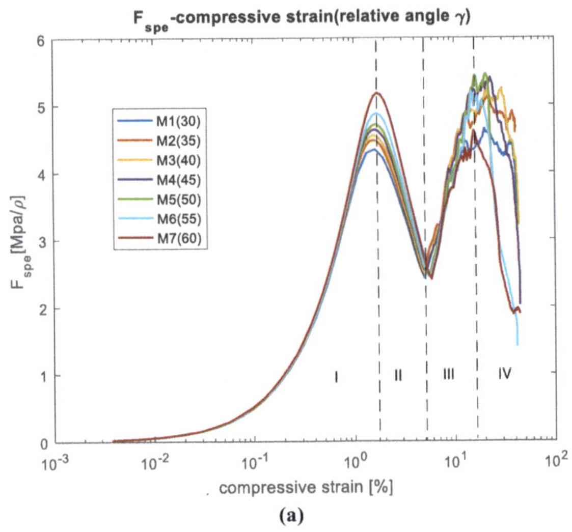

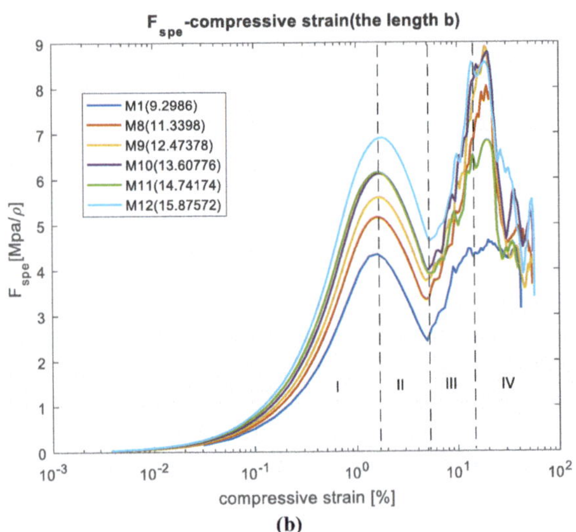

(b)

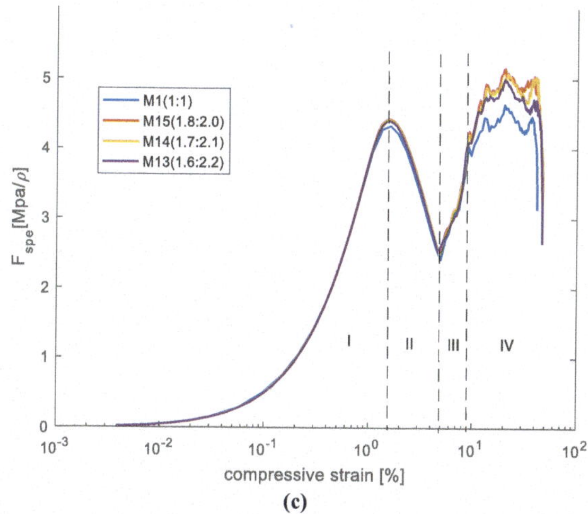



***Figure 6:*** *Specific compressive stress versus strain curves: (a) M1-M7, (b) M8-M12, (c) M13-M15.*

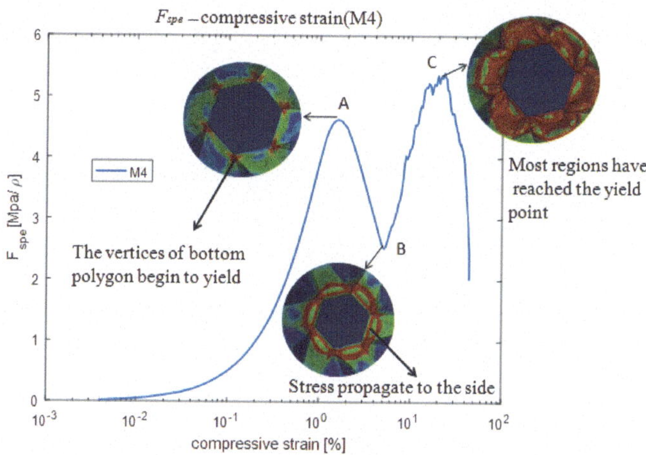



***Figure 7:*** *The compaction process of model M4*

The changes in the specific strength of the fifteen models with respect to the three parameters are plotted in Fig. 8. It is noted that the specific strength increases with the increase in $\gamma$ or $b$, and is more sensitive to the change of $b$ than that of $\gamma$. The location of the division point on the outer polygon has no obvious effect on the specific strength of hexagonal pattern.

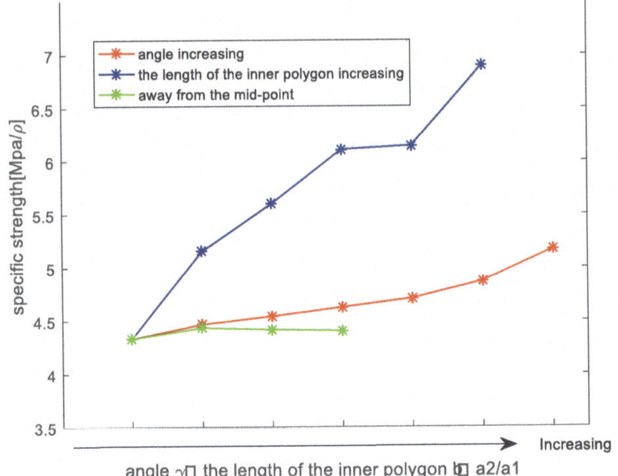


*Figure 8: The changing tendency of the specific strength in M1-M15 under compression*

Figure 9 compares the specific energy absorption of the fifteen models. We can reach the conclusion that the increase in the side length will significantly improve the energy-absorption capability. The other two parameters show no certain influences on the energy absorption capacity.

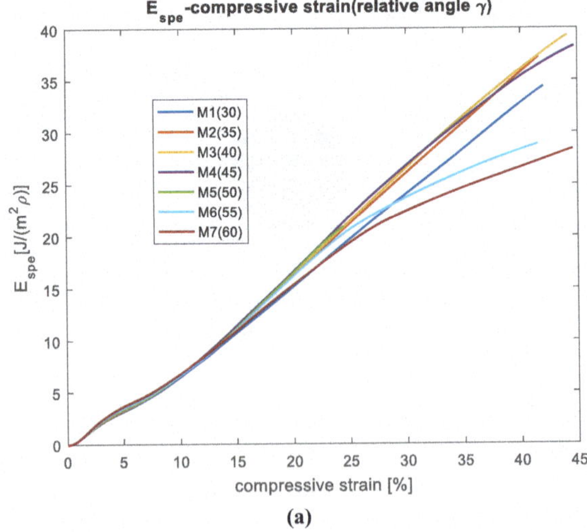


(a)

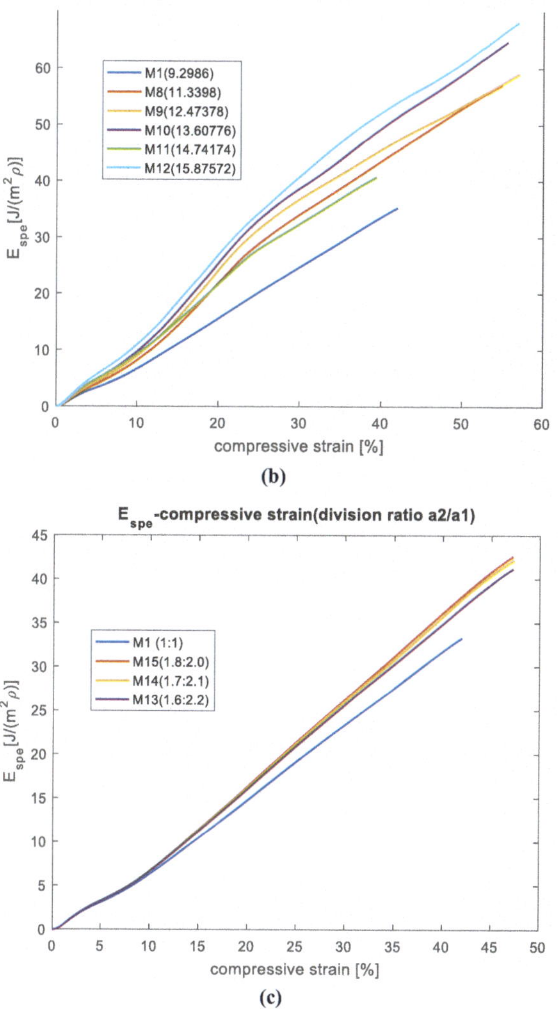

*Figure 9: Specific energy absorption versus strain curves: (a) M1-M7, (b) M8-M12, (c) M13-M15.*

The compressive stress versus strain curves and the energy absorption curves of models MF1-MF12 are similar to those of models M1-M15. Hence, they are not shown here for simplicity. The changes in the specific strength of model MF1-MF12 with respect to the three parameters are summarized in Fig. 10. Again, it is shown that the strength of the model increases as $\gamma$ and $b$ increase and the length of the inner polygon has the greatest influence on the strength of the models. The changing behaviour of the specific strength of models MT1-MT11 are shown in

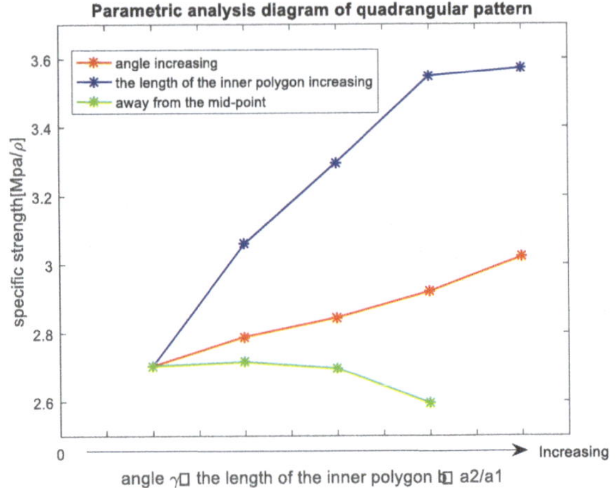


**Figure 10:** *The changing tendency of the specific strength in MF1-MF12 under compression*

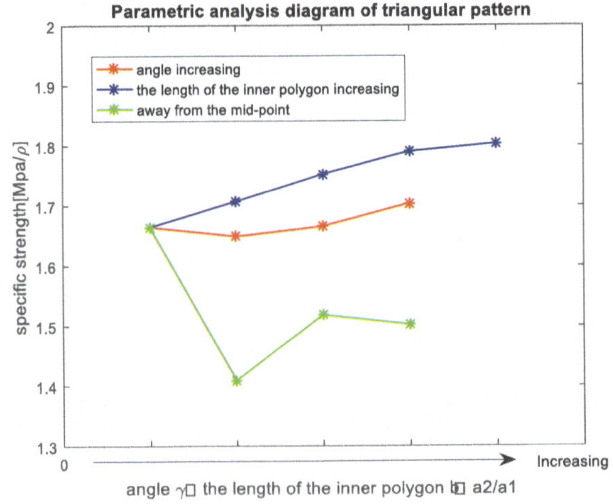


Figure 11: *The changing tendency of the specific strength in MT1-MT11 under compression*

## 5  Conclusions

A method for the virtual testing of foldcores structures using dynamic finite element simulations as an efficient alternative to experimental test series was presented in this paper. This study included different Resch Pattern foldcores of simulations showed that the hexagonal pattern has better mechanical properties than the quadrangular triangular patterns. It is notable that the compression stress process of the Resch Pattern has a distinctive reversing stage, which results from the fixation of the bottom. We also found that the specific strength of the Resch Pattern increases with an increase in $\gamma$ or $b$ and is more sensitive to changes in $b$ than in $\gamma$. The division point location on the outer polygon has no obvious effect on the specific strength.

## Acknowledgements

The financial support from National Natural Science Foundation of China (No. 51408357) is gratefully acknowledged.

## References

[Herrmann et.al 05] S. Herrmann, C. Zahlen, I. Zuardy. "Sandwich structures technology in commercial aviation: present application and future trends, in: Sandwich Structures 7: Advancing with Sandwich Structures and Materials", Springer, (2005), 13–26.

[Tachi et.al 13] T. Tachi. "Designing freeform origami tessellations by generalizing Resch's patterns", Journal. of Mechanical Design, 135, (2013),V06BT07A025.

[Klett et.al 07] Y. Klett, K. Drechsler. "Design of multifunctional folded core structures for aerospace sandwich applications", Deutscher Luft- und Raumfahrt Kongress; 2007.

[Hachenberg et.al 03] D. Hachenberg, C. Mudra, M. Nguyen. "Folded structures an alternative sandwich core material for future aircraft concepts", München: Deutscher Luftund Raumfahrt Kongress, 2003.

[Heimbs et.al 07] S. Heimbs, T. Mehrens, P.Middendorf, M. Maier, A. Schumacher, "Numerical determination of the nonlinear effective mechanical properties of folded core structures for aircraft sandwich panels", the 6th LS-DYNA Users Conference, Gothenburg, Sweden, 2007.

---

Xinying Lv
School of Aeronautics and Astronautics, Shanghai Jiao Tong University, 800 Dongchuan road, Shanghai, China, e-mail: xinying.lv@sjtu.edu.cn

Xiang Zhou
School of Aeronautics and Astronautics, Shanghai Jiao Tong University, 800 Dongchuan road, Shanghai, China, e-mail:xiangzhou@ sjtu.edu.cn

# The Mechanics of Metallic Folds

*M.G. Walker, K.A. Seffen*

**Abstract**: *We investigate the mechanical properties of metallic folds. In contrast to the common singular hinge assumption, we assume a finite geometry for the fold cross-section and consider the influence of the geometry on the mechanical behaviour. We derive the strain energy due to fold opening and show that reducing the size of the fold increases the energy required for fold opening and vice versa. In theory, as the fold size approaches the singular limit, the opening stiffness approaches infinity, but in practice this is limited by manufacturing constraints. Due to the large opening resistance of such folds, the deformation of origami structures made from folded metal will depend to a large extent on deformation of the panels. We consider a folded metallic strip with flexible panels manipulated by deformation at the panel edges, and identify the boundary between fold opening and panel bending dominated behaviour, which is confirmed by finite element analysis.*

## 1   Introduction

The increasing use of origami techniques in engineering applications has necessitated a shift away from paper-based materials to more robust metals and laminated composites. Folding paper locally tears fibres and, with repeated folding, gradually weakens the fold line, leading to eventual failure. In contrast, metal folding involves plastic deformation and work-hardening, where yield stresses in the fold robustly rise due to the forming process. Metal fold lines exhibit a typically high stiffness, which can subtract from the overall Origami character of the structure [Francis et al. 13].

The analysis of Origami structures has been dominated by two main techniques. The first assumes rigid plates connected by discrete fold lines, or 'rigid origami' [Evans et al. 15]. The folds themselves remain straight but accommodate relative rotations across them. Rigid folding this way can be described entirely by a kinematic analysis even if the fold lines are in some way stiff in practice; adding torsional springs along them does not affect the kinematic description. Example applications include deployable membranes for spacecraft [Miura 85], Origami-based robots [Felton et al. 14], metamaterials [Hawkes et al. 10], and medical stents [Kuribayashi et al. 06].

The second technique accounts for flexibility beyond the folds. For generality's sake, the fold line is treated as a general space curve connecting thin inextensible surfaces that can only bend [Duncan and Duncan 82, Dias and Audoly 14]. Differential geometry is then used to obtain the increase in strain energy due to deformation of the panels, as well as the change in fold angle, if it has been modelled elastically. This technique is more complex which prevents its immediate application to large-scale origami structures with multiple folds.

In either case, the fold itself is modelled as a singular feature. The assump-tion that an elastic fold can be represented by a singular hinge with a well-behaved torsional spring has not been fully investigated. In this work, we investigate the me-chanical properties of metallic folds for incorporation into the analysis of Origami structures.

Typical sheet metal folding processes create a constant curvature fold, and the formed radius and sector angle can be predicted using established formulae [Schey 00]. For this study we assume a finite geometry for the fold cross-section consisting of a uniform curvature cylindrical segment, in contrast to the common singular assumption. Using a linear elastic material model, we develop strain energy expressions for the opening of the fold and investigate the effect of the fold geometry on the opening behaviour of the cross-section.

Out-of-plane effects, such as flexure of the crease perpendicular to its axis, are not included in our analysis. We also assume that residual strains due to the forming process do not influence the elastic r esponse. O ur m odel i s then u sed to fi nd the fold opening elastic limit, which may be affected by residual strains, but we assume the cross-section is free from these effects here.

Due to the typically high opening stiffness of metallic folds, bending of the panels will have a significant i nfluence on th e be haviour of a fo lded sh eet. We investigate how the fold and panel geometry influence the transition between fold opening dominated behaviour and panel bending dominated behaviour. Finally, we consider the analysis of a folded sheet with deformable panels opened by imposed displacements at the panel edges and compare the results to finite element analysis.

This study is principally concerned with folds in metallic materials. However, the results are also applicable to Origami structures made from other linear-elastic materials across a range of length scales, for example: nanoscale silicon [Malachowski et al. 14] and graphene [Miskin et al. 18]. The results will contribute to the understanding and design of Origami structures made using elastic materials.

## 2   Constant Curvature Thin Shell Fold

In the first instance, the cross-section of a fold with infinite axial depth is modelled as a constant curvature cylindrical segment of radius, $r$, and sector angle $\beta$, as shown in Fig. 1a. Because the depth is infinite, w e m ay a ssume u niform axial behaviour governed by how the cross-section deforms. The fold has a uniform thickness, $t$, Young's modulus $E$, and Poisson's ratio $v$. The fold thickness-to-radius ratio is assumed to be small enough for thin shell theory to apply.

To derive the strain energy per unit depth of fold due to compatible panel rotations, $\theta$, on its edges, the deformed geometry, shown in Fig. 1b, is considered. The deformed fold is assumed to maintain a constant curvature, $\kappa$, since pure rotation is applied along its edges. This curvature is related to the edge rotation, $\theta$, by ensuring compatibility of slope between the fold and the panel:

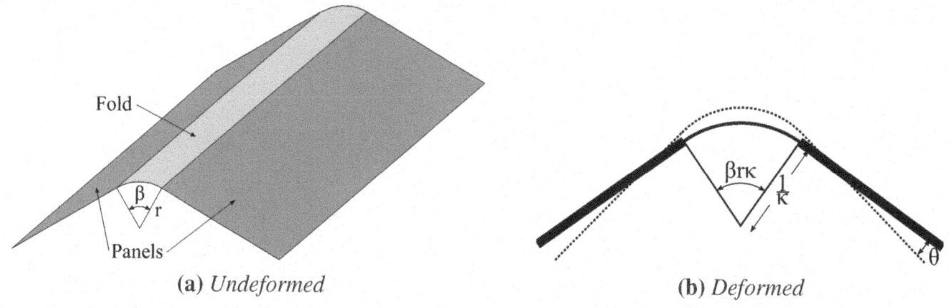



(a) *Undeformed*      (b) *Deformed*

**Figure 1:** *Fold cross-section geometry in the undeformed (a) and deformed (b) states. The panels are considered as rigid plates; therefore only the fold region deforms.*

$$\kappa = \frac{2\theta}{\beta r} + \frac{1}{r} \tag{1}$$

The strain energy due to the edge rotation is obtained from the corresponding change in curvature and neglecting any axial deformation:

$$U_{Thin} = \frac{\beta r D}{2}\left(\kappa - \frac{1}{r}\right)^2 = \frac{2D}{\beta r}\theta^2 \tag{2}$$

where $D = Et^3/12(1-v^2)$ is the shell flexural rigidity. This only depends on the initial geometry of the fold and the edge rotations. As $r$ approaches zero this energy approaches infinity; in reality this is limited by the onset of plastic deformations, as discussed in Section 4.

Large-radius folds (relative to thickness) can be used to accomodate the panel thickness in 'thick' Origami [Chen et al. 15, Ku and Demaine 16]. In theory, when the fold radius $r > t/2$, the panels can be folded flat without interference at the fold. However, large-radius folds can result in additional degrees of freedom and more complex behaviour in practice [Peraza Hernandez et al. 17]. In this study only the opening behaviour of the fold is considered. Metallic folds are commonly made with a radius roughly equal to the thickness, where thin shell theory does not apply. To investigate this violation we consider a "thick" fold next.

## 3   Constant Curvature Thick Fold

For thick folds, the through-thickness radial stresses are no longer negligible. We resort to a continuum mechanics approach, where the fold is treated as a thick curved beam. The corresponding stress state from end moments (applied in our case by each panel to the fold edges) was derived by Timoshenko [Timoshenko and Goodier 51]. Assuming uniform deformed curvature, the end moments can be related to the change in fold curvature by: $M = Et^3(1/r - \kappa)/12$. The stress state

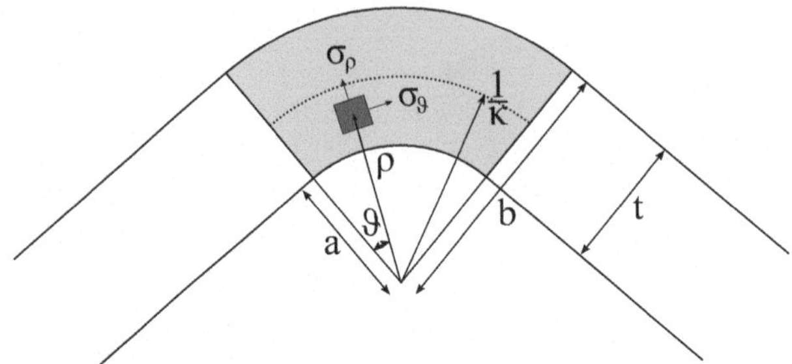



**Figure 2:** *Geometry of the thick fold model in addition to the geometry defined in Fig. 1, which applies along the neutral axis (shown dotted).*

can therefore be written as:

$$\sigma_\rho = \frac{Et^3}{3N}\left(\frac{1}{r} - \kappa\right)\left(\frac{a^2 b^2 \log\frac{b}{a}}{\rho^2} + a^2 \log\frac{a}{\rho} + b^2 \log\frac{\rho}{b}\right) \tag{3}$$

$$\sigma_\vartheta = \frac{Et^3}{3N}\left(\frac{1}{r} - \kappa\right)\left(-\frac{a^2 b^2 \log\frac{b}{a}}{\rho^2} + a^2 \log\frac{a}{\rho} - a^2 + b^2 \log\frac{\rho}{b} + b^2\right) \tag{4}$$

$$\tau_{\rho\vartheta} = 0 \tag{5}$$

where $\rho$ is a radial coordinate, $\vartheta$ is a circumferential coordinate, $a$ is the inner radius, $b$ is the outer radius, and $\kappa$ is the deformed curvature of the fold at the neutral axis, see Fig. 2. Also,

$$N = \left(b^2 - a^2\right)^2 - 4a^2 b^2 \log^2\left(\frac{b}{a}\right) \tag{6}$$

The strain energy of bending per unit depth into the page is obtained by integrating the strain energy density over the cross section, noting that the stresses do not depend on the circumferential coordinate $\vartheta$:

$$U_{Thick} = \frac{\beta r \kappa}{2E} \int_a^b \left[\sigma_\vartheta^2 + \sigma_\rho^2 - 2\nu\sigma_\vartheta\sigma_\rho\right]\rho\,d\rho \tag{7}$$

$$= \frac{\beta r \kappa E t^6 (b-a)(a+b)(\kappa - \frac{1}{r})^2}{36(1 - \nu^2)N} \tag{8}$$

Substituting $a = 1/\kappa - t/2$, $b = 1/\kappa + t/2$, and Eqn 1 into Eqn 8, yields the strain energy due to an edge rotation, $\theta$:

$$U_{Thick} = \frac{8Et^2\Phi^4\theta^2}{9\lambda(1 - \nu^2)(\Phi^2 - 4\lambda^2)^2 \log^2\left(\frac{2\lambda + \Phi}{2\lambda - \Phi}\right) - 144\lambda^3\Phi^2} \tag{9}$$

where $\Phi = \beta + 2\theta$ is the deformed sector angle of the fold, and $\lambda = \beta r/t$ is the sector length-to-thickness ratio. In contrast to Eqn 2 for thin shells, the fold stiffness

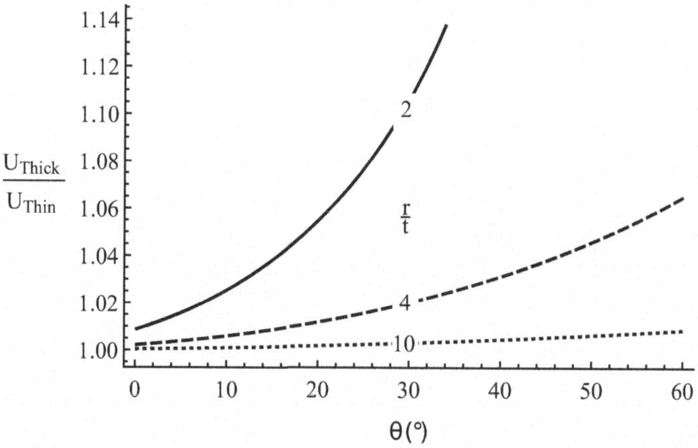


**Figure 3:** *Ratio of thick and thin fold strain energies, Eqn 9 and Eqn 2, for a fold with $\beta = 30°$ and $r/t$ ratios of 2, 4, and 10. For small values of $r/t$ the thick fold model requires more energy to deform (it is stiffer). As the $r/t$ ratio increases, the energy ratio approaches unity.*

here depends not only on the initial geometry of the fold, but also on the deformed fold sector angle, $\Phi$. The logarithmic term in Eqn 9 imposes a restriction on the change in fold sector angle for the thick model:

$$\theta < \beta \left( \frac{r}{t} - \frac{1}{2} \right) \tag{10}$$

which is ultimately a restriction on the maximum deformed curvature: $\kappa < 2/t$, due to Eqn 1. Physically this limit represents a sheet folded back onto itself; therefore, a larger curvature is not physically possible.

The ratio of the thick fold energy, Eqn 9, to the thin fold energy, Eqn 2, is plotted in Fig. 3, where it approaches unity as $r/t$ becomes large: a ratio of $r/t > 20$ is a commonly mentioned limit for thin shell behaviour, which works well in this case. The difference between the two energies also increases as the edge rotation, $\theta$, increases. By repeating this calculation over a range of edge rotations, the difference between the two models was found to be less than 20% for moderate rotations.

## 4   Rotation Limits for Elastic Behaviour

The maximum circumferential stress in the thick fold model is obtained using Eqn 4, with $\rho = a$ for tension and $\rho = b$ for compression. These turn out to be equal in magnitude; setting the maximum tensile stress equal to the yield stress, $\sigma_y$, and solving for the edge rotation angle, we find at the onset of yielding:

$$\theta_y = \frac{3\beta\gamma\left(\left(1-4\gamma^2\right)^2\log^2\left(\frac{2\gamma-1}{2\gamma+1}\right)-16\gamma^2\right)}{16\gamma-4(2\gamma+1)^2\log\left(\frac{2\gamma+1}{2\gamma-1}\right)}\frac{\sigma_y}{E} \qquad (11)$$

where $\gamma = r/t$, and the angle $\theta_y$ is assumed to be small. To test this assumption, consider a mild steel fold with $\gamma = 2$, a yield stress $\sigma_y = 200$ MPa, and elastic modulus $E = 200$ GPa. Using Eqn 11 results in $\theta_y = 0.0017\beta$, which is much less than unity for sector angles, $\beta < \pi$. For large $\gamma$, a good approximation of Eqn 11 is obtained using Eqn 1. Since the change in curvature is $\kappa - 1/r$, the fold rotation at yield for the thin shell model is: $\theta_y = \gamma\beta\sigma_y/E$; this is precisely the limit of Eqn 11 when $\gamma$ becomes large.

The edge rotation angle at the onset of yielding is generally small for metallic materials. Therefore, elastic behaviour of the fold only occurs over a small range of rotations. Further opening of the fold will result in permanent plastic deformations.

The strain energy, and consequently the stiffness, of a fold increases as the size of the fold is reduced and, in the absence of plastic effects, theoretically approaches infinity as the fold radius is reduced to zero. As a result, deformation of the panel regions will have an important contribution to the mechanical behaviour of a folded sheet. The transition between fold opening and panel bending dominated behaviour is investigated in the next section.

## 5 Fold Geometry Limits for Origami Behaviour

Since the bending strain energy of either side panel is proportional to the shell flexural rigidity, $D$ (units of N-m), and the fold opening energy is proportional to the fold opening stiffness (units of N), Lechenault et al. [F. et al. 14] propose that a natural length scale emerges from the ratio of these energies. They define this ratio as an Origami length scale, $L^* = 2D/k$, where $k$ is the stiffness of the fold: for panel lengths much less than $L^*$, fold opening will dominate; while for lengths much larger than $L^*$, panel bending will dominate. Between these extremes both deformation mechanisms will be present.

The strain energy of opening was derived for thin shell folds in Section 2, and thick folds in Section 3. An equivalent, singular, torsional spring stiffness representing the opening behaviour of the discrete fold is obtained by comparing these energy expressions to the potential energy of a torsional spring: $\frac{1}{2}k\theta^2$, where $k$ is the spring stiffness. Considering the thin shell energy in Eqn 2, the corresponding stiffness is therefore $4D/\beta r$. The Origami length scale of a thin shell fold is thus: $L^* = \beta r/2$. Similarly, using Eqn 9 for a thick fold:

$$L^* = \frac{3}{16\Phi^4}\left[16\lambda^2\Phi^2 - \left(\Phi^2 - 4\lambda^2\right)^2\log^2\left(\frac{2\lambda+\Phi}{2\lambda-\Phi}\right)\right]\frac{\beta r}{2} \qquad (12)$$

which approaches the thin shell Origami length scale, $L^* = \beta r/2$, for large $\gamma$ (note: $\lambda = \beta r/t = \beta\gamma$). On the other hand, Eqn 12 reduces to $L^* = 3\beta r/8$ for the minimum value of $\gamma$ equal to one half.

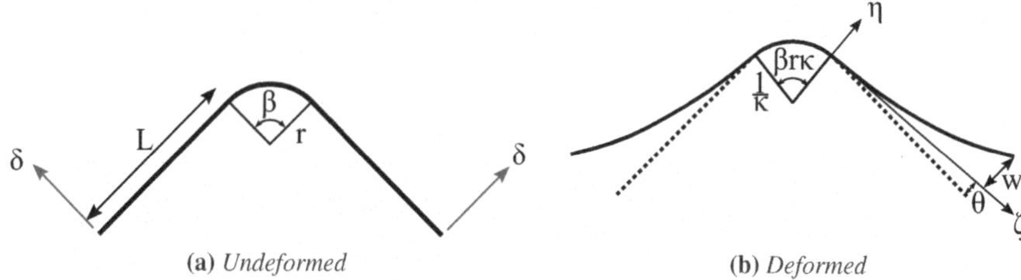

(a) Undeformed        (b) Deformed

**Figure 4:** *Cross-section geometry for a folded strip with deformable panels in the undeformed (a) and deformed (b) states. A displacement of $\delta = w + L\sin\theta$ is imposed perpendicular to the panel which opens the strip by a combination of panel bending and fold opening.*

## 6   Analysis of a Folded Strip with Flexible Panels

The undeformed cross-section of a folded strip is shown in Fig. 4a. The fold has the same geometry as shown in Fig. 1 and the panels each have a length of $L$. The folded strip is assumed to be infinite in depth as before. We consider the opening action caused by a displacement $\delta$ imposed at the free panel edges, without reference to the forces which cause such a displacement. Both the panel and the fold deform as a result of this displacement. The $\eta$-$\zeta$ coordinate system of the panel is located at the connection between the fold and the panel, as shown in Fig. 4b. Under deformation the fold edges rotate by $\theta$, and the $\zeta$ axis remains oriented along the fold tangent at the fold-panel interface. The panel itself deforms relative to the $\zeta$ axis by a displacement, $w$, at the free panel edge. These deformations sum to to the total imposed displacement $\delta = L\sin\theta + w$.

Considering the deformation of the panels, the governing equation of the deformation of a thin plate is given by:

$$D\frac{d^4\eta}{d\zeta^4} = 0 \tag{13}$$

The boundary conditions for the panel enforce compatibility at the fold-panel interface through:

$$\frac{d\eta}{d\zeta} = 0 \qquad\qquad \eta = 0 \tag{14}$$

and the boundary conditions at the free edges are:

$$\frac{d^2\eta}{d\zeta^2} = 0 \qquad\qquad \eta = w \tag{15}$$

The deformed panel shape is thus solved as:

$$\eta(\zeta) = \frac{w\zeta^2(3L - \zeta)}{2L^3} \tag{16}$$

and the total strain energy per unit depth due to bending of a single panel is:

$$U_P = \frac{D}{2} \int_0^L \left(\frac{d^2\eta}{d\zeta^2}\right)^2 d\zeta = D\frac{3w^2}{2L^3} \tag{17}$$

The energy of deformation of the fold region, $U_B$, is given in Eqns 2 or 9, depending on the $r/t$ ratio, resulting in a total energy of deformation of:

$$U = 2U_P + U_B \tag{18}$$

which depends on two parameters, $w$ and $\theta$, and the initial geometry.

For a particular displacement $\delta = L\sin\theta + w$, we find after substitution, for the thin shell fold:

$$U = D\left[\frac{3(\delta - L\sin\theta)^2}{2L^3} + \frac{2\theta^2}{\beta R}\right] \tag{19}$$

The angle, $\theta$, corresponding to the displacement, $\delta$, is found by minimising the total strain energy. For the thin shell fold:

$$\theta_{\text{Thin}} = \frac{3\Delta\mu}{3\mu + 2} \tag{20}$$

where $\mu = \beta r/L$ and $\Delta = \delta/L$. For the thick fold we find:

$$\theta_{\text{Thick}} = \frac{9\Delta\mu\left[(1 - 4\gamma^2)^2\log^2\left(\frac{2\gamma+1}{2\gamma-1}\right) - 16\gamma^2\right]}{9\mu(1 - 4\gamma^2)^2\log^2\left(\frac{2\gamma+1}{2\gamma-1}\right) - 16(9\gamma^2\mu + 2)} \tag{21}$$

where it can be confirmed that as $\gamma$ becomes very large, $\theta_{\text{Thick}}$ approaches $\theta_{\text{Thin}}$. The ratio of the fold energy to the panel energy, for the thin shell fold, is now:

$$\frac{U_{\text{Thin}}}{U_{\text{Panel}}} = \frac{6\Delta^2\mu}{(3\mu + 2)^2\left[\Delta - \sin\left(\frac{3\Delta\mu}{3\mu+2}\right)\right]^2} \tag{22}$$

Equation 22 has a weak dependence on $\Delta$ for realistic values of $\mu$ and $\Delta$ where $\mu\Delta \ll 1$. An equivalent, but lengthy, expression is obtained in the same manner for the thick fold case, which depends on the ratios, $r/t$ and $L/t$. As these ratios become very large the fold-to-panel energy ratio approaches Eqn 22. The ratio of the thick fold to panel energies is a minimum when $r/t = 0.5$ and $L/t \to \infty$, which returns after some algebra:

$$\frac{72\Delta^2\mu}{(9\mu + 8)^2\left[\Delta - \sin\left(\frac{9\Delta\mu}{9\mu+8}\right)\right]^2} < \frac{U_{\text{Fold}}}{U_{\text{Panel}}} < \frac{6\Delta^2\mu}{(3\mu + 2)^2\left[\Delta - \sin\left(\frac{3\Delta\mu}{3\mu+2}\right)\right]^2} \tag{23}$$

Equation 23 is plotted in Fig. 5. When $L/\beta r$ lies below $L^*/\beta r = 0.5$, the fold deformation energy increases rapidly in comparison to the panel energy. Above $L^*/\beta r = 0.5$, the rate of decrease of the energy ratio decreases. As $L/\beta r \to \infty$ the panel energy dominates and the energy ratio approaches zero.

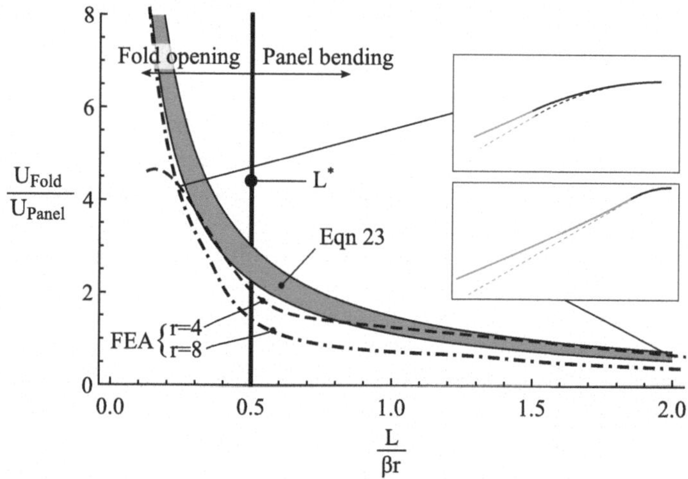


**Figure 5:** *Ratio of fold strain energy to panel strain energy for the boundary cases in Eqn 23, for $\Delta = \delta/L = 0.01$. Results from the finite element analysis study are also shown for strips with $\beta = 60°$, $t = 0.5$, $r = 4$ and $r = 8$. The insets show the deformed cross-section (solid) compared to the initial cross-section (dashed) for the finite element model. The origami length scale, $L^*$, provides a good measure of the transition between fold bending dominated behaviour and panel bending dominated behaviour.*

# 7   Finite Element Analysis

A finite element study was performed using ABAQUS [Systèmes 14] to evaluate the analytical prediction. The cross-section was modelled as shown in Fig. 4a, with a sector angle $\beta = 60°$, a thickness of 0.5 mm, and a linear elastic material model with Young's modulus $E = 200$ GPa and Poisson's ratio $v = 0.3$. Reduced integration, four-noded thin shell elements, S4R, were used, all degrees of freedom at the centreline of the fold were fixed and a non-linear static analysis was performed in half symmetry. A displacement of $\delta = 0.01L$ was imposed at the panel edge perpendicular to the initial panel inclination, an angle of $\beta/2$ to the vertical.

The fold to panel energy ratios, for fold radii of 4 mm and 8 mm, are included in Fig. 5. The shape of the curves follow Eqn 23 well but lie slightly below. One reason is Eqn 23 assumes a constant fold curvature; the stretching energy is also neglected. The finite element results also show a small dependence on the initial fold radius, $r$, which is not present in Eqn 23.

The predicted Origami length, $L^*$, provides a good indication of the transition between fold opening and panel bending dominated behaviour for both the analytical model as well as the finite element results. For a constant thickness strip, the deformation behaviour is a combination of panel bending and fold opening, except when the panel length is very long, or very short, in comparison to the fold sector arc-length.

# 8   Conclusions

This study considered the cross-sectional deformation mechanics of a fold with a discrete shape consisting of a constant curvature cylindrical segment. The strain energy due to the opening of this fold was derived using both a thin shell approach and a continuum mechanics approach. The thin shell approach is applicable when the fold radius-to-thickness ratio is greater than about 20. Below this limit, the thin shell assumptions are increasingly violated and the latter approach is appropriate. In either case, if plasticity is neglected, the stiffness (and energy) approaches infinity as the fold radius is reduced to zero.

Simple design rules were developed for the fabrication of metallic Origami structures. The fold opening limit for elastic behaviour was derived and found to be $\theta_y = r\beta\sigma_y/Et$ for thin folds. For thicker folds, the more complex expression in Eqn 11 must be used. Because folds in metals are typically small, they have a high bending stiffness in comparison to the panels. The panel length where the opening of a folded sheet transitions between fold opening dominated to panel bending dominated behaviour was shown to be one half of the fold sector length, $L^* = \beta r/2$, for thin folds. For thick folds this decreases to $L^* = 3\beta r/8$.

The relative deformation of the fold and panel regions, under imposed displacement at the panel edges, was considered in detail using an energy approach and compared to finite element analysis, showing excellent agreement. The results also showed that the Origami length, $L^*$, provides a good indication of the dominant deformation mode. The results of this study will be useful for the design and analysis of Origami structures made from folded metal and other linear-elastic materials.

# References

[Chen et al. 15] Y. Chen, R. Peng, and Z. You. "Origami of thick panels." *Science* 349:6246 (2015), 396–400. doi:10.1126/science.aab2870.

[Dias and Audoly 14] M.A. Dias and B. Audoly. "A non-linear rod model for folded elastic strips." *Journal of the Mechanics and Physics of Solids* 62 (2014), 57–80. doi:10.1016/j.jmps.2013.08.012.

[Duncan and Duncan 82] J.P. Duncan and J.L. Duncan. "Folded developables." *Proceedings of the Royal Society of London A* 383 (1982), 191–205.

[Evans et al. 15] T.A. Evans, R.J. Lang, S.P. Magleby, and L.L. Howell. "Rigidly foldable origami gadgets and tessellations." *Royal Society Open Science* 2:9 (2015), 150067. doi:10.1098/rsos.150067.

[F. et al. 14] Lechenault F., B Thiria, and M. Adda-Bedia. "Mechanical response of a creased sheet." *Physical Review Letters* 112:24 (2014), 244301. doi:10.1103/PhysRevLett.112.244301.

[Felton et al. 14] S. Felton, M. Tolley, E. Demaine, D. Rus, and R. Wood. "A method for building self-folding machines." *Science* 345:6197 (2014), 644–646. doi:10.1126/science.1252610.

[Francis et al. 13] K.C. Francis, J.E. Blanch, S.P. Magleby, and L.L. Howell. "Origami-like creases in sheet materials for compliant mechanism design." *Mechanical Sciences* 4:2 (2013), 371–380. doi:10.5194/ms-4-371-2013.

[Hawkes et al. 10] E. Hawkes, B. An, N. M. Benbernou, H. Tanaka, S. Kim, E. D. Demaine, D. Rus, and R. J. Wood. "Programmable matter by folding." *Proceedings of the National Academy of Sciences* 107:28 (2010), 12441–12445. doi:10.1073/pnas.0914069107.

[Ku and Demaine 16] Jason S. Ku and Erik D. Demaine. "Folding flat crease patterns with thick materials." 8:June 2016 (2016), 1–6. doi:10.1115/1.4031954.

[Kuribayashi et al. 06] Kaori Kuribayashi, Koichi Tsuchiya, Zhong You, Dacian Tomus, Minoru Umemoto, Takahiro Ito, and Masahiro Sasaki. "Self-deployable origami stent grafts as a biomedical application of Ni-rich TiNi shape memory alloy foil." *Materials Science and Engineering A* 419:1-2 (2006), 131–137. doi:10.1016/j.msea.2005.12.016.

[Malachowski et al. 14] Kate Malachowski, Mustapha Jamal, Qianru Jin, Beril Polat, Christopher J. Morris, and David H. Gracias. "Self-folding single cell grippers." *Nano Letters* 14:7 (2014), 4164–4170. doi:10.1021/nl500136a.

[Miskin et al. 18] Marc Z. Miskin, Kyle J. Dorsey, Baris Bircan, Yimo Han, David A. Muller, Paul L. McEuen, and Itai Cohen. "Graphene-based bimorphs for micron-sized, autonomous origami machines." *Proceedings of the National Academy of Sciences* 115:3 (2018), 201712889. doi:10.1073/pnas.1712889115.

[Miura 85] K. Miura. "Method of packaging and deployment of large membranes in space." Technical report, Institute of Space and Astronautical Science, 1985.

[Peraza Hernandez et al. 17] E.A. Peraza Hernandez, D.J. Hartl, and D.C. Lagoudas. "Design and simulation of origami structures with smooth folds." *Proceedings of the Royal Society A: Mathematical, Physical and Engineering Science* 473:2200 (2017), 20160716. doi:10.1098/rspa.2016.0716.

[Schey 00] J.A. Schey. *Introduction to Manufacturing Processes*, Third edition. McGraw-Hill, 2000.

[Systèmes 14] Dassault Systèmes. *Abaqus Theory Guide*, 2014.

[Timoshenko and Goodier 51] S. Timoshenko and J.N. Goodier. *Theory of Elasticity*, Second edition. London: McGraw-Hill, 1951.

---

Martin G. Walker

Mary Ewart Junior Research Fellow, Somerville College and Department of Engineering Science, University of Oxford, Woodstock Road, Oxford, OX2 6HD

e-mail: martin.walker@some.ox.ac.uk

Keith A. Seffen

Reader in Structural Mechanics, Department of Engineering, University of Cambridge, Trum-pington Street, Cambridge, CB2 1PZ

e-mail: kas14@eng.cam.ac.uk

# Crease Pattern Simplification for Automatic Folding

*J. A. Romero, L. A. Diago, C. Nara, J. Shinoda, and I. Hagiwara*

**Abstract**: *Origami, the ancient art of paper-folding has recently attracted the attention of the scientific community, due to its applicability in several fields, such as: aerospatial, medicine, architecture, and packaging. Being able to do automatic paper-folding using a robot has been a challenge for almost 20 years. The origami patterns developed in previous works are intended to be assembled by hand, and have folds that are very difficult to execute with a robot due to handling problems and manipulations that require multiple movements at the same time. Here, two methodologies are proposed to create crease patterns that can be automatically folded by simplifying the difficult procedures using simple folds and gluing segments to create 3D shapes. A methodology based on surface of revolution is introduced to produce several 3D shapes. Later, this methodology is extended to shapes with irregular shapes, that not necessary need to be created by surface of revolution.*

## 1   Introduction

Origami, the ancient art of folding a two-dimensional flat materials such as paper into three-dimensional (3D) objects. It has recently gained popularity among scientists and engineers because the technique can be used to create interesting-looking packages [Dai and Cannella 08], shape-changing structures [Felton et al. 14], or be applied in robotics [Balkcom and Mason 08].

Several folding robots have been developed in the past to automatically transform a 2D flat sheet of paper into a 3D object. However, it is extremely difficult to introduce automation in robots to properly handle flexible objects such as paper. The difficulty in handling flexible objects has been an on-going problem in the robotic community for over 20 years [Nakazawa 98]. There are two main problems associated with paper handling. One of them is controlling deformation of objects that has infinite degrees of freedom using a finite number of manipulated variables. The other problem is how to handle fragile material such as paper without exerting excessive stress, i.e. how to handle it safely and reliably.

Balkcom et al. provide a fundamental basis for robotic paper manipulation [Balkcom and Mason 08]. However, they used a robot that was specially designed for simple folds. Moreover, the generalization to more complex folds, such as squash and petal folds that required multiple manipulations simultaneously was not possible. Yao and Dai, focuses on the dexterous manipulation of origami cartons using robotic fingers based on the Interactive Configuration Space (ICS). Furthermore, ICS includes a geometric configuration of control vectors associated with a spherical five-bar mechanism allocated in the four corners of the cartons. Trajectories for the manipulations using four robotic fingers are generated simultaneously using the ICS. However, their control strategy is very limited to an specific folding pattern (i.e. a carton box). Kihara and Yokokohji, developed a folding robot able to fold a "tadpole" pattern. This pattern includes a squash fold in the folding sequence, that is considered to be a more complex type of folding, requiring to perform multiple manipulations at the same time. Their method synthesized a desired trajectory and sensory feedback control based on direct teaching data from a person. Although canonical correlations between forces and velocities of human trajectories are used to correct the motion deviation from nominal trajectories, complex direct teaching (e.g. many repetitions) and no implicit manipulations are included in order to avoid fluctuations of the paper.

Namiki et al. [Namiki et al. 03], use real-time modeling and visual tracking in their robot. The measured data of the paper is used to create the trajectories of the robot in real-time operation. The method by Namiki applies a visual recognition by camera and impedance control to produce softness in the trajectories of the fingers. The required kinematics adjustments are applied according to the data measured by the small sensors in the fingers. Although this robot can produce some interesting results, the complexity in the systems and high cost make it not suitable to be used. Elbrechter et al. [Elbrechter et al. 12], model the bending of a sheet of paper, and paper crease lines in order to monitor deformations. Tactile-based and vision-based closed loop controllers are introduced in an anthropomorphic robot to fold paper. Although high-speed manipulations of flexible objects can be achieved, this implementation is still very complex to be implemented and very expensive.

There have been several methodologies to create computer-aided folding-crease patterns. Robert. J. Lang and his software "Tree Maker" [Lang 17] is one of the examples that can be found on internet. This program creates the crease pattern based on the internal stick structure of the target figure (known as the t ree). The target shape can have any type of complicated structure such as insects or any other type of animal. However, the final c rease p attern c reated w ith t his s oftware can be extremely complicated to fold, making it very difficult to perform patterns created with this methodology using a robot. Another interesting program that can be found online is the "Origamizer" created by Tomohiro Tachi [Tachi 09]. This program can generate crease patterns from an arbitrary polygonal model that possesses the topology of a disk. This program has a wide applicability as most of the 3D shapes can be modified to match the topology of a disk by adding some cut lines to the final crease pattern. Although "Origamizer" can be applied to any type of shape, the final crease pattern could be over-complicated to fold. The software "ORI-REVO" [Mitani 17] designed by Jun Mitani uses a rotational sweep to create the crease pattern from a 2D poly-line. This program can be used to create a vast number of 3D shapes that can be used in different type of applications. ORI-REVO can create crease patterns less complex than Tree Maker or Origamizer, but is limited to figures using only surface of revolution. Despite the crease pattern is not very complicated to fold by hand, proper adaptations of this methodology must be carried out in order to be implemented in our robot. Lately new methodologies to create crease patterns for rigid origami applications have been in development. The methodology presented in [Miyamoto et al. 17] uses the information of the poly-lines of the cross section of the polygon to create crease patterns of irregular shapes. Although this methodology can be used to create irregular shapes, it is think to built flat-foldable structures and not for robotic folding.

In our previous works [Romero et al. 16], a paper folding and gluing robot has been designed and developed using LEGO MINDSTORMS NXT® to simplify the design process and test its effectiveness and applicability in a mass-production procedure. In order to simplify the folding process and reduce the number of manipulations, a novel methodology to produce three-dimensional (3D) objects based on surfaces of revolution (SOR) [Mitani 09] was proposed in [Romero et al. 17]. The idea in this methodology is to use a 2D poly-line (see Fig.1a), and rotate it $2\pi/$ $Krad$, where $K$ is a integer number greater than 2, and represent the number of sub-segments in the 3D shape (see Fig.1b). The resulting crease pattern is composed by stack of trapeze and triangles, with some extra areas that represent the places where the glue has to be applied (see Fig.1b). The shape of these gluing areas is necessary to create a stamp that later is used to perform the automatic folding; for deep explanation about the robot manipulations and the functionality of this stamp please refer to [Romero et al. 16].

Although several 3D shapes can be made using the SOR methodology, this is limited to these particular structures. The main idea in this paper is to start from the SOR methodology and extract the required features, necessary to design a crease pattern, and extended to irregular shapes. This enhanced methodology, must have the necessary requirement to be automatically folded and glued using a robot. However, due to the irregularity in these shapes, multiple stamps are necessary to generate the desired 3D shape.

## 2   Creation of crease patterns using the SOR methodology

The methodology used to create crease patterns in our robot is based on the SOR methodology presented in [Mitani 09]. Proper modification have been done into the procedure to incorporate the gluing areas, and correct some problems with the allocation of the resulting flaps (i.e gluing areas in the 3D shape). In order to perform the folding process using a robot two requirements have to be taken in account:

- The folding pattern is a series of intercallation between valley and mountain folds with gluing segments in between to generate the desired shape after the glue is dried. Thus, the distance between these folds has to be always the same to allow the paper been pasted in the final stage of the folding and gluing process in our robot [Romero et al. 16].

- The gluing areas have to be symmetrical and congruent (i.e. vertically mirror) to

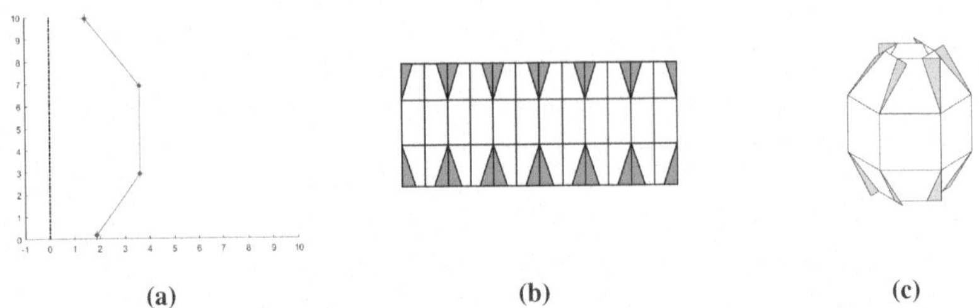

**Figure 1:** *Creation of a 3D shape with the SOR methodology. (a) Poly-line, (b) Crease pattern, (c) 3D shape. The gray lines represent the gluing areas*

fit perfectly when two slides of paper are pasted together.

Each pair of points in the poly-line forms a stage. In the cases where there is only one stage in the poly-line (see Fig.2a), the flaps can rotate freely because there is no physical connection with other flaps (see Fig.2c). However, adding new stages into the poly-line like in Fig.3a, generates connections between flaps, limiting their movement. Furthermore, if the changes in the $y$ axis of the poly-line are not monotonically increasing or decreasing, i.e. $P_{(n-1,y)} \leq P_{(n,y)}(1 \leq n < N)$ or $P_{(n-1,y)} \geq P_{(n,y)}(1 \leq n < N)$, extra considerations have to be included to generate the desired 3D shape.

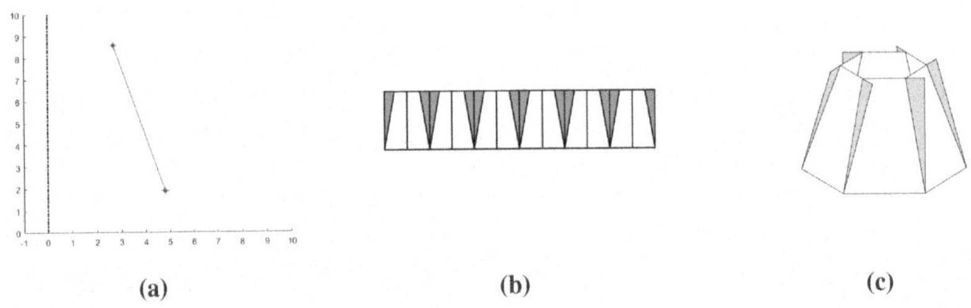

**Figure 2:** *Creation of a 3D shape with one stage. (a) Poly-line, (b) Crease pattern, (c) 3D shape. The gray lines represent the gluing areas*

We propose as solution to this problem to perform cut lines allocated in the crese line in between stages in the gluing areas (see Fig.3b, these lines are colored in red). These cutting lines can be added to allow the free movement in these flaps and avoid crushing between them and the main body or other flaps.

In figure 4a, $P_{(n,x)}$, and $P_{(n,y)}$ denote the $x$ and $y$ coordinates respectively of the vertex $P_n$ in the input poly-line. For each point $P_n$ in the input profile, two points ($P1'_n$ and $P2'_n$) are created in the crease pattern. In Fig. 4a, $P'_{(n,x)}$, and $P'_{(n,y)}$ denote the $x$, and $y$ coordinates respectively of the vertex $P'_n$ in the crease pattern. The new points $P1'_n$ and $P2'_n$ have the same $y$ coordinate but different $x$ coordinates. Using

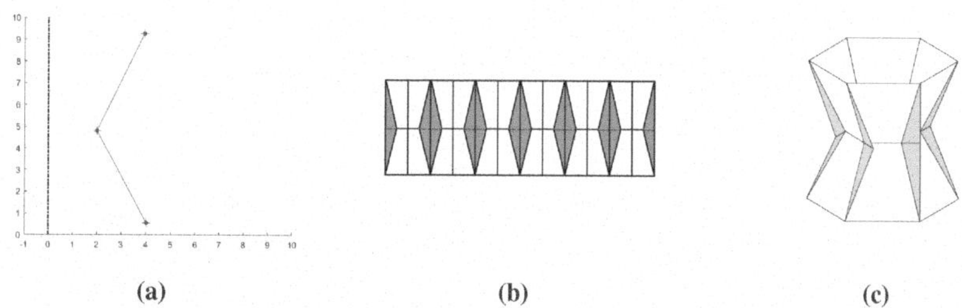

**Figure 3:** *Creation of a 3D shape with two stages. (a) Poly-line, (b) Crease pattern, (c) 3D shape. The gray lines represent the gluing areas. The red lines indicate the location of the cut lines*

the information in Fig.4a, the values of Fig.4b can be calculated as follows:

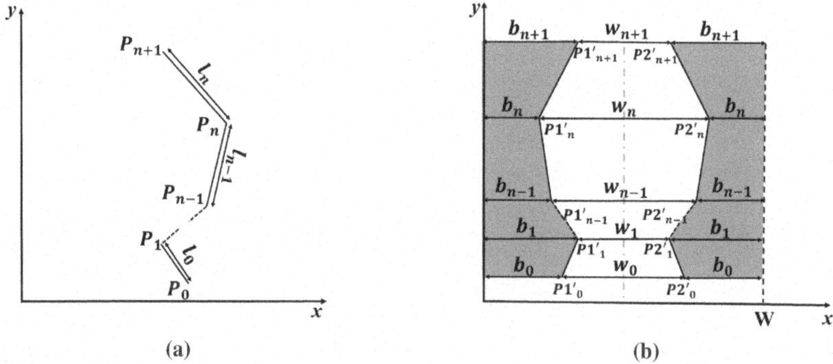

**Figure 4:** *Input profile and crease points defining a sequence of segments of N stages $(n = 0 \cdots N)$. (a) input profile, (b) segment of the crease pattern. The dashed gray line is a middle valley fold. The gray area is the gluing area and the white area is the area in the surface of the 3D object (non-glued area)*

$$w_n = 2P_{(n,x)} \sin \tfrac{\pi}{K}, \quad W = max(w_n). \tag{1}$$

$$b_n = \frac{W - w_n}{2}. \tag{2}$$

$$l_n = ||P_n - P_{(n-1)}||. \tag{3}$$

$$P1'_{(n,x)} = \frac{W - w_n}{2}, \quad P2'_{(n,x)} = \frac{W + w_n}{2}. \tag{4}$$

$$P1'_{(n,y)} = P2'_{(n,y)} = P'_{(n,y)} = \begin{cases} 0, & if(n=0), \\ P'_{(n-1,y)} + \sqrt{l_n^2 - \left(\frac{w_n - w_{n-1}}{2}\right)^2}, & if(n>0). \end{cases} \quad (5)$$

where $w_n$ and $b_n$ represent the width of the non-glued and glued areas respectively of each segment at point $P_n$ of the profile, and $W$ represents the size of each segment in the crease pattern.

## 3    Adaptations of the SOR methodology for irregular shapes

The methodology explained in section 2, can be used to create 3D shapes that are based on surface of revolutions (SOR), able to be built by a robot. Although this methodology expands the applicability of the robot from [Romero et al. 16], there are some shapes that are not based on SOR, because they have irregular shapes and do not posses a clear pivot axis. The resulting crease patterns for these complex shapes are even more complicated to build than the SOR counterparts, requiring to perform difficult folds where multiple manipulations have to be carried out at the same time, in combination with holding techniques that make them extremely difficult to be assembled using a robot.

In the SOR methodology, a single profile is rotated in sections with an angle of $2\pi/K$. These sections generate $N$ equal stages, side by side to each other. The idea with the second methodology proposed in this paper is to create a 2D pattern of shapes that have two or more different profiles. This new methodology can only be applied to figures with star-shaped-projected polyhedrons.[1] An example of a star-shaped polygon is shown in fig. 5.

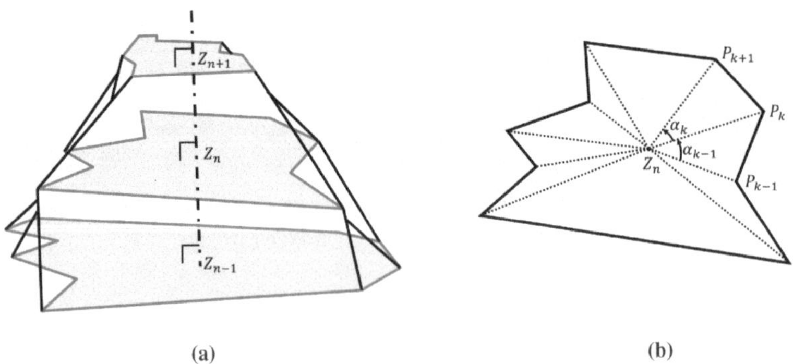


**Figure 5:** *Example of a star-shaped polygon. (a) Star-shaped polyhedron. (b) Star-shaped polygon extracted from the polyhedron.*

---

[1]A star-shaped-projected polyhedron is a structure, that seeing from above, and performing vertical slices, all the slices have a star-shaped polygon form.

A star-shaped-polygon is an irregular 2D polygon that contains a point $Z_n$, from which the entire boundary of the polygon is reachable, i.e. all vertex $P_{0,n}, \cdots,$ $P_{k-1,n}, P_{k,n}, P_{k+1,n}, \cdots, P_{K,n}$, where $K$ represents the total of profiles, and $n$ the current stage [Preparata and Shamos 85].

The art-gallery theorem states that $\frac{K}{3}$ points are necessary to cover an $K$-gon, and in particular, this shows that polygons for which $3 \leq K \leq 5$ are necessarily star-shaped [Aggarwal 84, Avis and Toussaint 81]. There are many ways to calculate the center $Z_n$. One of the most efficient algorithms is the one proposed by Lee and Preparata [Lee and Preparata 79]. However, in the tests performed in this paper, we consider the points $Z_n$ to be previously known.

As previously said, this new methodology translates and performs modification into the SOR methodology explained in section 2. One difference that can be noticed analyzing irregular shapes like the one in Fig.5a, is that they have two or more different profiles, instead of the single profile of the SOR co unterpart. Therefore, a novel procedure has been proposed to extract the required profiles from a 3D model, and create the 2D crease pattern using this information.

## 3.1 Extraction of the poly-lines from a 3D model

The first step in our methodology consists in extraction and arranging all the possible poly-lines in a 3D model. These poly-lines must have the same number of points, and all these points must be at the same height (i.e must have the same $z$). Using the information of the vertices from the 3D model, the heights ($P_{k,n,z} = Z_{n,z}$) and rotation angles $\alpha_k$ can be obtained using the algorithm in Table 1.

In these poly-lines the maximum height $Z_{max}$ and minimum height $Z_{min}$ must be the same for all poly-lines. After all the heights and rotational angles have been extracted from the 3D model, the next step is to extract the poly-lines. To obtain these poly-lines, the intersection between two planes is used. A cutting plane is created for each one of the poly-lines using the information of $Z_{max}$, $Z_{min}$, and $\alpha_k$. Figure 6a shows the location of the first three cutting planes, and Fig.6b shows an approximated poly-lines of the 3D model in Fig.5a if all poly-lines are aligned and represented in a 2D plane. This procedure has to be done for all $\alpha_k$ angles, and all the information of the resulting vertices is used later to create the 2D crease pattern.

## 3.2 Creation of the crease pattern using the poly-lines

After the information of the poly-lines is known the next step is to use this information to create the 2D crease pattern. To do this, we suppose two different poly-lines with three stages as the one in Fig.7a. Both poly-lines have the same number of points and their projections form star-shaped polygons.

For all 3D shapes there are multiple possible crease pattern solutions. This occur due to the irregularity in the trapezoids that compose each stage of the crease pattern. In the SOR methodology, all of the trapezoids on each stage form regular isosceles trapezes. However, with irregular shapes these trapezoids could form non-isosceles polygons which could be divided into two irregular triangles tracing a diagonal line inside the trapezoid. Depending on the allocation of this diagonal

**Table 1:** *Algorithm to obtain the heights and rotation angles from a 3D model*

---

Having a 3D model as Fig. 5a.
For each point $P_m$ in the 3D model:
   if ($Z_{n,z}$ and $\alpha_k$ are empty ):
     $Z_{n,z} = P_{m,z}$;

$$\alpha_k = tan^{-1} \frac{P_{m,y}-Z_{n,y}}{P_{m,x}-Z_{n,x}};$$
     if ($\alpha_k < 0$): $\alpha_k = \alpha_k + 2\pi$;
     if ($\alpha_k \geq 2\pi$): $\alpha_k = \alpha_k - 2\pi$;
     $n = 0$;
     $k = 0$;
  else:
    if ($P_{m,z} \notin Z_{n,z}$):
     $Z_{n,z} = P_{m,z}$;
     $n++$;

$$\beta = tan^{-1} \frac{P_{m,y}-Z_{n,y}}{P_{m,x}-Z_{n,x}};$$
    if ($\beta < 0$): $\beta = \beta + 2\pi$;
    if ($\beta \geq 2\pi$): $\beta = \beta - 2\pi$;
    if ($\beta \notin \alpha_{0 \cdots k-1}$):
     $\alpha_k = \beta$;
     $k++$;

---

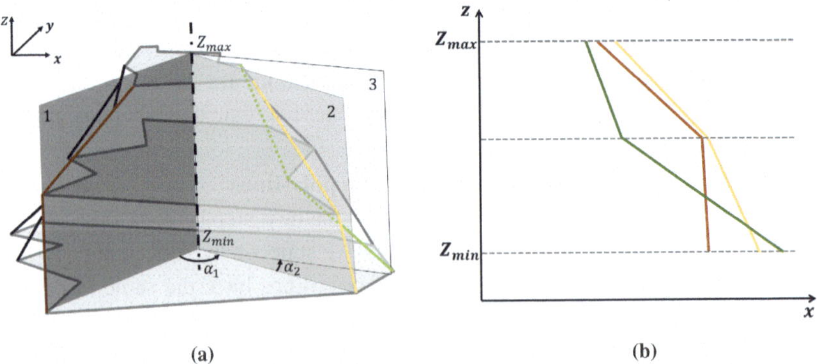


**Figure 6:** *Extraction of the poly-lines from a 3D model. (a) Cutting planes representation. (b) Poly-lines aligned in a 2D plane. Red: First poly-line. Yellow: Second poly-line. Green: Third poly-line.*

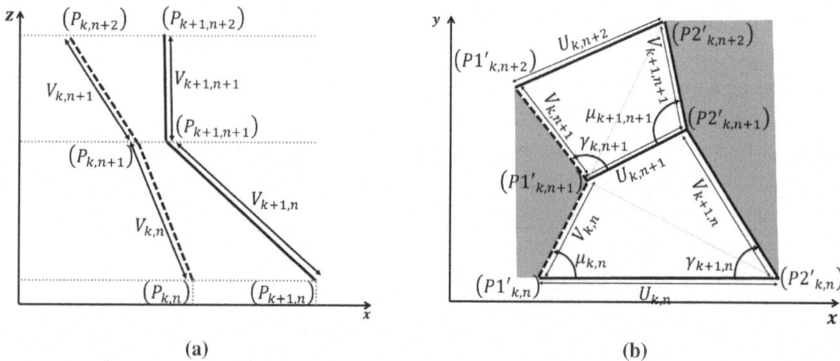



**Figure 7:** *Calculation of the crease pattern of a single sub-segment of a irregular shape. (a) Poly-lines (b) Crease pattern. k represents the poly-lines, and n represents the stages. In this case the diagonal of the first stage starts even, and then starts odd in the second stage*

line (gray dashed lines in Fig.7b) one of the two possible calculations angles (i.e $\gamma_{k,n}$ or $\mu_{k,n}$) is used to determinate the position of the points in the crease pattern. The value of these angles can be obtained using the following equations:

$$\mu_{k,n} = cos^{-1}\left(\frac{\mathbf{V_{k,n}} \cdot \mathbf{U_{k,n}}}{||\mathbf{V_{k,n}}|| \cdot ||\mathbf{U_{k,n}}||}\right). \tag{6}$$

$$\gamma_{k,n} = cos^{-1}\left(\frac{\mathbf{V_{k,n}} \cdot \mathbf{D_{k,n}}}{||\mathbf{V_{k,n}}|| \cdot ||\mathbf{D_{k,n}}||}\right) + cos^{-1}\left(\frac{\mathbf{D_{k,n}} \cdot \mathbf{U_{k,n}}}{||\mathbf{D_{k,n}}|| \cdot ||\mathbf{U_{k,n}}||}\right). \tag{7}$$

where $\mathbf{V_{k,n}}$ is the longitudinal vector, $\mathbf{U_{k,n}}$ is the width vector, and $\mathbf{D_{k,n}}$ is the diagonal vector, that can be calculated as follows:

$$\begin{aligned} \mathbf{V_{k,n}} &= \left[v_{k,n,x}, v_{k,n,y}, v_{k,n,z}\right] \\ v_{k,n,x} &= P_{k,n+1,x}cos(\alpha_k) - P_{k,n,x}cos(\alpha_k), \\ v_{k,n,y} &= P_{k,n+1,x}sin(\alpha_k) - P_{k,n,x}sin(\alpha_k), \\ v_{k,n,z} &= P_{k,n+1,y} - P_{k,n,y}. \end{aligned} \tag{8}$$

$$\begin{aligned} \mathbf{U_{k,n}} &= \left[u_{k,n,x}, u_{k,n,y}, u_{k,n,z}\right] \\ u_{k,n,x} &= P_{k+1,n,x}cos(\alpha_k) - P_{k,n,x}cos(\alpha_k), \\ u_{k,n,y} &= P_{k+1,n,x}sin(\alpha_k) - P_{k,n,x}sin(\alpha_k), \\ u_{k,n,z} &= P_{k+1,n,y} - P_{k,n,y}. \end{aligned} \tag{9}$$

$$\begin{aligned} \mathbf{D_{k,n}} &= \left[d_{k,n,x}, d_{k,n,y}, d_{k,n,z}\right] \\ d_{k,n,x} &= P_{k+1,n+1,x}cos(\alpha_k) - P_{k,n,x}cos(\alpha_k), \\ d_{k,n,y} &= P_{k+1,n+1,x}sin(\alpha_k) - P_{k,n,x}sin(\alpha_k), \\ d_{k,n,z} &= P_{k+1,n+1,y} - P_{k,n,y}. \end{aligned} \tag{10}$$

To maintain the symmetry in the gluing area, required by the robot to perform the gluing process correctly, the angles $\mu$ and $\gamma$ have to be intercallated on each stage and stages as can be observed in Fig.7b. This also means that the number of stages $N$ must be a even number to keep this symmetry to be able to folded with the robot. Also, it has to be guarantied that $\mu + \gamma < \pi$ or else the vertical vectors will cross each other. To prevent this is recommended to be sure that the differences in $x$ between adjacent poly-lines is not very extreme.

Finally, using the values and vectors calculated in equations (6) - (10), the points that represent the crease pattern can be calculated as follows:

$$P1'_{k,n,x} = \begin{cases} ||V_{k,n}||\cos(\mu_{k,n}) + d1_k + d2_k, & \textit{if the diagonal starts even,} \\ ||V_{k,n}||\cos(\gamma_{k,n}) + d1_k + d2_k, & \textit{if the diagonal starts odd.} \end{cases} \tag{11}$$

$$P2'_{k,n,x} = \begin{cases} ||V_{k,n}||\cos(\gamma_{k,n}) + d1_k + d2_k, & \textit{if the diagonal starts even,} \\ ||V_{k,n}||\cos(\mu_{k,n}) + d1_k + d2_k, & \textit{if the diagonal starts odd.} \end{cases} \tag{12}$$

$$P1'_{k,n,y} = \begin{cases} ||V_{k,n}||\sin(\mu_{k,n}), & \textit{if the diagonal starts even,} \\ ||V_{k,n}||\sin(\gamma_{k,n}), & \textit{if the diagonal starts odd.} \end{cases} \tag{13}$$

$$P2'_{k,n,y} = \begin{cases} ||V_{k,n}||\sin(\gamma_{k,n}), & \textit{if the diagonal starts even,} \\ ||V_{k,n}||\sin(\mu_{k,n}), & \textit{if the diagonal starts odd.} \end{cases} \tag{14}$$

$$d1_k = |min(P1'_{(k,x)})| \tag{15}$$

$$d2_k = \begin{cases} 0, & if(k=0) \\ d2_{k-1} + max(P2'_{k-1,x}) - min(P1'_{(k-1,x)}), & if(k>0). \end{cases} \tag{16}$$

## 4   Results

This section shows the results of two experiments in order to test the proposed methodologies. Two shapes were created using each one of the two methodologies, the SOR and the enhanced version explained in the previous section. The first two shapes were folded and glued using the robot from [Romero et al. 16]. The other two shapes were built by hand. However, the crease patterns of the shapes made by the enhanced version meet the requirements to be folded by the robot except the equality in the distances (i.e simmetry and congruency). These shapes can be observed in Fig.8 - 12.

The shapes made using the SOR methodology in Fig.8 and 9 are an approximation of a paper dome and a hexagonal box. The flaps were allocated outside the main structure in both cases. This means that no cut lines were required because the poly-lines are monotonically increasing. However, if the flaps are desired to be allocated inside the main structure the cut lines are required for the dome case.

For the examples made by using the enhanced version of the SOR methodology exposed in figures 11 and 12, the poly-lines are the same for both cases, however, in the first case, the rotation angle $\alpha_k$ is constant (see Fig.10b), and in the second case

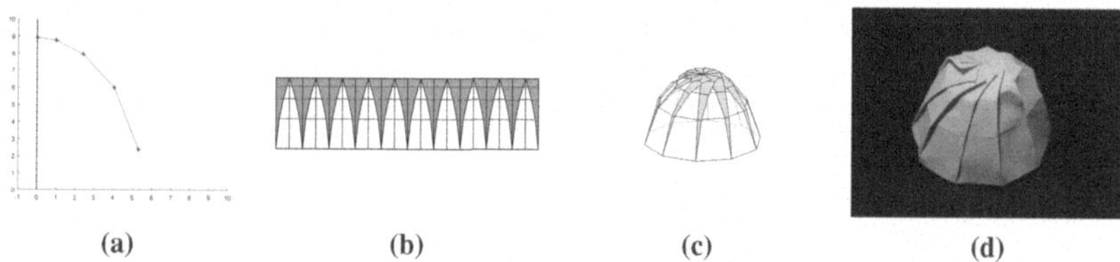

**Figure 8:** *Example of a paper dome build with the SOR methodology. (a) Poly-line. (b) Crease Pattern. (c) 3D model approximation. (d) Paper shape. N = 5, K = 10*

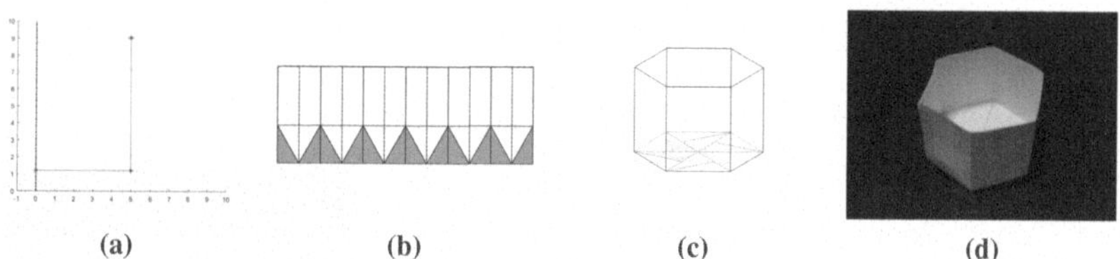

**Figure 9:** *Example of a hexagonal box build with the SOR methodology. (a) Poly-line. (b) Crease Pattern. (c) 3D model approximation. (d) Paper shape. N = 3, K = 6*

these angles are different (see Fig.10c). It can be observed that the resulting flaps have different shapes and sizes. Thus make necessary a gluing stamp per poly-line, and it is recommended to always perform the cutting lines to allow movement in the flaps and be able to complete and open the 3D shapes.

It can be observed from the results from Fig.11c and 12c, that changing only the angle between poly-lines the final shape change, however, using the proposed methodology the pattern can be created in both cases. In both examples the final shape was closed (i.e join the first and final gluing areas together) and opened by hand. This procedure still very complicated to be performed automatically due to the differences in shapes that can be created with these two methodologies, and the dexterous manipulations involved in this process. On the other hand, the robot can fold and glue a shape in 225 seconds, while a person could take more than 10 minutes to perform these procesures.

# 5   Conclusions

In this paper, two methodologies to create 3D shapes that can be automatically folded using the robot system from [Romero et al. 16] were proposed. The first methodology is based in the SOR methodology similar to the one in [Mitani 09], with some modifications to allow free movement in the resulting flaps, and be able to implement the designs in the robot.

The use of simple folding patterns in combination of gluing areas was proposed

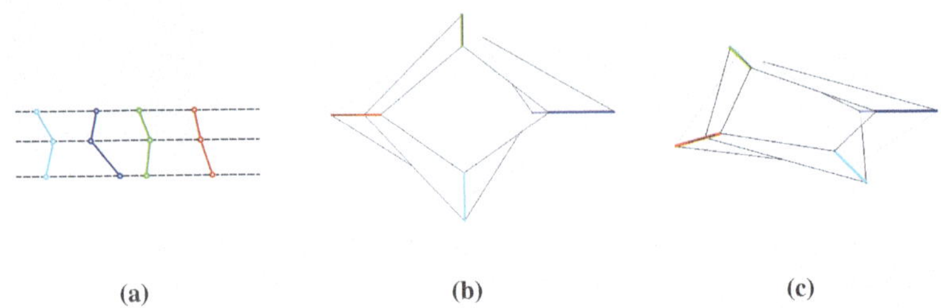

(a)        (b)        (c)

**Figure 10:** *Poly-lines of the polygon corners and top views. (a) Poly-lines of the corners. (b) Top view of the polygon with $\alpha_k$ constant. (c) Top view of the polygon with $\alpha_k$ variable.*

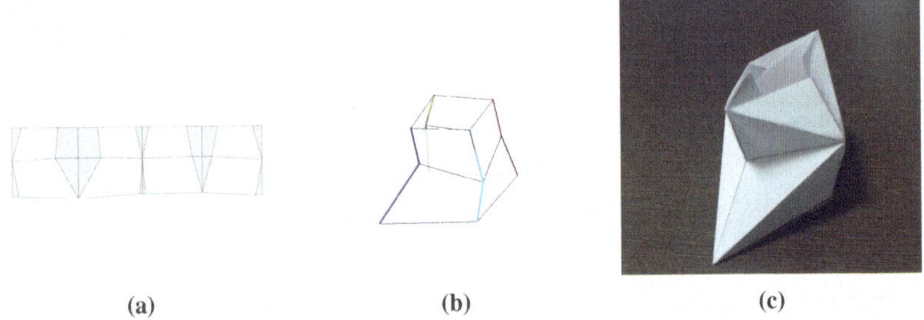

(a)        (b)        (c)

**Figure 11:** *Example of a 3D shape build with the enhanced SOR methodology. (a) Poly-lines. (b) Crease Pattern. (c) 3D model approximation. (d) Paper shape. $N = 3$, $K = 4$, $\alpha_k$ is the same in all segments*

in [Romero et al. 16] as a solution to automatic paper-folding. A series of requirements (i.e symmetrical gluing areas, intercalating between mountain and valley folds ) were explained to be able to perform the automatic folding.

Based in the SOR methodology, and taking in account the requirements to automatic folding, an enhanced version of the SOR methodology was also proposed in this paper. Using this methodology, 2D crease patterns from irregular shapes can be designed and folded by hand. These designs contain the requirements to be automatically folded using a modified version of the robot proposed in [Romero et al. 16]. This new version of the robot must have the possibility to use several stamps that can be designed using the methodology proposed in this work.

## 6  Future Works

The gluing areas from the enhanced version of the SOR methodology could have different shapes and sizes. An optimization procedure has to be done to calculate the optimal inclination of the trapezoids in order to reduce the differences between

1324

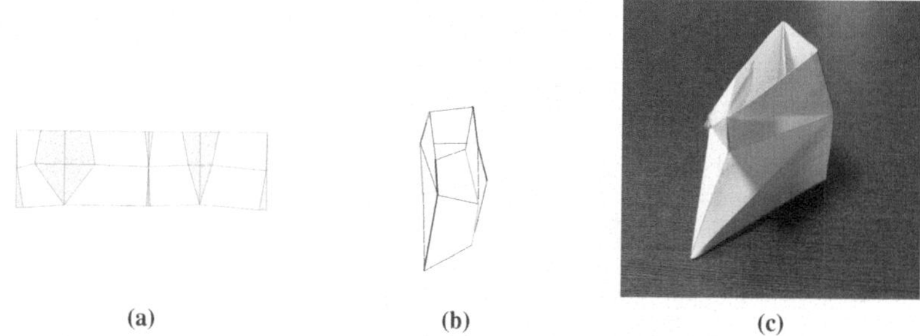

**Figure 12:** *Example of a 3D shape build with the enhanced SOR methodology. (a) Crease Pattern. (b) 3D model approximation. (c) Paper shape.* $N = 3$, $K = 4$, $\alpha_k = \pi/4, 3\pi/4, \pi/3, 2\pi/3$.

all gluing areas.

A new version of the robot proposed in [**Romero et al. 16**] has to be designed to be able to use multiple stamps, and be able to automatically fold irregular shapes, designed using the methodology explained in this work.

The methodology of multiple poly-lines parts from the premise of having a star-shaped projected polyhedron. However, a vast number of shapes with irregular shapes usually do not have this type of form. As a solution to this problem, we propose to search for the skeleton of of the irregular shape [**Chuang et al. 00**], and use the resulting bones as independent pivot axis. The resulting structure will be a group of crease patterns that have to be assembled forming the final shape.

# References

[Aggarwal 84] Alok Aggarwal. "The art gallery theorem: its variations, applications and algorithmic aspects."

[Avis and Toussaint 81] David Avis and Godfried T Toussaint. "An efficient algorithm for decomposing a polygon into star-shaped polygons." *Pattern Recognition* 13:6 (1981), 395–398.

[Balkcom and Mason 08] Devin J. Balkcom and Matthew T. Mason. "Robotic origami folding." *The International Journal of Robotics Research* 27:5 (2008), 613–627. doi:10.1177/0278364908090235.

[Chuang et al. 00] Jen-Hui Chuang, Chi-Hao Tsai, and Min-Chi Ko. "Skeletonisation of three-dimensional object using generalized potential field." *IEEE Transactions on Pattern Analysis and Machine Intelligence* 22:11 (2000), 1241–1251.

[Dai and Cannella 08] Jian S Dai and Ferdinando Cannella. "Stiffness characteristics of carton folds for packaging." *Journal of mechanical design* 130:2 (2008), 022305.

[Elbrechter et al. 12] C. Elbrechter, R. Haschke, and H. Ritter. "Folding paper with anthropomorphic robot hands using real-time physics-based modeling." In *2012 12th IEEE-RAS International Conference on Humanoid Robots (Humanoids 2012)*, pp. 210–215, 2012. doi:10.1109/HUMANOIDS.2012.6651522.

[Felton et al. 14] S. Felton, M. Tolley, E. Demaine, D. Rus, and R. Wood. "A method for building self-folding machines." *Science* 345:6197 (2014), 644–646. doi:10.1126/science.1252610.

[Lang 17] Robert J Lang. "TreeMaker V5.0." `http://www.langorigami.com/article/treemaker`, 2005 (accessed January 9, 2017).

[Lee and Preparata 79] Der-Tsai Lee and Franco P Preparata. "An optimal algorithm for finding the kernel of a polygon." *Journal of the ACM (JACM)* 26:3 (1979), 415–421.

[Mitani 09] Jun Mitani. "A Design Method for 3D Origami Based on Rotational Sweep." *Computer-Aided Design and Applications* 6:1 (2009), 69–79. doi:10.3722/cadaps.2009.69-79.

[Mitani 17] Jun Mitani. "ORI-REVO v1.01." `http://mitani.cs.tsukuba.ac.jp/ori_revo/#install`, 2011 (accessed January 2, 2017).

[Miyamoto et al. 17] Emi Miyamoto, Yuki Endo, Yoshihiro Kanamori, and Jun Mitani. "Semi-Automatic Conversion of 3D Shape into Flat-Foldable Polygonal Model." In *Computer Graphics Forum*, 36, 36, pp. 41–50. Wiley Online Library, 2017.

[Nakazawa 98] Masaru Nakazawa. "Special Issue on Handling of Flexible Object." *Journal of Robotics and Mechatronics* 10:03 (1998), 167–169. doi:10.20965/jrm.1998.p0167.

[Namiki et al. 03] Akio Namiki, Koichi Hashimoto, and Masatoshi Ishikawa. "A Hierarchical Control Architecture for High-Speed Visual Servoing." *The International Journal of Robotics Research* 22:10-11 (2003), 873–888. doi:10.1177/0278364490302210006.

[Preparata and Shamos 85] Franco P Preparata and Michael Ian Shamos. "Introduction." In *Computational Geometry*, pp. 1–35. Springer, 1985.

[Romero et al. 16] Julian A. Romero, Luis A. Diago, Chie Nara, Junichi Shinoda, and Ichiro Hagiwara. "Norigami Folding Machines for Complex 3D Shapes." In *ASME. International Design Engineering Technical Conferences and Computers and Information in Engineering Conference*, 5B: 40th Mechanisms and Robotics Conference, 5B: 40th Mechanisms and Robotics Conference, 2016. doi:10.1115/DETC2016-60580.

[Romero et al. 17] JA Romero, LA Diago, and I Hagiwara. "Norigami Crease Pattern Model Design Based on Surfaces of Revolution." In *ASME 2017 International Design Engineering Technical Conferences and Computers and Information in Engineering Conference*, pp. V05BT08A047–V05BT08A047. American Society of Mechanical Engineers, 2017.

[Tachi 09] Tomohiro Tachi. "Simulation of rigid origami." *Origami* 4 (2009), 175–187.

Julian Andres Romero

Meiji University, Graduate School of Advanced Mathematical Sciences, 4-21-1 Nakano, Nakano-ku, Tokyo, Japan 164-8525, e-mail: jaromero_18@hotmail.com

Luis Ariel Diago

Interlocus Inc., Tokyo Tech YVP W305, 4259-3 Nagatsuta-Cho, Midori-ku, Yokohama City 226-8510, e-mail: ldiago@i-locus.com

Chie Nara

Meiji University, Graduate School of Advanced Mathematical Sciences, 4-21-1 Nakano, Nakano-ku, Tokyo, Japan 164-8525, e-mail: cnara@jeans.ocn.ne.jp

Junichi Shinoda

Interlocus Inc., Tokyo Tech YVP W305, 4259-3 Nagatsuta-Cho, Midori-ku, Yokohama City 226-8510, e-mail: jshinoda@i-locus.com

Ichiro Hagiwara

Meiji University, Graduate School of Advanced Mathematical Sciences, 4-21-1 Nakano, Nakano-ku, Tokyo, Japan 164-8525, e-mail: ihagi@meiji.ac.jp

# Folding Fabrication of Curved-Crease Origami Spindle Beams

*W.Q. Cui, T. Gfeller, D. Fernando, M.T. Heitzmann, J.M. Gattas*

**Abstract**: *Curved-crease origami have been utilised as inspiration for numerous engineering applications, however the accuracy of employed manufacturing methods is largely unknown. This paper presents manufacturing methods for two structural sheet materials suitable for use in curved-crease thin-walled structures: typical structural steel and a novel hybrid fibre-reinforced polymer timber material. For both materials, a range of curved-crease 'spindle' beam geometries are manufactured and verified through 3D scanning and surface error analysis. The manufacturing methods are shown to be reliably accurate to within ±50% of plate thickness, which is in conformance with requirements for acceptable surface imperfections in thin-walled structures.*

## 1 Introduction

Curved-crease origami is characterised by curved fold lines and component panels which possess a non-zero principal curvature once folded. The striking aesthetic and potential mechanical advantages of such geometries has led to their adaption for many engineering and architectural purposes. Applications include deployable structures [Tachi 13], thin-walled structures [Buri et al. 11, Hansen et al. 16], sandwich panels [Miura 72, Gattas and You 14, Klett et al. 17], crash boxes [Garrett et al. 16], propellers [Miyashita et al. 15], pipelines [Guo et al. 16], and static and adaptive facades [Vergauwen et al. 13, Epps and Verma 13].

The manufacturing methods utilised for these applications are correspondingly diverse and include connection of discrete panel components [Buri et al. 11], robotic folding about predefined crease lines [Epps and Verma 13], and deformation of an uncreased sheet using press dies or vacuum thermoforming [Gattas and You 14, Zhao et al. 15]. The accuracy of these manufacturing methods is largely unknown, despite the fact that the performance of thin-walled elements is highly sensitive to surface imperfections for certain applications. Quantification of manufacturing error is not straightforward. Curved-crease origami possesses component panels that deform during folding, yet consideration of material deformation behaviours during manufacture is often excluded from the geometric design process. The few studies that have measured as-built curved-crease origami suggest that the correspondence between designed and manufactured geometry is approximate only, with discrepancies arising from spring-back or inaccurate design assumptions about surface non-zero principal curvature [Lee and Gattas 16, Vergauwen et al. 17].

In this paper, we aim to assess the accuracy of two folding fabrication methods suitable for production of curved-crease origami thin-walled structures. Section 2 first presents the origami geometry of interest, a closed and bi-symmetric 'spindle' beam generated using the elastica surface extrusion method. Section 3 then presents the two manufacturing methods of interest: one for use with a typical steel sheet material and the other for use with a novel hybrid FRP-timber sheet material. Section 4 conducts a surface error analysis with 3D scanning of manufactured prototypes, followed by a discussion of the manufacturing efficacy.

## 2  Geometry

The specific geometry considered in this paper is a closed and bi-symmetric curved-crease origami form as shown in Figure 1. The form tapers from a large depth at mid-span to a small depth at each end and is so termed a spindle. The spindle possesses a non-uniform structural capacity akin to the non-uniform bending moment distribution of a simply-supported beam under point or uniform distributed loading conditions.   This kind of profile is thus well-suited for application as a thin-walled structural beam, with a geometric similarity to beams with a cross-sectional profile optimised for material efficiency under bending loads [Mayencourt and Mueller 17, Samyn 04].

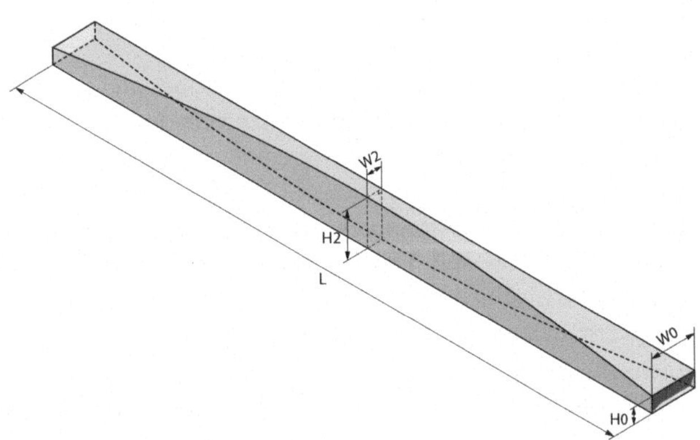


**Figure 1:** *Form and parameters of the curved-crease origami spindle beam.*

Figure 1 shows a parametric definition of the beam geometry. Five parameters are sufficient to determine a unique beam configuration: beam length $L$, end height $H0$, end width $W0$, mid-span height $H2$, and mid-span width $W2$. These parameters are sufficient to determine the volumetric boundary but do not determine the non-zero principal curvature of the curved surface.

Several methods exist to characterise surface curvature, with several assuming an approximate, inexact curvature for the sake of computational efficiency,

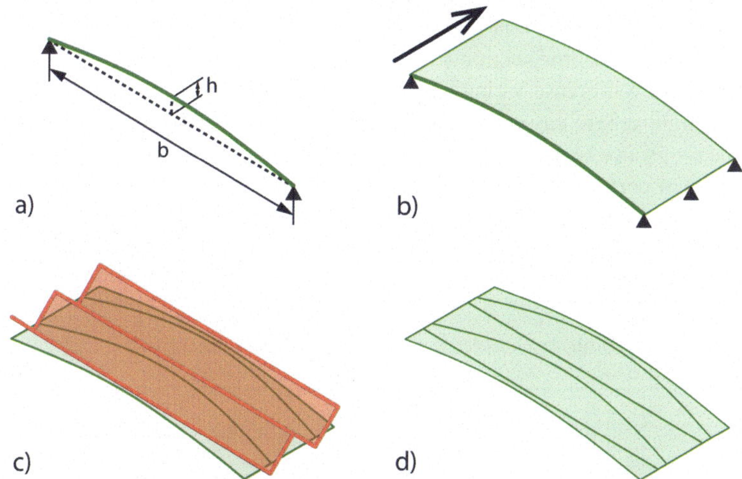

**Figure 2:** *Geometry Modelling Process. a)Elastica Curve Generation. b) Linear Extrusion. c) Sequential Mirror Truncation Planes. d) 3D Curved-Crease Pattern.*

for example, elliptic [**Gattas and You 14**] or sinusoidal [**Edison 11**] profiles. This paper shall adopt the recently-developed elastica surface generation method [**Lee et al. 18**]. This method adopts the 'elastica' as the non-zero principal curvature of a 3D curved-crease origami surface. The elastica is a minimum elastic bending curve for a slender 1D rod and adopting it gives a concise and exact formulation for curved-crease surfaces which are folded such that only elastic deformation occurs within component panels.

An elastica surface profile can be determined from beam parameters using the geometric modelling process shown in Figure 2. A 1D rod of some length greater than $L$ is deformed and pinned at each end. After deformation, support distance is $b = L$ and mid-span beam height is $h = (H2 - H0)/2$, as shown in Figure 2a. Parameters $b$ and $h$ are elastica parameters and are sufficient to determine a unique elastica profile. The profile can be extruded to a surface and divided with mirror reflection planes with relative inclinations corresponding to the spindle edge angle $90°$ and offset with alternating distances $W0$ and $H0$. The final divided surface, shown in Figure 2d, can be reflected about mirror planes to form the target spindle shape and can be unrolled to 2D crease pattern to facilitate manufacture.

## 3   Manufacture

An experimental study was undertaken to assess the accuracy of curved-crease spindle geometries manufactured from folded structural materials.   Eight geometric cases were selected for investigation, with parameters summarised in Table 1.  These cases were designed on the basis of 400 mm fixed perimeter, which gave an approximately uniform weight for all specimens. Cases 6 and 7 were designed as 'straight-crease' cases, that is they are a traditional rectangular hollow section (RHS) and square hollow section (SHS), respectively. Remaining cases were then designed as curved-crease variants by changing either mid-span or end cross section of straight-crease cases. With reference to Fig. 3, cases 0, 3, 6 and 2, 5, 7 are two case groups with a common mid-span cross-section and were intended for comparison of performance through a curved-to-straight profile transition. Cases 0, 1, 2, and 3, 4, 5 then form another two case groups with a fixed end cross-section and with a steep-to-shallow panel curvature transition, that is they have decreasing $h/b$ elastica parameters.

**Table 1:** *Dimension of spindle and RHS beam cases*

| Case | Dimensions [mm] | | | | Elastica |
| # | $W0$ | $H0$ | $W2$ | $H2$ | Parameter $h/b$ |
| --- | --- | --- | --- | --- | --- |
| 0 | 150 | 50 | 50 | 150 | 0.0238 |
| 1 | 150 | 50 | 75 | 125 | 0.0179 |
| 2 | 150 | 50 | 100 | 100 | 0.0120 |
| 3 | 120 | 80 | 50 | 150 | 0.0167 |
| 4 | 120 | 80 | 60 | 140 | 0.0143 |
| 5 | 120 | 80 | 100 | 100 | 0.0050 |
| 6 | 50 | 150 | 50 | 150 | 0 |
| 7 | 100 | 100 | 100 | 100 | 0 |

## 3.1 Manufacture of FRP-timber Spindle Beams

Manufacturing methods were explored for two disparate structural sheet materials: typical structural steel and a novel hybrid fibre-reinforced polymer timber (FRP-timber) material. All cases were manufactured for FRP-timber beams, while only cases 0, 3 and 6 were manufactured for steel beams. Example final manufactured samples for case 3 are shown in Figure 4.

Hybrid FRP-timber spindles were folded using a cured-in-place manufacturing process which utilises a differential curing time between two types of resins impregnated in a glass-fibre sheet. A 24-hour slow-cure resin is used on crease lines and a 3-hour fast-cure resin is used on panel regions. The 2D sheet is folded into a 3D section between the fast-cure and slow-cure periods, when panel regions are rigid while hinge regions remain flexible.

The fabrication procedure is shown in Figure 5 and is described as follows. A 295 gsm bidirectional glass fibre sheet was first marked with crease lines and timber panels were manufactured on a CNC router from 6.5 mm thick plywood. Plywood parts allow precise folding on the target crease line, possess an edge chamfer to control the maximum fold angle, and when used with FRP, create a hybrid material with excellent structural characteristics [Gattas et al. 18].

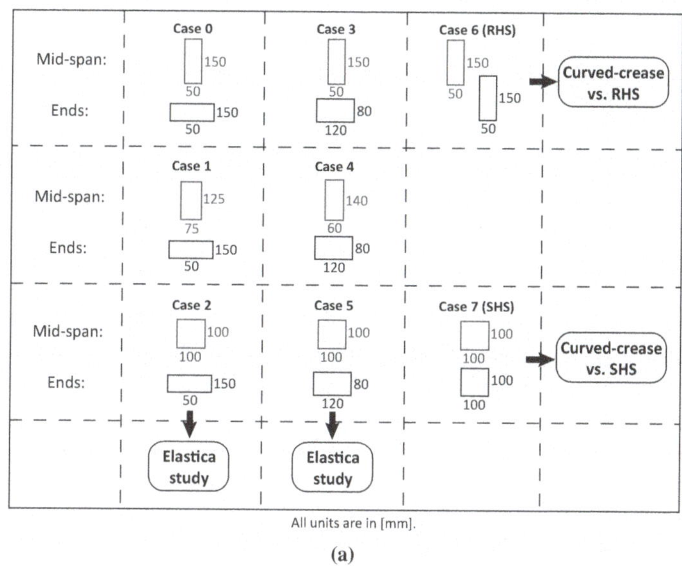


(a)

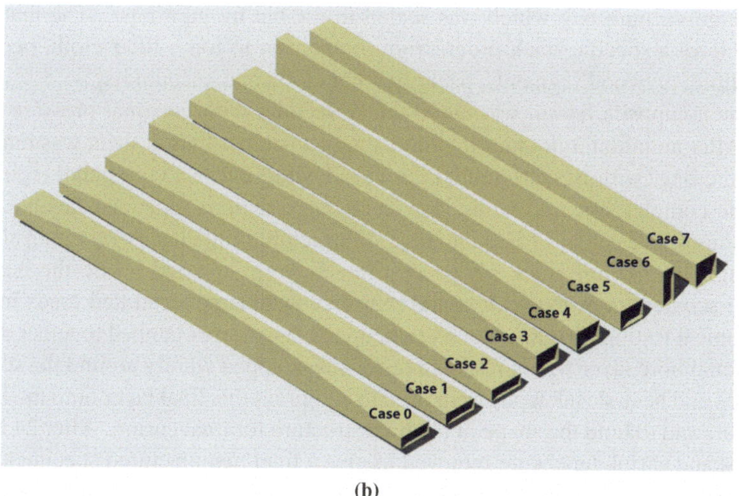


(b)

**Figure 3:** *Manufactured cases. a) Cross-sectional design. b) Illustration of all cases.*

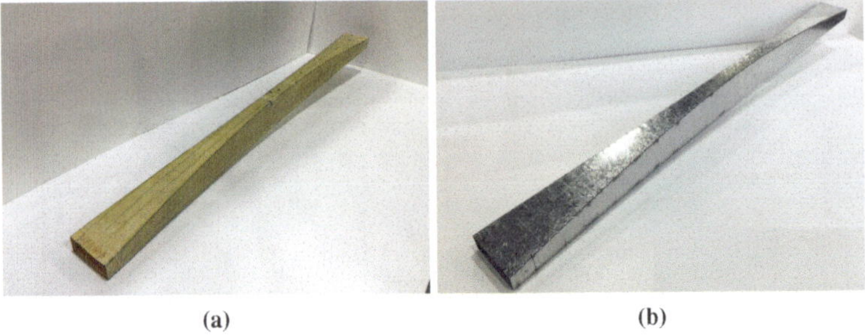

(a)         (b)

**Figure 4:** *Final folded specimens. a) Steel specimen. b) FRP-timber specimen.*

The glass fibre sheet was wet laminated with a Gurit AMPREG 22 epoxy, with fast hardener (3-hour demould time) applied on panel regions and slow hardener (25-hour demould time) applied on hinge regions. The volume of resin usage was controlled for each manufacture, with 189g slow-cure resin (150g epoxy resin plus 39g slow cure hardener) and 320g fast cure resin (250g epoxy resin plus 70g fast cure hardener) used for all specimens. Some auxiliary materials were used to help with the lamination process. A thixotropic silica powder was added to the slow cure resin to limit resin bleed, filter cloth absorbed extra resin, and peel ply enabled easy separation of layers after curing. All materials for wet lamination were sealed in a large vacuum bag which was sealed to a table by gum tape. The materials inside have a specific stack order, from the bottom to top – filter cloth, peel ply, glass fibre, plywood segments, peel ply, filter cloth, and vacuum bag.

The composite lay-up was cured under vacuum with maximal pressure of 90 kPa. After an initial 3-hour curing time, the FRP-timber hybrid sheet was removed from the bag, with the differential curing allowing the fast-cure panel regions to then be completely cured while slow-cure hinge regions remained flexible. The 2D composite sheet was folded up to a 3D structure immediately on removal from the bag. Four 9 mm-thick plywood stiffeners were epoxied inside the beam to ensure squareness of the section and reinforce against concentrated loads in later mechanical testing. To close the section, open edges were clamped together and an additional fibre layer with a fast cure resin was wrapped wholly around the section. Clamps and heat-shrink tapes were used to compress this final layer onto the folded laminate and to hold the shape of the final structure for final curing. After 24 hours, clamps and shrink tape were removed to give a final manufactured specimen. Sixteen beams were manufactured in total, two per case, with an example completed specimen shown in Figure 4a.

## 3.2 Manufacture of Steel Spindle Beams

Steel spindle beams are made with a computer numerical control (CNC) manufacturing technique developed by Industrial Origami [Durney and Pendley 05], in which curved slits are CNC-cut into the sheet metal along a notional geometric fold

**Figure 5:** *FRP-timber spindle manufacture process. a) Application of slow-cure resin to hinge regions. b) Application of fast-cure resin to panel regions. c) 3-hour curing in vacuum. d) Application of slow-cure resin. e) Folding and wrapping with clamps and shrink tapes. f) 24-hour curing.*

line. Adjacent panels remain connected with 'straps' of material, which can plastically deform at a low load so as to allow the steel spindle to be folded manually.

The modelling of slits was achieved with the Digital Fabrication Toolbox [Gattas 14], a Rhino/Grasshopper parametric CAD tool which includes a parameterisation algorithm for design of connection details and a superposition algorithm for placement of connections along target 1D or 2D crease lines [Gattas and You 16]. Two parametric connection details were used. A slits-and-straps detail was used along internal fold lines and followed the parametrisation developed in [Shi et al. 17], with $a_s = 25$ mm, $a_r = 5$ mm and $a_\theta = 90°$. A lapped rivet connection was placed along the edges of the sheet, which when folded allowed the spindle to be closed along the centreline of one panel. The connection was simply a 10 mm edge flap which overlapped in the folded configuration and then rivet holes which were placed at a spacing of approximately 190 mm. The generated crease pattern with integral connection details is shown in Figure 6a.

Patterns were manufactured on a CNC waterjet cutting machine from a 1 mm thick galvanised steel sheet, with a manufacturing time of approximately 30 minutes per pattern. The straps-and-slits connection allowed manual folding by two people in approximately 30 minutes, with only a pair of pliers and a rivet gun. The folding process is illustrated in Figure 6b-d and required all creases to be folded gradually and concurrently until the final position is reached, at which point the lapped edge was riveted together to close and fix the section. Four internal stiffener plates were again inserted inside the beam during folding, to ensure squareness of the section and reinforce against concentrated loads in later mechanical testing. These were also CNC-cut from 1 mm thick steel and were riveted to the four walls of the spindle beam. Six beams were manufactured in total, two each of case 0, 3 and 6, with an example completed specimen shown in Figure 4b.

## 4 Imperfection Analysis and Discussion

### 4.1 Method

Manufacturing accuracy can be assessed in terms of surface imperfections, that is the difference between what is designed and what is manufactured. It is of primary importance in determining whether a manufacturing process is feasible or not and in the case of thin-walled structures, it is also of primary importance due to the high influence imperfections have on the static response of thin-walled structures [Godoy 96].

The surface imperfections of manufactured spindle prototypes was assessed using a 3D scanning method. A FARO Edge ScanArm was used to digitally capture the shape of each spindle, with the data captured as a 3D point cloud with a measurement accuracy of 0.041mm maximum volumetric deviation. The point cloud was converted into a 3D surface mesh using the commercial software Geomagic Studio and then exported to a Rhino/Grasshopper CAD environment for analysis. A previously developed optimisation routine was applied to locate the design surface relative to the scanned surface such that a minimum of surface error was achieved [Lee et al. 18]. The measure of surface error is here taken as the average

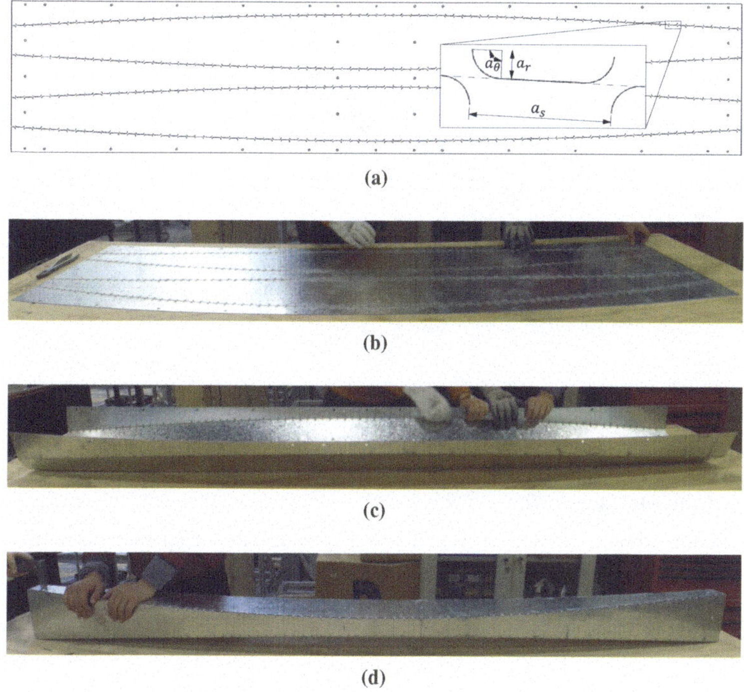

**Figure 6:** *Steel spindle manufacturing process. a) CNC-input pattern. b) CNC-cut unfolded steel sheet. c) Partially-folded sheet. d) Rivets used to close the section.*

absolute surface error, measured as the average of the absolute distance between a set of approximately 40,000 sampling points on the design surface and the corresponding closest points on the scanned surface. Results are displayed in Figure 7 as coloured lines between the coupled sampling points on designed and scanned surfaces, with a color legend set for a range of -50% to 50% of plate thickness $t$. Analysis data for FRP-timber and steel specimens is summarised in Table 2 and Table 3, respectively.

## 4.2   FRP-Timber Spindle Beams

A first analysis was undertaken to determine the efficacy of the differential curing folding fabrication method, based on the consistency of error between manufactured samples of the same type and on the surface error of all manufactured samples. Figure 8a shows the average absolute error for all specimens and it can been seen that samples of the same type have similar levels of error, with largest difference being only 0.4 mm for  case 7.  For all cases, the average absolute surface error is well below a threshold of 50% plate thickness, or 3.6 mm for the case of FRP-timber. This is the permissible maximum centreline deviation of a thin-walled section and imperfection levels below this threshold will only manifest as typical

(a)         (b)         (c)

**Figure 7:** *Surface error measurement process. a) Designed surface. b) Best-fit scanned surface. c) Surface error result.*

structural buckling behaviours, that is the manufactured specimens are sufficiently accurate for structural design. The error obtained from averaging all cases is very low at only approximately 1.58 mm, which indicates high manufacturing accuracy of FRP-timber beams.

A second analysis was undertaken to assess whether the surface curvatures employed for different spindle cases had any correlation with measured surface error. Figure 8b plots the mean surface errors for each case against elastica parameter $h/b$. A low elastica parameter indicates low or no surface curvature, for example the RHS cases with $h/b = 0$. A higher elastica parameter indicates a higher surface curvature. There is no obvious trend of surface error variation with the increase of $h/b$, with the variation between types is small compared with the baseline level of error seen across all samples.

Three conclusions can be drawn from these results. First, the FRP-timber folding fabrication process is effective, with the manufacturing accuracy consistently high across a range of curved-crease spindle configurations. Second, the elastica parameter has no obvious effect on manufacturing accuracy, with relatively small variation seen with different spindle types and certain types being more accurate than comparable RHS beams with no surface curvature. Third, the elastica surface generation method can be concluded to be an effective surface design method, capable of specifying the actual shape of an elastically-deformed curved-crease surface.

## 4.3   Steel Spindle Beams

Similar to FRP-timber specimens, a first analysis for steel specimens was undertaken to observe the surface error of all manufactured samples and investigate the consistency of error between samples of the same type. According to figure 9a, the overall average absolute error for all steel cases is about 0.56 mm, slightly exceeding 0.5 mm, the threshold of 50% steel thickness. However, the manufacturing accuracy can still be concluded acceptable considering the inevitable surface error from overlapped connection area with double thickness. Comparing between samples of the same type, the largest difference occurs at case 0, being just 0.16 mm, which indicates the manufacturing process was consistent.

**Table 2:** *Surface error measurement data for FRP-timber beams.*

| Surface Error | Case Number | | | | | | | | | | | | | | | |
|---|---|---|---|---|---|---|---|---|---|---|---|---|---|---|---|---|
| | 0-1 | 0-2 | 1-1 | 1-2 | 2-1 | 2-2 | 3-1 | 3-2 | 4-1 | 4-2 | 5-1 | 5-2 | 6-1 | 6-2 | 7-1 | 7-2 |
| Avg. Error | 1.23 | 1.49 | 1.31 | 1.20 | 1.13 | 1.24 | 1.36 | 1.49 | 1.93 | 2.02 | 1.74 | 1.68 | 2.02 | 1.81 | 1.55 | 1.95 |
| Avg. Abs. Error | 1.24 | 1.50 | 1.31 | 1.20 | 1.15 | 1.29 | 1.36 | 1.50 | 1.93 | 2.02 | 1.74 | 1.68 | 2.03 | 1.81 | 1.56 | 1.96 |
| Max. Error | 4.20 | 5.79 | 3.87 | 3.72 | 6.03 | 4.78 | 4.59 | 5.49 | 5.29 | 5.48 | 5.30 | 4.49 | 6.92 | 5.52 | 5.26 | 6.18 |
| Min. Error | -1.41 | -2.96 | -2.96 | -0.47 | -3.07 | -2.57 | -1.93 | -1.92 | -0.78 | -0.83 | -1.39 | -0.05 | -3.84 | -2.41 | -1.83 | -3.68 |

| Subcase Average | Case 0 | Case 1 | Case 2 | Case 3 | Case 4 | Case 5 | Case 6 | Case 7 |
|---|---|---|---|---|---|---|---|---|
| Avg. Error | 1.36 | 1.26 | 1.19 | 1.43 | 1.98 | 1.71 | 1.92 | 1.75 |
| Avg. Abs. Error | 1.37 | 1.26 | 1.22 | 1.43 | 1.98 | 1.71 | 1.92 | 1.76 |
| Max. Error | 5.00 | 3.80 | 5.41 | 5.04 | 5.39 | 4.89 | 6.22 | 5.72 |
| Min. Error | -2.18 | -1.71 | -2.82 | -1.92 | -0.80 | -0.72 | -3.12 | -2.76 |
| $h/b$ (%) | 2.38 | 1.79 | 1.19 | 1.67 | 1.43 | 0.48 | 0.00 | 0.00 |

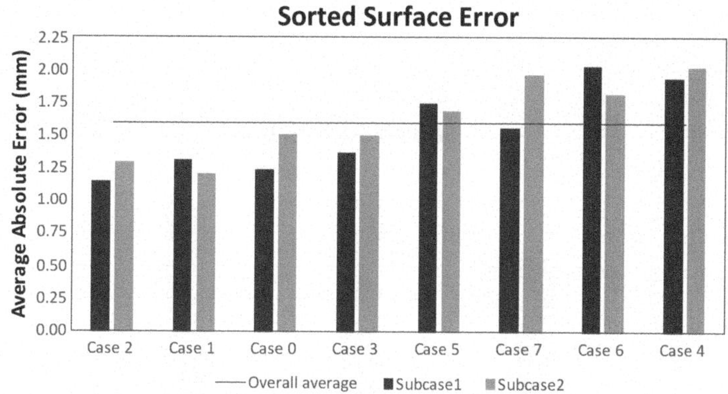


(a) *Surface error overview for all FRP-timber specimens*

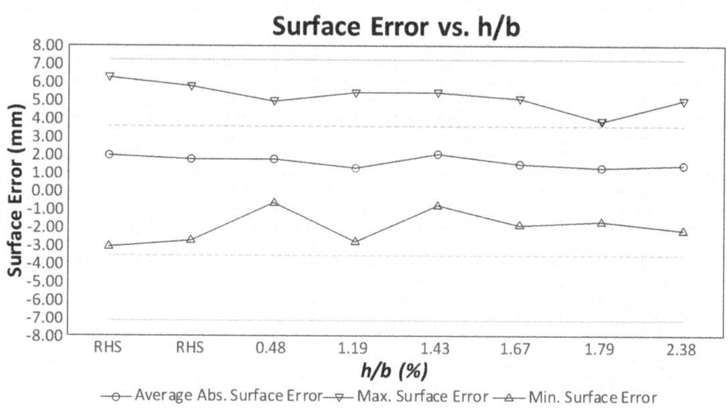


(b) *Surface error comparison between FRP-timber cases*

**Figure 8:** *Surface error analysis for FRP-timber beams.*

**Table 3:** *Surface error measurement data for steel beams.*

| Surface Error | Case Number | | | | | |
|---|---|---|---|---|---|---|
| | 0-1 | 0-2 | 3-1 | 3-2 | 6-1 | 6-2 |
| Avg. Error | -0.39 | -0.11 | -0.21 | -0.34 | 0.02 | -0.18 |
| Avg. Abs. Error | 0.67 | 0.51 | 0.47 | 0.54 | 0.62 | 0.56 |
| Max. Error | 3.51 | 3.70 | 2.45 | 4.02 | 3.84 | 2.63 |
| Min. Error | -3.15 | -3.04 | -2.58 | -4.15 | -2.49 | 0.00 |
| Subcase Average | Case 0 | | Case 3 | | Case 6 | |
| Avg. Error | -0.25 | | -0.27 | | -0.08 | |
| Avg. Abs. Error | 0.59 | | 0.51 | | 0.59 | |
| Max. Error | 3.60 | | 3.24 | | 3.23 | |
| Min. Error | -3.09 | | -3.37 | | -1.24 | |
| $h/b$ (%) | 2.38 | | 1.67 | | 0.00 | |

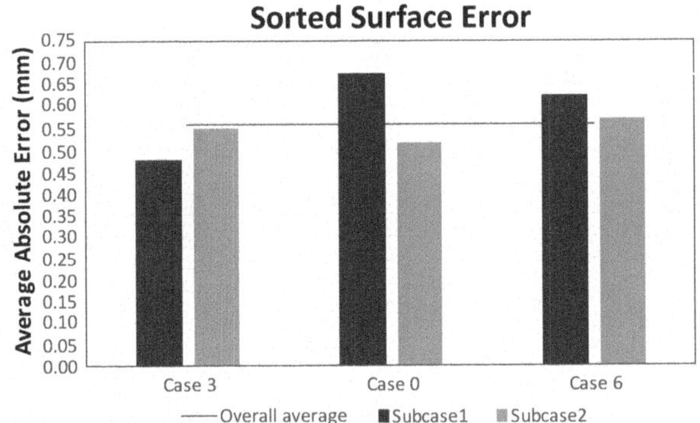

(a) *Surface error overview for all steel specimens*

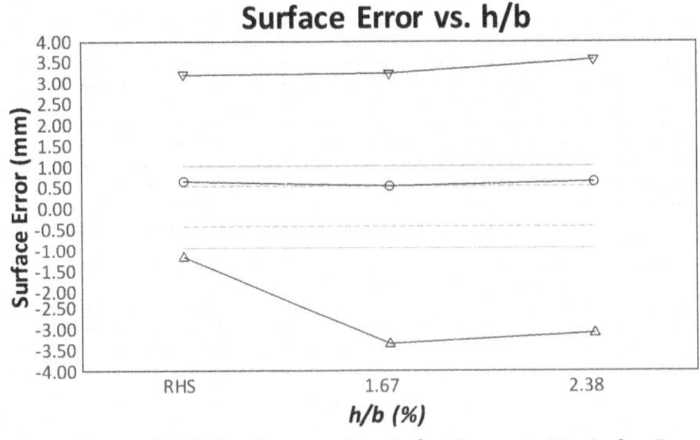

(b) *Surface error comparison between steel cases*

**Figure 9:** *Surface error analysis for steel beams.*

A second analysis was undertaken to assess whether the error-curvature variation trend for steel cases is similar to that for FRP-timber cases. Figure 9b shows the plot of mean surface errors for each case against elastica parameter $h/b$. It can be seen that case 3 has lower average absolute error than case 6, which is similar to the trend obtained in FRP-timber cases. However, all three cases have very similar mean surface error and so again there is no clear correlation between surface error and surface curvature. From figure 9b, excessive maximum and minimum errors were observed for all cases, which can mostly be attributed to the sheet overlap at folding joints.

Three conclusions can be drawn from these results. First, the steel folding fabrication process is effective, with acceptably low error for all cases and consistent error seen between different cases of the same type. Second, the elastica parameter has no obvious effect on manufacturing accuracy, with extremely small variation seen with different spindle types. Third, the elastica surface generation method is effective, with accurate spindle products obtained from both steel and FRP-timber manufacturing processes.

## 5   Conclusion

This paper applied two innovative fabrication methods to the manufacture of curved-crease folded structures. Previously, the cured-in-place method was developed as a concept proof for building thin-walled FRP-timber structures, while the slits-and-straps technique has only been applied for folding straight-crease folded structures. This study has shown that both methods can be effectively extended for the manufacture of large-scale curved-crease structures. For FRP-timber spindles, the fabrication process has been optimised with fibre wrapping and shrink-tape finishing. For steel spindles, fast fabrication has been achieved with a parametric fold-line detail and CNC waterjet cutting. Both methods produced curved-crease spindle structures with surface imperfections that were sufficiently low to enable the future use as curved-crease thin-walled structural elements.

## Acknowledgement

This paper has been awarded the 7OSME Gabriella & Paul Rosenbaum Foundation Travel Award. The authors also gratefully acknowledge the financial support provided by the Australian Research Council Discovery Project DP160103279.

## References

[Buri et al. 11]  Hans Ulrich Buri, Ivo Stotz, and Yves Weinand. "Curved Folded Plate Timber Structures." In *IABSE-IASS 2011 London Symposium*. IABSE-IASS, 2011.

[Durney and Pendley 05]  Max W. Durney and Alan D. Pendley.  "Method for precision bending of sheet of materials, slit sheets fabrication process.", 2005.  Patent US6877349.

[Edison 11] Christine E. Edison. "Compression and Rotational Limitations of Curved Corrugations." In *Origami 5: Fifth International Meeting of Origami Science, Mathematics, and Education*, pp. 69–79. CRC Press, 2011.

[Epps and Verma 13] Gregory Epps and Sushant Verma. "Curved Folding: Design to Fabrication Process of Robofold." In *HYPERSEEING Special Issue on SMI/ISAMA 2013*, pp. 75–83. ISAMA, 2013.

[Garrett et al. 16] Daniel Garrett, Zhong You, and Joseph M. Gattas. "Curved Crease Tube Structures as an Energy Absorbing Crash Box." In *ASME 2016 International Design Engineering Technical Conferences and Computers and Information in Engineering Conference*, 2016.

[Gattas and You 14] Joseph M. Gattas and Zhong You. "Miura-Base Rigid Origami: Parametrizations of Curved-Crease Geometries." *Journal of Mechanical Design* 136:12 (2014), 121404.

[Gattas and You 16] Joseph M. Gattas and Zhong You. "Design and digital fabrication of folded sandwich structures." *Automation in Construction* 63 (2016), 79–87.

[Gattas et al. 18] Joseph M. Gattas, Mitchell L. O'Dwyer, Michael T. Heitzmann, Dilum Fernando, and Jin-Guang Teng. "Folded hybrid FRP-timber sections: concept, geometric design and experimental behaviour." *Thin-Walled Structures* 122 (2018), 182–192.

[Gattas 14] Joseph M. Gattas. "Digital Fabrication Toolbox.", 2014. Available at: `http://joegattas.com/digital-fabrication/`.

[Godoy 96] Luis A. Godoy. *Thin-walled structures with structural imperfections.* Elsevier,
1996.

[Guo et al. 16] Zhiming Guo, Joseph M. Gattas, and Hassan Baji. "Load Analysis and Buckling Behaviour of Curved-Crease Origami Pipelines." In *Twelfth (2016) Pacific-Asia Offshore Mechanics Symposium*. International Society of Offshore and Polar Engineers, 2016.

[Hansen et al. 16] Benjamin J. Hansen, Jonathan Tan, Joseph M. Gattas, Dilum Fernando, and Michael Heitzmann. "Folded fabrication of FRP-timber thin-walled beams with novel non-uniform cross-sections." In *WCTE 2016 World Conference on Timber Engineering*. Vienna University of Technology, 2016.

[Klett et al. 17] Yves Klett, Carla Zeger, and Peter Middendorf. "Experimental characterization of pressure loss caused by flow through foldcore sandwich structures." In *ASME 2017 International Design Engineering Technical Conferences and Computers and Information in Engineering Conference*, p. V05BT08A051. American Society of Mechanical Engineers, 2017.

[Lee and Gattas 16] Ting-Uei Lee and Joseph M. Gattas. "Folded fabrication of composite curved-crease components." In *8th International Conference on Fibre-Reinforced Polymer (FRP) Composites in Civil Engineering (CICE 2016)*, 2016.

[Lee et al. 18]  Ting-Uei Lee, Zhong You, and Joseph M. Gattas. "Elastica surface generation of curved-crease origami." *International Journal of Solids and Structures* 136-137 (2018), 13–27.

[Mayencourt and Mueller 17]  Paul Mayencourt and Caitlin Mueller. "Structural Optimization and Digital Fabrication of Timber Beams." In *IABSE Symposium Report*, 108, 108, pp. 153–154. International Association for Bridge and Structural Engineering, 2017.

[Miura 72]  Koryo Miura. "Zeta-Core Sandwich-Its Concept and Realization." *ISAS report/Institute of Space and Aeronautical Science, University of Tokyo* 37:6 (1972), 137–164.

[Miyashita et al. 15]  Shuhei Miyashita, Isabella DiDio, Ishwarya Ananthabhotla, Byoungkwon An, Cynthia Sung, Slava Arabagi, and Daniela Rus. "Folding angle regulation by curved crease design for self-assembling origami propellers." *Journal of Mechanisms and Robotics* 7:2 (2015), 021013.

[Samyn 04]  Philippe Samyn. *Étude de la morphologie des structure: à l'aide des indicateurs de volume et de dépalcement*, 2004.

[Shi et al. 17]  Quan Shi, Xiaoqiang Shi, Joseph M. Gattas, and Sritawat Kitipornchai. "Folded assembly methods for thin-walled steel structures." *Journal of Constructional Steel Research* 138 (2017), 235–245.

[Tachi 13]  Tomohiro Tachi. "Composite Rigid-Foldable Curved Origami Structure." In *The First Transformables 2013*. School of Architecture, Seville, 2013.

[Vergauwen et al. 13]  Aline Vergauwen, Lara Alegria Mira, Kelvin Roovers, and Niels De Temmerman. "Parametric design of adaptive shading elements based on Curved-line Folding." In *First Conference Transformables 2013*. School of Architecture, Seville, 2013.

[Vergauwen et al. 17]  Aline Vergauwen, Lars De Laet, and Niels De Temmerman. "Computational modelling methods for pliable structures based on curved-line folding." *Computer-Aided Design* 83 (2017), 51–63.

[Zhao et al. 15]  Danyang Zhao, Yujie Li, Minjie Wang, Chunzheng Duan, and Zhong You. "Fabrication of Polymer Origami-Based V-Type Folded Core." In *ASME 2015 International Design Engineering Technical Conferences and Computers and Information in Engineering Conference*, pp. V05BT08A038–V05BT08A038. American Society of Mechanical Engineers, 2015.

---

W.Q. Cui
School of Civil Engineering, University of Queensland, Australia

T. Gfeller
Institute of Structural Engineering (IBK) — ETH Zurich, Switzerland

D. Fernando
School of Civil Engineering, University of Queensland, Australia

M.T. Heitzmann
School of Mechanical Engineering, University of Queensland, Australia

J.M. Gattas
School of Civil Engineering, University of Queensland, Australia, corresponding author
e-mail: j.gattas@uq.edu.au

# Fold Printing: Using Digital Fabrication of Multi-Materials for Advanced Origami Prototyping

*M. Gardiner, R. Aigner, H. Ogawa, E. Reitböck, R. Hanlon*

**Abstract**: *Computational Origami invents a new kind of complexity: the highly irregular crease pattern, posing a significant challenge to foldability. We design a method for digital fabrication of multi-material composites to overcome issues of foldability and durability faced when producing functional prototypes with paper. Our fabrication method, called Fold Printing, uses off-the-shelf and customised 3D printing technology, heat pressing, and elastomers to control fold memory in Folded Polymer Textile Elastomer Composites (FPTEC). Our results show that Fold-Printing affords foldability of high-irregularity crease patterns, and can produce durable advanced origami prototypes.*

## 1   Introduction

We present a method to increase foldability and durability of complex folding patterns by digital fabrication of multi-material composites with rigid plates separated by perfect hinges. Using a process we call *Fold Printing* (FP), that produces Fold Printed Textiles (FPT), and Fold Printed Textile Elastomer Composites (FPTEC). Similar in function to textile electroplating technique [De-Ruysser 09], and a cement textile process [van der Woerd et al. 15] with the advantages that no pre-tooling or chemistry beyond polymer filaments, and textiles are required, and due to the digital fabrication workflow, designs can be altered with each iteration. We also present a custom-built cartesian robot, designed for prototyping sheets up to 105x105cm to accommodate an increased complexity in the number of folds per model, and increased utility by producing prototypes at scale. We evaluate our methods by proof-of-concept in a variety of design targets, featuring regular and irregular software generated folding patterns. We compare foldability in terms of time-to-fold laser-scored paper with FPTs, compare processing techniques, and critically evaluate physical and aesthetic properties of the completed prototypes.

## 2   Related Work

Physical prototyping is used to validate folded designs, and often, the first choice of material is paper. It is strong, thin, retains an elastic memory of folds, and can be pre-scored by machine to increase foldability of complex patterns. However, properties that make paper readily accept creases, also make it vulnerable to corruption. Unwanted creases impact the aesthetic of a design, functionally they can impact the accuracy of subsequent folds, and can introduce buckling that corrupts a collapsing pattern. Paper prototypes can be fragile and delicate, and even more so for low-degree-of-foldability origami, and kinetic origami structures.

In recent years, the new tool in the origami studio isn't a folding bone, its a machine to aid creasing, as can be seen in the way digital fabrication techniques have extended the origami toolkit: from early work with plotting [Resch 92], laser scoring [Gardiner 13, p.21], perforation of paper, polymers, and textiles [Sallas and Zscheckel 10], etching of metals [Kuribayashi et al. 06], electroplating [De-Ruysser 09], as well as pleating by pressure and heat [Gardiner 15] are commonly known methods in industry and practices of origami artists. A prime example of the rigid plate/perfect hinge technique can be seen in the design object BAO BAO by Issey Miyake [Miyake 10]. A simple pattern of rigid polymer panels heat-fixed to a flexible polymer netting creates an immediately foldable structure and compelling aesthetic object. Combinations of origami principles with elastomers, polymers and composite approaches to robotics have been explored by a number of researchers such as [Felton et al. 13], [Martinez et al. 12], and [Felton et al. 14].

Our contribution is a method for prototyping origami patterns in durable, low-cost, easily acquired materials, by using an off-the-shelf FDM (Fuse Deposit Modelling) 3D Printer to increase foldability of regular and irregular designs. The key features of our method are to 3D print rigid polymer plates onto textiles, and to treat with heat pressing. Followed by optional application of elastomers to introduce elastic memory, and pneumatic sealing qualities, for per-instance setting of the folded-shape memory of the Folded-Polymer-Textile-Elastomer-Composite (FPTEC) and Folded-Polymer-Textile (FPT).

# 3 Method

## 3.1 Fabricating Foldable Polymer Textile Elastomer Composites

We call *Fold Printing* the process of using Fuse Deposit Modelling (FDM) to print polymers onto textiles. The polymer forms semi-rigid plates separated by the near-perfect hinge of the textile to form a flexible plate structure. The FPTEC process extends the basic step of printing onto textiles, by heat pressing to enforce the polymer-textile bond, and by application of elastomer to create elastic memory in the composite. FPTEC (see Fig. 1) is the result of trial-and-error experimentation designed to find a digital fabrication method that produces durable, flexible, fold-programmable materials, for application in contexts requiring robust rapid-prototyping and rapid-folding, i.e kinetic origami, design proof-of-concept for advanced origami structures. The following section details a step by step overview of the process.

### 3.1.1   FTPEC Process Overview

The following list details the general guidelines for creating an FPTEC. Steps 1–2 are related to textile and polymer processing used in Fold Printing and should be read and followed in conjunction with the instructions given in Section 3.2. Steps 3–8 relate the application of Elastomer to an FPT, and related notes are given in Section 3.3. Application of the elastomer listed in Step 3. requires a skilled origamist, due to the short gel time of the elastomer.

1. Fold Print polymer geometry onto a natural textile: the coarse microstructure of natural fibres is ideal for adhesion to the polymer. Autoignition of cotton is above 400°C, we print PLA at higher than normal ranges: 210-230°C Fig. 2. shows deeper penetration of the polymer through the weave at 230°C.

2. Heat press at the printing temperature 200°C for PLA. We used a t-shirt transfer heat press with adjustable temperature and timer, with sheets of Teflon or grease-proof baking paper to prevent adhesion of the polymer to the heat press.

3. Apply a thin coat of elastomer, pressing the elastomer into the textile weave with a flat tool such as a spatula. We use Dow Corning 734, a strong, low-viscosity flowable silicone, with excellent adhesion to a wide range of materials including cotton and plastics, 315% elongation at break, a 3.0kN/m tear strength [Corning 11].

4. Working quickly, cover the wet elastomer with a non-stick film such as PE (Polyethylene), this layer is critical as it protects adjacent surfaces from bonding as the elastomers cure.

5. Quickly vacuum press the composite to eliminate bubbles between the elastomer and PE film, and remove from the vacuum. We used one layer of vacuum bagging material over a flat bench to ensure a tight pull onto a registered flat surface.

6. Working as quickly as possible, fold the composite into the desired shape, and leave to cure, under light pressure or clamps to retain shape.

7. After curing, remove the non-stick film, and trim off any excess polymer.


### 3.2   Fold Printing 101

The following notes for Fold Printing are useful for use with off-the-shelf 3D-printers, our experiments were conducted on Ultimaker 2[1] and Mankati Fullscale XT[2]. The following general notes are to aid first-time printer set up and calibration for Fold Printing.

The first step is to calibrate the printer head to be in direct contact with the textile (See Fig. 2). Calibration is made by adjusting the nozzle height as low as possible, and while printing, incrementally move upwards until the extrusion is visible. Approximately 50% coverage of the textile is ideal, no extrusion is a sign of the nozzle being set too low. Material feed rate can be reduced by up to

---

[1]https://ultimaker.com/en/products
[2]http://www.mankati.com/fullscale-xt-plus

**Figure 1:** *Left: FPTEC Miura-ORI, in a closed configuration. Right: Close up of open Yoshimura-ORI. Both sheets were coated on the non-printed side and pressed under vacuum at 0.3 Pa.*

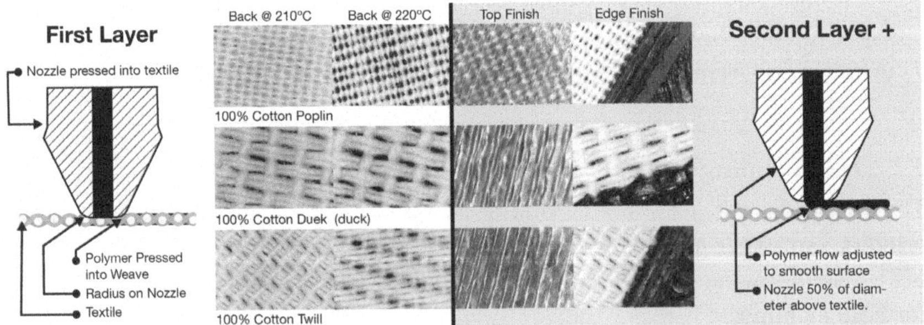


**Figure 2:** *L: first layer setup guide showing nozzle in contact with textile; CL: 500x magnification shows the effect of temperature on polymer penetration into the textile weave. CR: Top and edge finish on different cotton weaves. R: Second layer nozzle setup.*

50% to preventing material excess, visible as a raised bump around the edge of the extrusion noodle, as the excess is pushed outwards from the nozzle point.

Set up your slicer to print a minimum of 2-3 layers. For a 0.4mm nozzle, set the layer height at 0.2mm. the first layer temperature 220°C for PLA. The second or third layers can be reduced to a lower print temperature between 190-210°C, this varies between filaments and requires per-instance tuning.

Bed levelling is crucial, we identified that a constant distance of the printing head relative to the printer bed, $\pm$ 0.1mm greatly improved the consistency of both adhesion and finish. 75% of calibration time was spent bed levelling and adjusting nozzle height, the last 25% was spent on material flow and temperature adjustments. Glass and aluminium build platforms work equivalently, as the print bed does not require heating. We use *bulldog* paper clamps, and for loose textile weaves, we also apply double-sided tape, to affix the textile to the build platform.

### 3.3 Setting Fold Memory: Water, Elastomers, Resins

FPTs have either slight or no elastic memory, and thus an unfolded FPT can *forget* its folds. *Fold Memory* can be set using a variety of methods: by water for a more paper-like memory; by elastomer for a muscle-like elasticity; and by resin for a rigid component. Each application is by liquid and requires drying or curing time. Water is the simplest, our experiments used 60°C (water from a domestic hot tap) to saturate the FPT, and after folding was held in shape until dry. We experimented with elastomers to enable the folded form to retract into a set configuration when expanded or contracted, see Fig. 1. We tested a range of industrial RTV silicones and two component silicones. Dow Corning 734 Flowable Silicone Sealant was selected as the best candidate, due to its adhesion to many materials, including cotton and plastics, its excellent viscosity for application onto textiles, and the performance of its elastic memory in our results. The higher viscosity Dow Corning 732 Silicone Adhesive did not penetrate the cotton and felt weaves, and resulted in thicker layers of silicone, creating an over-stiff elastic memory that impaired movement of the folded pattern. In our experiments with resin, we used polyester and epoxy casting and fibreglassing resins, to permanently harden the sculpture in a set shape. This method of origami sculpture produces resilient, lightweight folded structures. To facilitate accurate curing positions of more open patterns, we fabricated negative moulds to set the target fold geometry.

### 3.4 Large Format Fold Printer

Our large format Fold Printer, playfully dubbed *NIWASHI* from Japanese meaning *gardener*, was developed to produce FPTs and FPTEC artefacts at scale, and to allow more folds per sheet to afford complex geometries due to increased size sheet. The design of the printer (cartesian robot), was based on pick-and-place gantry robots, we built a 125 x 125cm XY gantry locating a retractable 7.5cm Z-axis, to print onto 110cm wide textile rolls.

The machine build raised many design issues particular to our design requirements. Our build recommendations are as follows:

**Figure 3:** *Building NIWASHI, from bare bones, gantry structure, prototype in progress, to functional printer*

- **Choice of hardware and firmware**: we chose open-source hardware RUMBA[3], running open-source software Marlin[4] for controlling the printer, as Marlin provides various bed levelling functions for continuous adjustment of the Z-axis. This compensates for bed height variations, based on a matrix of pre-measured positions.

- **Gantry system**: for NIWASHI-I we machined custom connecting plates to fit IGUS drylin[5] belt-driven linear rails. A more rigid, ball-bearing guided, CCM[6] linear rail system was used on the dual-head NIWASHI-II, we recommend the latter due to improved rigidity.

- **Fine layer height calibration**: a 0.5mm pitch screw-drive spindle was installed as the Z axis.

- **High powered extrusion to push the material into the textile, and material transport**: E3D Titan Extruder[7], fed with 1.75mm filament through a long flexible Teflon *Bowden* tube[8], fed through the e-chain to the material spool, mounted at the end of the X carriage.

- **Holding the fabric in place during printing**: a martensitic stainless steel 1.4000 bed allows magnets to be used to hold textiles in place.

- **Print levelling calibration due to deflection and bed levelness**: using bed levelling functions built into Marlin, at the maximum grid resolution of 9 x 9, the printer was calibrated to compensate for the deflection, measuring Z offsets with an error tolerance of 0.01mm, with the result of substantially improved print quality across the bed.

- **Material deposition rates and calibration**: our experience using CURA slicer for the Ultimaker and Mankati showed that print settings, especially material feed rate, and Z-height adjustments were critical for print quality. Simplify3D[9] generates G-code (RS-274), from user-defined material profiles, and includes a

---

[3]RepRap RUMBA http://reprap.org/wiki/RUMBA

[4]Marlin Open Source 3D Printer Firmware http://marlinfw.org/

[5]IGUS drylin http://www.igus.com/drylin

[6]CCM linear motion rails https://www.ccmrails.com/

[7]E3D Titan Extruder http://e3d-online.com/Titan-Extruder

[8]Erik's Bowden Extruder http://reprap.org/wiki/Erik%27s_Bowden_Extruder

[9]Simplify 3D http://simplify3d.com/

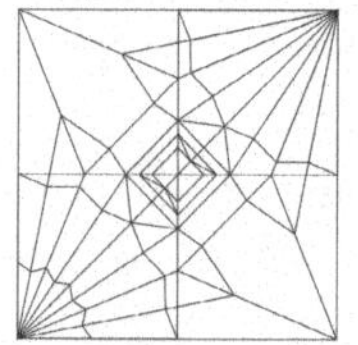   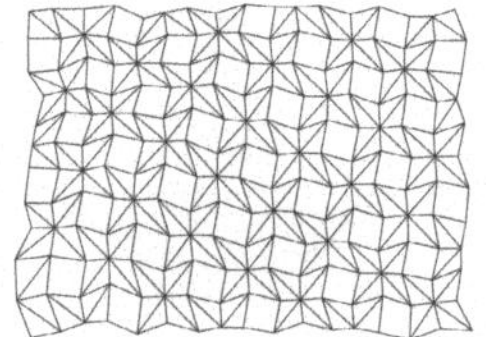

**Figure 4:** *FPTEC Experiment Crease Patterns. From left: Orizuru, Miura45, FF-Dome, Resch4i*

**Figure 5:** *FPTEC Experiment Print Patterns. From left: Orizuru with variable (1.5–6mm) width creases, FFDome with 2.5mm width creases. Black shows the extruded section printed onto the textile.*

soft machine control panel to support tweaking of print settings such as temperature and material feed rate during print calibration.

# 4   Results

## 4.1   FPTEC experiments

**Orizuru** is an example of the additional work required to using FPTs for multi-layered folding, where the complete folding pattern can be resolved in the print. **Miura45** is a Miura-ori, or herringbone pattern, with high regularity due to the 45° chevrons. Collapsing this pattern causes a cascade of folds to fall into place with gentle prodding. **FFDome** is a pattern output from *FreeForm Origami*, a dome of subdivided icosahedron geometry was imported and Origamized on default settings. The repeated generalised Resch structure makes the collapse of this piece come together quickly in the FPT, in paper it was difficult to pre-crease the acute angles, resulting in a longer paper fold time. **Resch4i** is a deliberately irregular

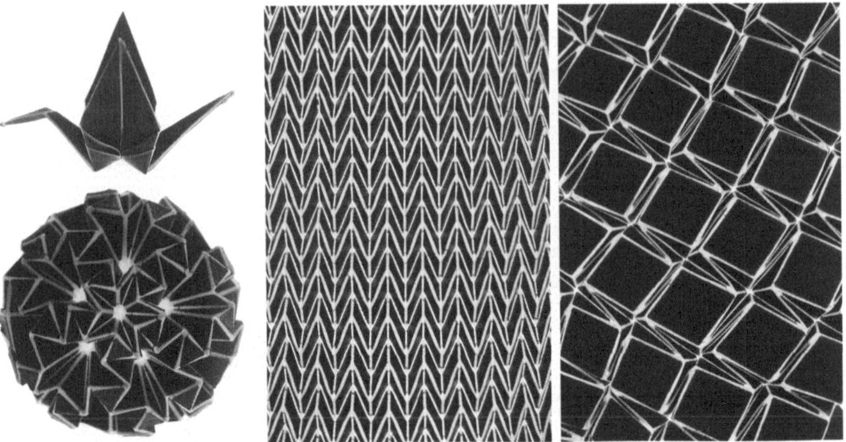

**Figure 6:** ***FPTEC Experiments*** *paper vs FPTEC from left: Orizuru (paper crane), FFDome from FreeForm Origami Origamizer algorithm, Miura45, and Resch4i from ORI*gh*

arrangement of the standard Resch-4 pattern mapped onto an uneven surface, causing the squares to become rhomboids of various dimensions. The Resch geometry, even in irregular arrangement, was simple to actuate with FPT.

## 4.2　Folding time comparison

In general, we observed that folding FPTs was unquestionably easier than folding any other medium we have experienced. It requires some gentle compression on the pattern and with gentle manipulation the crease patterns *pop* into place. We conducted timed folding experiments to evaluate the benefit, starting a stopwatch prior to folding, and stopping immediately on completion. The same author timed a selection of crease patterns in FPTs and laser-cut machine-made Washi or Elephant Hide. Results in Table 1 show the clear improvement of foldability between paper and FPTs. Speed advantages are produced by the semi-rigid plates and perfect hinges, resulting in 1.7 to 11.5 times faster folding times. In addition, as visible in Fig. 6, there is little or no fold induced deformations or imperfections in the FPTs, but the paper samples (not shown), being folded as quickly as possible, suffered minor crumpling.

**Table 1:** *Time to fold Laser Scored Paper vs FPTEC, folds: number of Folds. Folding time in seconds for Paper, Seconds per Fold for Paper. Time in seconds for FPTEC, Seconds per fold for FPTEC, and Percentage Difference between paper and FPTEC. Shows significant improvement between paper and FTPEC in %diff column.*

| Model | Folds | PAPER | s/f | FPTEC | s/f | % diff |
|---|---|---|---|---|---|---|
| Orizuru | 28 | 50 s | 1.79 | 28 s | 1.00 | **179%** |
| Miura45 | 1188 | 1981 s | 1.67 | 330 s | 3.60 | **600%** |
| FFDome | 405 | 2339 s | 5.78 | 202 s | 2.00 | **1158%** |
| Resch4i | 555 | 1635 s | 2.95 | 308 s | 1.80 | **531%** |

## 4.3 General Design Considerations

A key issue for developmental work with FPTs is the issue of generating suitable geometry. Our experimental results indicated the following general design consideration that can be applied by researchers applying this technique:

- **Tesselations and Corrugations**: showed high applicability for the FPT & FPTEC process. Designers working with complex patterns can effortless achieve a folded result with an unmodified extruded crease pattern.

- **Irregular and complex designs**: are exceptionally easy to fold, despite the irregular positioning of folds, angles, and lack of parallelity, FPTs and FPTECs outperform other methods of folding a sheet with respect to time-to-fold.

- **Fold Density**: is proportional to crease width, which necessarily occupies a larger than normal area when compared to creases in paper. Ultimately, the fold count is limited by the print size, and combined material thickness.

- **Heat Pressing**: changes the geometry, especially crease width. The loss was approximately 1mm less than the printed crease width.

- **Multi-layered designs**: require creases to be adjusted width dependently, to allow for overlapping folds (see variable width creases in Figure 5). The maximum factors for assigning crease width are shown in Figure 7. The pliability of textiles is such that the radial fold width can effectively be up to 50% less than these maximums. Therefore, with T in mm, we adjusted our width calculation to be:

$$f = \frac{T\pi}{2} + 1mm \tag{1}$$

## 5 Evaluation and Further Work

### 5.1 On Textile selection

Fig. 2 shows experimental results of calibrated prints on three cotton weaves: Poplin, Duck and Twill. Top and edge finish are consistent across each weave, having similar visual qualities. Poplin's top finish is smoother, due to the low profile and variance of the weave. Poplin proved to be preferable for fine folding

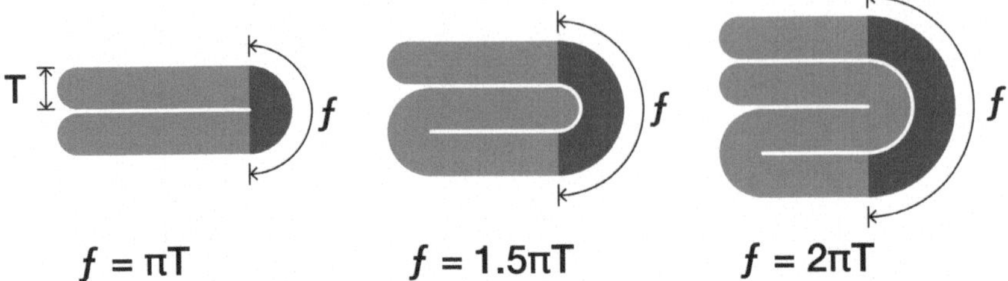



**Figure 7:** *Calculating fold radial width from left: f=Tπ ; f=1.5Tπ ; f=2Tπ ; Where f is radial width of the fold, and T is the thickness of the material. This value can be up to 50% less for pliable textiles.*

applications with a multi-material thickness of 0.25 mm. Duck has extra thickness, more body, and retains fold direction memory, it also functions better for larger folded plate structures with a thickness of 0.5 mm. Further experiments included felts, ranging from pure wool to synthetic blends, concluding with a preference for Offitex[®][10], a textile designed for under-collar application and consists of approximately 70 % wool and 30 % rayon. For printing onto Offitex[®], the initial printer head position (Z0) had to be adjusted to 80 % of the material height, such that the nozzle creates an indent when traversing the felt. Due to the material's porosity, the extruded polymer is absorbed into the fibres. We achieved excellent adhesion with felt, with attempts to remove the printed layer destroying the material

## 5.2   On FPTs and FPTECs

Our general observations and measured results show marked improvement of foldability, material properties and application domains over paper. The application of polymers as semi-rigid plates allows FPTECs and FPTs to be folded up to 11.5 times faster than laser-scored paper, enabling the researcher to rapid-prototype and rapid-fold new designs of increased complexity. The increased durability, water resistance, and tear resistance of textiles over paper, and the ability of textiles to accept coatings of water, elastomers or resins, affords a wider range of applications, and material customisations. Elastomers have previously been successfully deployed in origami soft-robotic actuators, and we expect that research-based experimentation into FPTECs will lead to new developments in advanced origami applications such as new soft-robotic actuators, and soft interfaces in tangible human-computer interaction.

---

[10]`http://www.bwf-group.de/en/bwf-feltec/products/undercollar_felts.html`

## 6   Conclusion

FPTs and FPTECs as fabrication methods are shown to have higher foldability
than paper in regular and irregular geometries. Affording potential for researchers
to investigate materials in new areas of advanced origami prototyping, and new aes-
thetics due to freedom of geometry. Further work suggests an increased application
domain for origami prototypes through experimental multi-material combinations,
including variant 3D printing filaments such as conductive carbon, flexible nylons,
with an assortment of textiles and other sheet materials. We speculate that this
technique could realise complex folding structures for applications in robotics, especially soft robotics, where programmability of structure, and material properties
such as fold-memory elasticity are required.

## 7   Acknowledgements

PEEK Project AR272-G21. Funded through the FWF, PEEK Program, Program
Management: Dr Eugen Banauch. Special thanks to Ars Electronica Futurelab for
their continued support of this research.

## References

[Corning 11]  Dow Corning.  "Dow Corning® 734 Flowable Adhesive Sealant - Product
Information Sheet.", 2011.

[De-Ruysser 09]  Tine De-Ruysser.  "'Wearable Metal Origami'?: The Design and Manufacture of Metallised Folding Textiles." Ph.D. thesis, Royal College of Art, 2009.

[Felton et al. 13]  Samuel M. Felton, Michael T. Tolley, ByungHyun Shin, Cagdas D. Onal,
Erik D. Demaine, Daniela Rus, and Robert J. Wood. "Self-Folding with Shape Memory Composites." *Soft Matter* 9:32 (2013), 7688–7694. doi:10.1039/C3SM51003D.

[Felton et al. 14]  Samuel Felton, Michael Tolley, Erik Demaine, Daniela Rus, and Robert
Wood. "A Method for Building Self-Folding Machines." *Science* 345:6197 (2014),
644–646.

[Gardiner 13]  Matthew Gardiner. *Designer Origami*. Melbourne, Australia: Hinkler Books,
2013.

[Gardiner 15]  Matthew Gardiner. "Folding and Unfolding a Million Times over: Longevity,
Origami, Robotics and Biomimetics as Material Thinking in Oribotics." *Symmetrion*
26:2 (2015), 189–202.

[Kuribayashi et al. 06]  Kaori Kuribayashi, Koichi Tsuchiya, Zhong You, Dacian Tomus,
Minoru Umemoto, Takahiro Ito, and Masahiro Sasaki. "Self-Deployable Origami
Stent Grafts as a Biomedical Application of Ni-Rich TiNi Shape Memory Alloy Foil."
*Materials Science and Engineering: A* 419:1 (2006), 131–137.

[Martinez et al. 12]  Ramses V. Martinez, Carina R. Fish, Xin Chen, and George M.
Whitesides. "Elastomeric Origami: Programmable Paper-Elastomer Composites as
Pneumatic Actuators." *Advanced Functional Materials* 22:7 (2012), 1376–1384.
doi:10.1002/adfm.201102978.

[Miyake 10] Issey Miyake. "BAO BAO ISSEY MIYAKE." http://www.baobaoisseymiyake.com/, 2010.

[Resch 92] Ron Resch. "The Ron Resch Paper and Stick Film on Vimeo." https://vimeo.com/36122966, 1992.

[Sallas and Zscheckel 10] Joan Sallas and Frohmut Zscheckel. *Gefaltete Schönheit: Die Kunst Des Serviettenbrechens*. J. Sallas, 2010.

[van der Woerd et al. 15] Jan Dirk van der Woerd, R Chudoba, and J Hegger. "Design and Construction of a Thin Barrel Vault by Folding." In *Proceedings of the IASS Symposium*, 2015.

---

Matthew Gardiner
Ars Electronica Futurelab, Ars Electronica Strasse 1, Linz 4040, Austria.
University of Newcastle, School of Creative Industries, Newcastle Australia.
e-mail: matthew.gardiner@aec.at

Roland Aigner, Hideaki Ogawa, Erwin Reitböck and Rachel Hanlon
Ars Electronica Futurelab, Ars Electronica Strasse 1, Linz 4040

# Kirigami Fabrication of Shaped, Flat-foldable Cellular Materials based on the Tachi-Miura Polyhedron

*Sam Calisch, Neil Gershenfeld*

**Abstract**: *The Tachi-Miura polyhedron possesses a number of useful properties, including bidirectional, rigid flat-foldability, tailorable Poisson ratio, and tileability. This work shows how to efficiently produce shaped volumes with these properties by modifying constructions for shaped kirigami honeycombs to replace the hexagonal cells with Tachi-Miura polyhedral cells. We analyze the rigid mechanics of these Tachi-Miura honeycombs, and experimentally test the elastic behavior subject to realistic materials and boundary conditions. Finally, we prototype, fabricate, and test a running shoe sole using these techniques.*

## 1  Introduction

In [Miura and Tachi 10] and [Tachi and Miura 12], several rigid-foldable origami cylinders are proposed and analyzed. One example, the Tachi-Miura polyhedra, is a rigid, bidirectionally flat-foldable cylinder which is tileable. Such foldable, space-filling cylinders open the possibility of constructing cellular materials with rigid folding mechanisms. The mechanical properties of cellular materials based on the Tachi-Miura polyhedron have been analyzed in several studies, including [Yasuda et al. 13], [Yasuda and Yang 15], and [Yasuda et al. 16]. As these materials are rigidly flat-foldable, they can be used to create objects that undergo large reversible strains, with Poisson ratios determined by the Miura angle used to create the underlying polyhedra. This capability is a powerful tool for engineering compliant structures.

Physically constructing such cellular materials by simply joining many cylinders is difficult to implement at scale, but we can adapt strategies that were developed to efficiently construct origami honeycombs [Nojima and Saito 06] [Saito et al. 11] [Saito et al. 14] [Wang et al. 17]. These works show how a single sheet can be cut and folded to create a straight-walled honeycomb filling the space between two bounding surfaces. The current research extends these methods to fabricate honeycombs with Tachi-Miura polyhedra instead of hexagonal prismatic cells. These cellular materials are efficiently constructed by cutting and scoring a flat sheet which rigidly folds into the cellular material. By specifying locations of cuts, these cellular materials acquire a desired shape, just as with hexagonal honeycombs. We call such structures *Tachi-Miura honeycombs*.

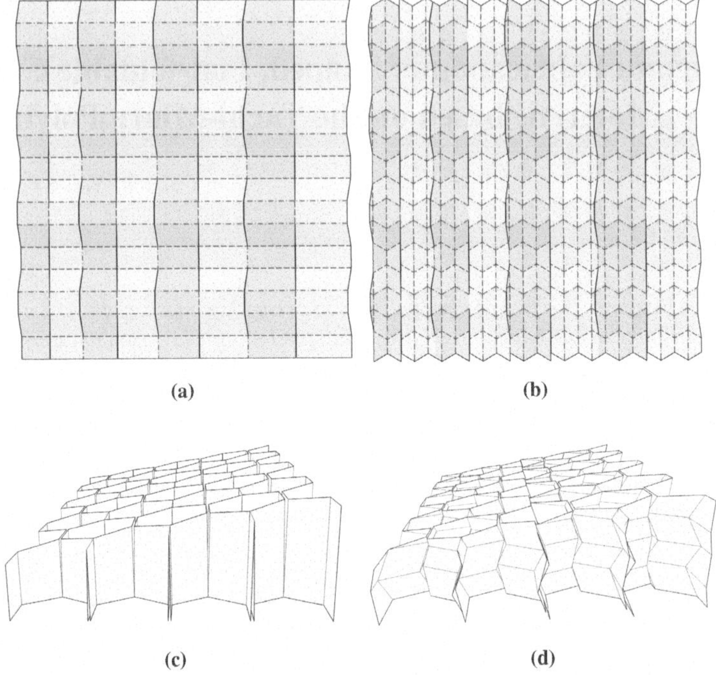

(a)

(b)

(c)

(d)

**Figure 1:** *A) Cutting and folding pattern for the straight-walled honeycomb (blue dashed lines denote a valley fold, red dash-dot lines denote a mountain fold, and black solid lines denote a cut), B) Cutting and folding pattern, for Tachi-Miura honeycomb, made replacing straight folds with zig-zag folds, C) Shaped straight-walled kirigami honeycomb, D) Shaped Tachi-Miura kirigami honeycomb.*

In this work, we first describe our construction for these cellular materials and characterize the rigid folding motions of the resulting cellular materials by calculating their Poisson ratios. To validate the application of these cellular materials to engineer compliant materials, we measure the mechanical response of samples under compressive loading. By varying the Miura angle of the underlying Tachi-Miura polyhedron, we can tune response to match a range of common engineering foams. We then design a prototype utilizing both the large compliance of the folding mechanism and the shape control offered by the honeycomb construction: a running shoe sole. We laser cut, fold, and test this prototype.

## 1.1  Pattern generation

We begin by describing our method of constructing an origami pattern for a Tachi-Miura honeycomb filling a given volume. The volume to be filled is defined as all

1358

points $(x,y,z) \in \mathbb{R}^3$ such that $0 < x < X$, $0 < y < Y$, and

$$u(x,y) < z < t(x,y). \tag{1}$$

In other words, the volume is the space bounded by two functions $u$, $t$, restricted to a rectangular region of the $x - y$ plane. For instance, in Figure 1c, $t$ and $u$ are taken as linear functions describing planes.

In [Saito et al. 14, Calisch and Gershenfeld 18], methods are derived for calculating the shape and positions of the slits in Figure 1a in order to produce the honeycomb shown in Figure 1c. These methods use evaluations of the functions $t$ and $u$ in recursive equations to calculate these parameters. For brevity, we do not reproduce these derivations here, but refer the reader to these articles for the details of these calculations.

Our method for producing Tachi-Miura honeycombs is a direct adaptation of the method described above, motivated by modifying the cutting and folding pattern shown in Figure 1. First, we introduce some nomenclature. We distinguish *corrugation folds*, the horizontally-running folds in Figure 1a, from *strip folds*, the vertically-running folds joining adjacent cuts. Then, we conceptually subdivide the pattern into *strips*, the regions of material bounded on left and right by cuts and strip folds. For example, Figure 1a is composed of eight strips.

Within each strip, we replace each corrugation fold line with a zig-zag fold of the same mountain/valley orientation. As shown in Figure 2, these zig-zag folds make an angle of $\pm\alpha$ with the vertical axis and has a period of $2d$. Vertical mountain and valley folds are added between each vertex of the zig-zag as shown to create a Tachi-Miura quarter cylinder [Tachi and Miura 12]. We apply this modification to each strip, taking care to adjust the phase of the zig-zag fold pattern so that adjacent strips align when the strip folds are actuated.

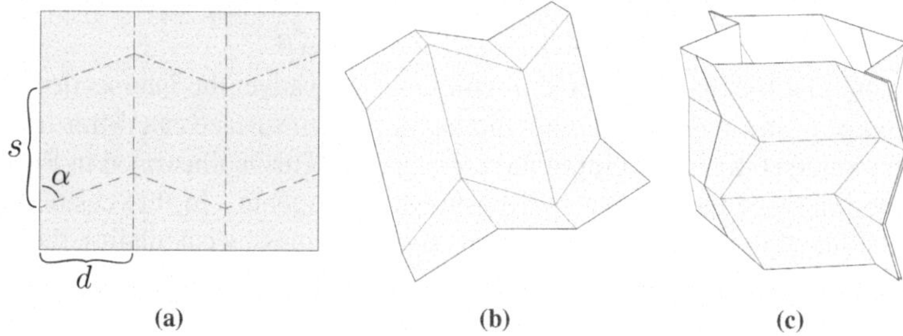

(a)  (b)  (c)

**Figure 2:** *A) Parameterized Tachi-Miura unit cell folding pattern, B) Tachi-Miura unit cell three dimensional geometry. C) The unit cell and its mirror image arranged into a Tachi-Miura polyhedron.*

This modification allows us to produce Tachi-Miura honeycombs from one sheet, but to accurately determine the shape, we must analyze the rigid folding

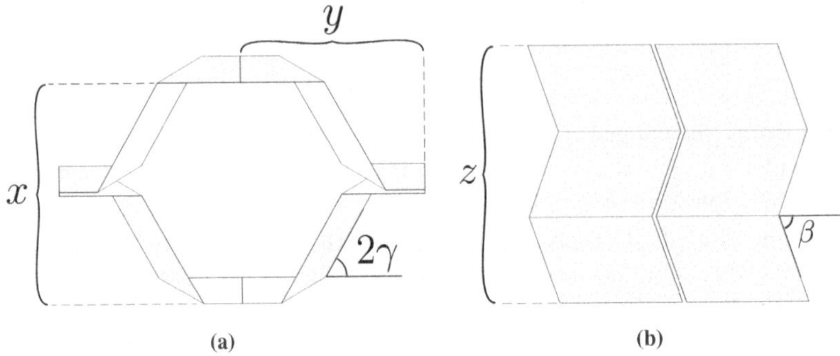

**Figure 3:** *A) Plan view, defining x, y, and γ, B) Altitude view, defining z and β.*

mechanism of the Tachi-Miura polyhedron. To do this, we use an angle parameterization, introduced and proven in [**Miura and Tachi 10**], given by the equation:

$$\tan\alpha\cos\beta = \tan\gamma \tag{2}$$

where the *Miura angle* $\alpha$ sets the relationship between the exterior angle of each cell ($2\gamma$) and the dihedral angle of the trapezoidal faces ($2\beta$). These angles are shown graphically in Figure 3a.

Using these angle definitions, we define variables $x$, $y$, $z$ for the spatial extents of a cell in each coordinate direction, shown in Figure 3b. We calculate:

$$x = 2s\sin 2\gamma \qquad y = s + s\cos 2\gamma \qquad z = Nd\sin\beta \tag{3}$$

where $N$ is the number of zig-zag half-periods in our strip. We see that the expressions for $x$ and $y$ are identical to those of a standard hexagonal honeycomb cell, while the expression for $z$ is simply scaled by a factor of $\sin\beta$. Therefore, to accurately reproduce the shape of functions $t$ and $u$, we must scale our generated patterns in the horizontal direction by a factor of $1/\sin\beta$.

We note that for functions $t$ and $u$ with large derivative, the intersection of the folded image of the zig-zag fold with the top or bottom surface can differ from the function's value at the base point of the zig-zag fold. This is illustrated in Figure 4, where the image of the zig-zag fold is shown in magenta. In this case, we can correct the function values used in the recursive equations by calculating the actual intersection of the zig-zag fold image with the bounding surfaces.

To do this, we define the linear segments of the magenta path by vectors $\mathbf{v}_+$ and $\mathbf{v}_-$ (shown in Figure 4b). Using the angle parameterization of the Tachi-Miura polyhedron and the parallelogram area interpretation of the cross product, we calculate

$$\mathbf{v}_+ = d\left[\cot\alpha, \cot\alpha - \frac{\cos\beta}{\sin 2\gamma}, \sin\beta\right]^{\mathsf{T}} \tag{4}$$

$$\mathbf{v}_- = d\left[-\cot\alpha, -\cot\alpha + \frac{\cos\beta}{\sin 2\gamma}, \sin\beta\right]^{\mathsf{T}} \tag{5}$$

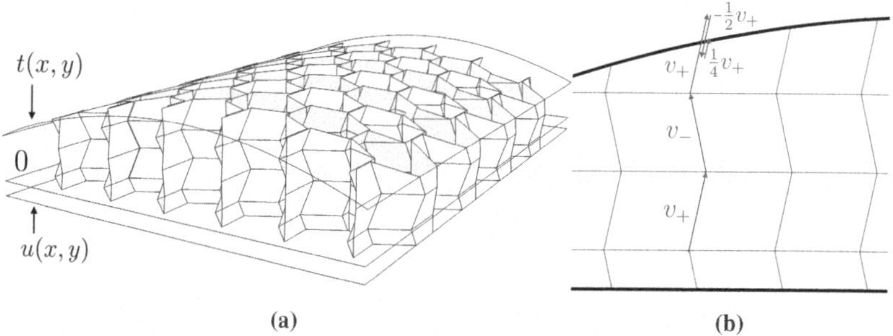

**Figure 4:** *A) Bounding surfaces t and u shown in reference to the $z = 0$ plane. B) Calculating the intersection of the zig-zag fold image curve with the bounding surfaces using binary search.*

Thus, to calculate the intersection point we start from a point known to lie between $t$ and $u$ (shown in Figure 4 as the $z = 0$ plane). We alternatively add $\mathbf{v}_+$ and $\mathbf{v}_-$ until we leave the valid region. At this point, we have bracketed the intersection point by the end points of one linear segment of the zig-zag fold image path. Hence, we can simply perform a root search (e.g. binary search) to approximate the intersection point as accurately as is desired. For example, in Figure 4, the intersection point is computed to be approximately $\mathbf{v}_+ + \mathbf{v}_- + \mathbf{v}_+ - \frac{1}{2}\mathbf{v}_+ + \frac{1}{4}\mathbf{v}_+$ away from the starting point.

Finally, we also note that for functions $t$ and $u$ with large curvature, self-intersections of the flat folding pattern can occur as a result of the zig-zag phase shift. In practice, these are small, and for the purposes of the present work, we ignore them. With these corrections and caveats, our modification of the kirigami honeycomb pattern produces Tachi-Miura honeycombs filling the same volume.

## 2 Results

### 2.1 Poisson Ratios

To analyze the mechanical behavior of these Tachi-Miura honeycombs, we first calculate their Poisson ratios (i.e., negative the ratio of strains between coordinate directions during the folding motion). We start from the expressions derived above for the $x$, $y$, and $z$ extents, calculating the Poisson ratios in terms of $\alpha$ and $\gamma$. Our expressions are similar to those derived in [Yasuda and Yang 15] but with a parameterization (shown in Figure 3) of cell extents that generalizes to volumes of adjacent cells (instead of applying to single cells).

To derive the poisson ratios of these structures, we must write expressions for the strains $\varepsilon_x = dx/x$, $\varepsilon_y = dy/y$, and $\varepsilon_z = dz/z$. Towards this end, we take derivatives of the spatial extents relative to $\gamma$:

$$\frac{dx}{d\gamma} = 4s\cos 2\gamma \tag{6}$$

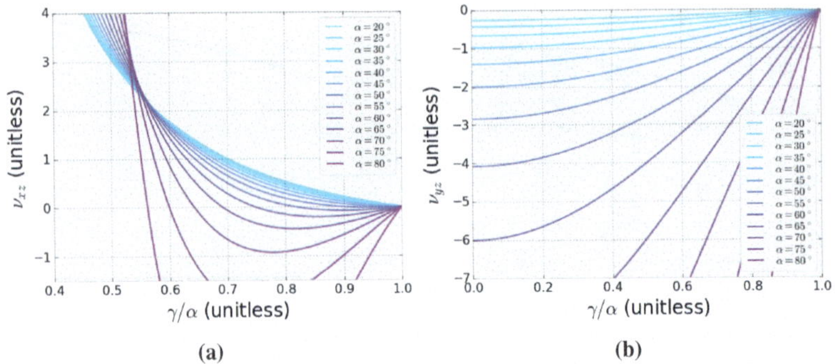

**Figure 5:** *Poisson ratios of Tachi-Miura honeycomb. A) $v_{xz} = -\varepsilon_x/\varepsilon_z$, B) $v_{yz} = -\varepsilon_y/\varepsilon_z$.*

$$\frac{dy}{d\gamma} = -2s\sin 2\gamma \tag{7}$$

$$\frac{dz}{d\gamma} = \frac{dz}{d\beta}\frac{d\beta}{d\gamma} = Nd\cos\beta\frac{d\beta}{d\gamma} = Nd\frac{-1}{\tan\beta\tan\alpha\cos^2\gamma} \tag{8}$$

where the final line follows from implicitly differentiating the governing equation. Now we calculate the strains:

$$\varepsilon_x = \frac{dx}{x} = \frac{dx}{d\gamma}\frac{d\gamma}{x} = \frac{d\gamma(1-\tan^2\gamma)}{\tan\gamma} \tag{9}$$

$$\varepsilon_y = \frac{dy}{y} = \frac{dy}{d\gamma}\frac{d\gamma}{y} = -2d\gamma\tan\gamma \tag{10}$$

$$\varepsilon_z = \frac{dz}{z} = \frac{dz}{d\gamma}\frac{d\gamma}{z} = \frac{-d\gamma\tan\gamma}{\tan^2\alpha\cos^2\gamma - \sin^2\gamma} \tag{11}$$

These expressions can be used directly to calculate the Poisson ratios, as the factor of $d\gamma$ cancels:

$$v_{xz} = -\frac{\varepsilon_x}{\varepsilon_z} \tag{12}$$

$$v_{yz} = -\frac{\varepsilon_y}{\varepsilon_z} \tag{13}$$

These values are shown in Figure 5. We see that the Poisson ratio depends on the choice of $\alpha$, but it is independent of $d$ and $N$. Further, for many choices of $\alpha$ and over much of the folding range, $v_{xz}$ is approximately zero, while $v_{yz}$ is very sensitive to $\alpha$ over this same range. Thus, by selecting $\alpha$, we can control the Poisson properties of the honeycomb.

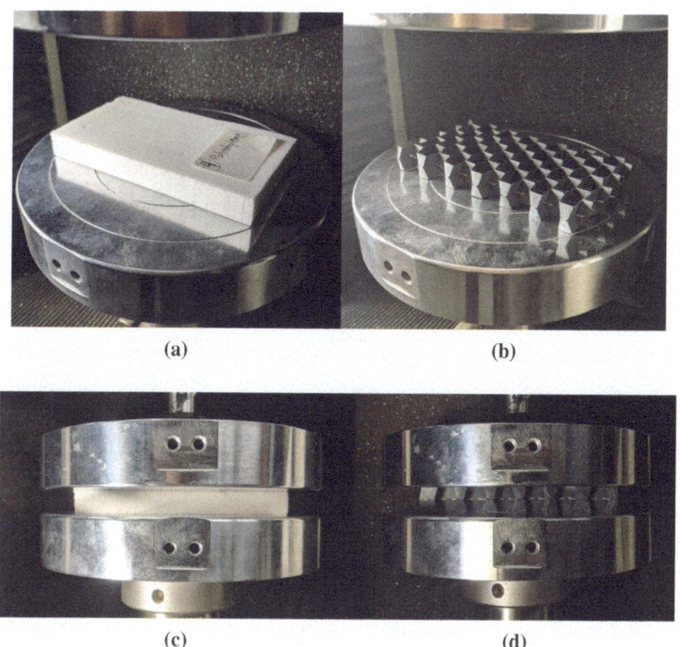

**Figure 6:** *A) Foam sample on test platen, B) Folded sample on test platen, C) Foam sample loaded by platens, D) Folded sample loaded by platens.*

## 2.2 Elastic behavior testing

The behavior of Tachi-Miura honeycombs under realistic boundary conditions is a combination of the rigid mechanism analyzed above and elastic deformations of the underlying material. A nonzero Poisson ratio means that fixed or frictional boundary conditions effectively resist the structure's tendency to follow its rigid mechanism under a compressive load in the $z$ direction [Yasuda et al. 13]. The larger the magnitude of the Poisson ratio, the more lateral movement is required by the rigid mechanism, and hence, the more resistance is offered from the boundary condition. As the average magnitude of the Poisson ratios $v_{xz}$ and $v_{yz}$ increases with $\alpha$, we expect the effective stiffness of Tachi-Miura honeycombs subject to fixed or frictional boundary conditions to also increase with $\alpha$.

To confirm this effect, we performed compression testing of Tachi-Miura honeycombs comparing the results to those of commercially available engineering foams. Using a constant base material and sheet thickness, we prepared samples with varying Miura angles. We stitched these samples to retain the folded state of the strip folds.

We applied a load of 100 N to the samples over their full area of 54 $cm^2$, cycling the load six times at a rate of 1 $mm$ per minute. We tested commercial foam samples of several materials having similar density to the pleated honeycomb samples with

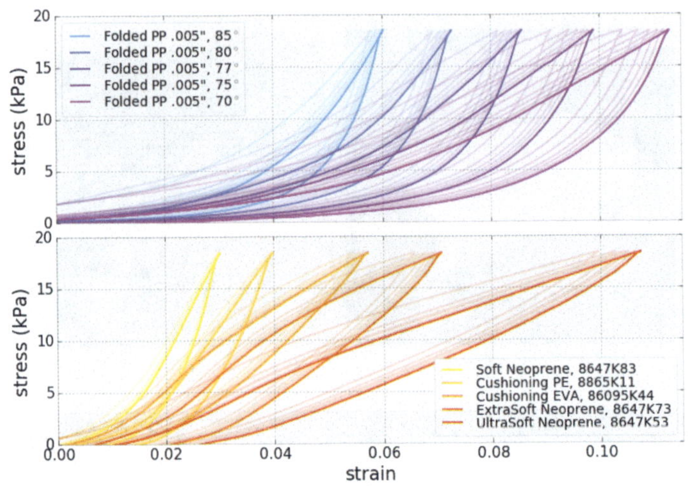

**Figure 7:** *Loading curves of Tachi-Miura polyhedra compared with those of engineering foams*

the same test parameters. The results (graphed in Figure 7) show the the pleated honeycombs span a range of average modulus from roughly 140 to 300 kPa, while the commercial foams range from roughly 170 to 500 kPa.

This shows that by using a single material and simply changing the Miura angle, we can produce a similar range of stiffness as is exhibited by a range of commercial foams. The conventional foams use a variety of materials and process parameters to achieve this range of stiffness. Thus, the reduction of material properties to ge-ometry has the potential to simplify the manufacturing supply chains for products using foam-like materials. Further, foams are commonly made in a heat-induced expansion process, giving no ability to spatially vary material properties. Tachi-Miura honeycombs can be constructed where Miura angle varies from cell to cell, effectively giving the ability to spatially map stiffness. While polyhedra with dif-fering Miura angle are not strictly tileable, similar methods as outlined above can work in practice to produce such honeycombs with only minor gaps between adja-cent cells. Quantifying the precise limits to this approach is an active area of future research.

## 2.3 Shoe sole prototype

Given that Tachi-Miura honeycombs can exhibit similar compliance as commercial foams while filling a prescribed shape, we now use our construction to prototype a running shoe sole. The thermal processes involved in creating a conventional sole are responsible for nearly half of the energy required to manufacture an athletic shoe [Cheah et al. 13], and so creating a sole using only cutting and folding processes is an attractive strategy for embodied energy reduction. Further, athletic soles are conventionally produced in a machined mold, requiring a large number of identical soles to be produced for economic feasibility. Specifying the sole geometry using the fold pattern instead of a mold has the potential to enable customized shoes without this overhead.

Figure 8 shows the steps of producing this folded running shoe sole. First, the geometry is digitally designed. In this case, the sole has a thickness of 20mm throughout the heel and tapers to 10mm through the forefoot. The footprint of cells which are populated with Tachi-Miura polyhedra are selected to fit inside the shape of common mens size 10 sole outline. The angle $\alpha$ is $50°$, while the lengths $d$ and $s$ are 2.8$mm$ and 8.8$mm$, respectively. The top three images at left of Figure 8 show the three-dimensional design and the corresponding folding and cutting pattern.

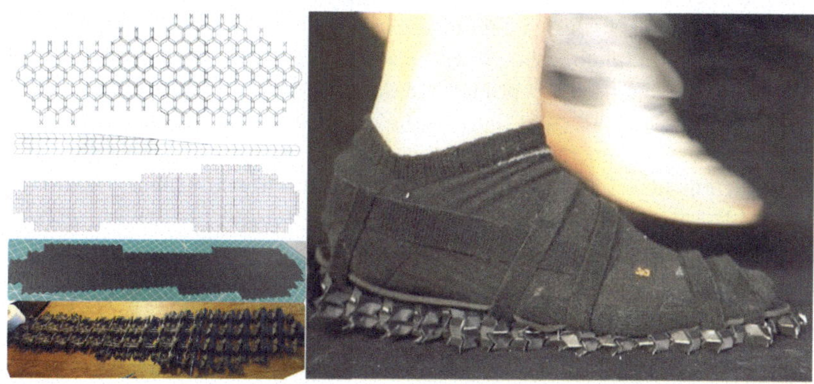

**Figure 8:** *A) Plan view of digitally designed geometry for shoe sole, B) Altitude view, C) Cut and fold pattern for shoe sole, D) Laser cut polypropylene prototype, E) After initial folding, F) Testing the shoe sole at running speed.*

This pattern is sent to a laser-cutter, where the outline and slits are through-cut and the creases are half-cut. The material used is 0.25$mm$ polypropylene, which produces robust creases with this laser processing technique. This sheet is then folded with the aid of 3D printed creasing dies of progressively larger fold angles $\gamma$. The flat and folded states are shown in the bottom two images at the left of Figure 8, and the folding motion is shown in this video. Finally, cotton thread is passed through laser cut holes and pulled tight in order to keep the strip folds held closed. The finished shoe is shown in Figure 8 at right during a running test.

When loaded by a footstep, the folded shoe sole compresses to levels similar to that of a conventional foam shoe, absorbing the impact of the foot strike. At this point, the forces arising from the Poisson effects and boundary conditions balances the force from the foot, and the Tachi-Miura honeycomb is limited from continuing its flat-folding mechanism. The deformation occurs chiefly at regions of highest force (for example, under the heel and metatarsal heads), effectively conforming to the underfoot shape. With a thin neoprene sock liner, the result was comfortable for a range of test athletes. This video shows high speed video of the shoe sole in use at running speeds.

# 3 Conclusions

We have shown how existing techniques for producing shaped hexagonal-celled honeycombs can be adapted to produce shaped honeycombs with Tachi-Miura polyhedra as cells. The bi-directional flat folding mechanism of the Tachi-Miura polyhedron endows the resulting honeycombs with the same flat foldings. In physical prototypes, finite-thickness materials and boundary conditions turn this flat folding into an elastic mechanism, a powerful tool for building robust structures with compliance.

Motivated by this, we investigated the mechanical properties of Tachi-Miura honeycombs, first calculating the rigid Poisson ratios. Next, we compared the mechanical behavior of the structures to common engineering foams, showing it is possible to create a similar range of compliance by changing the angle $\alpha$. Finally, we use these findings to design a prototype making use of both the compliance and the shape control: a running shoe sole.

In this work, we used 3D-printed creasing dies as a tool to fold the samples quickly. This was effective but required considerable manual labor. For future work, we will increase production rates by either automating the use of similar incremental forming dies, or using "pattern and gather" approaches [Schenk 12].

# References

[Calisch and Gershenfeld 18] Sam Calisch and Neil Gershenfeld. "Towards continuous production of shaped honeycombs." In *ASME 2018 Manufacturing Science and Engineering Conference*. American Society of Mechanical Engineers, 2018.

[Cheah et al. 13] Lynette Cheah, Natalia Duque Ciceri, Elsa Olivetti, Seiko Matsumura, Dai Forterre, Richard Roth, and Randolph Kirchain. "Manufacturing-focused emissions reductions in footwear production." *Journal of Cleaner Production* 44 (2013), 18 – 29. Available online (http://www.sciencedirect.com/science/article/pii/S0959652612006300).

[Miura and Tachi 10] Koryo Miura and Tomohiro Tachi. "Synthesis of Rigid-Foldable Cylindrical Polyhedral." *Symmetry: Art and Science, International Society for the Interdisciplinary Study of Symmetry, Gmuend.* Available online (http://www.tsg.ne.jp/TT/cg/FoldableCylinders_miura_tachi_ISISSymmetry2010.pdf).

[Nojima and Saito 06] Taketoshi Nojima and Kazuya Saito. "Development of newly designed ultra-light core structures." *JSME International Journal Series A Solid Mechanics and Material Engineering* 49:1 (2006), 38–42. Available online (http://adsabs.harvard.edu/abs/2006JSMEA..49...38N).

[Saito et al. 11] Kazuya Saito, Fabio Agnese, and Fabrizio Scarpa. "A cellular kirigami morphing wingbox concept." *Journal of intelligent material systems and structures* 22:9 (2011), 935–944. Available online (http://jim.sagepub.com/content/22/9/935.refs).

[Saito et al. 14] Kazuya Saito, Sergio Pellegrino, and Taketoshi Nojima. "Manufacture of arbitrary cross-section composite honeycomb cores based on origami techniques." *Journal of Mechanical Design* 136:5 (2014), 051011. Available online (http://pressurevesseltech.asmedigitalcollection.asme.org/article.aspx?articleid=1831332).

[Schenk 12] Mark Schenk. "Folded shell structures." Ph.D. thesis, University of Cambridge, 2012. Available online (http://www.markschenk.com/research/files/PhD%20thesis%20-%20Mark%20Schenk.pdf).

[Tachi and Miura 12] Tomohiro Tachi and Koryo Miura. "Rigid-foldable cylinders and cells." *J. Int. Assoc. Shell Spat. Struct* 53:4 (2012), 217–226. Available online (http://origami.c.u-tokyo.ac.jp/~tachi/cg/RigidFoldableCylindersCellsTachiMiuraJournalIASS2012.pdf).

[Wang et al. 17] Lijun Wang, Kazuya Saito, You Gotou, and Yoji Okabe. "Design and fabrication of aluminum honeycomb structures based on origami technology." *Journal of Sandwich Structures & Materials* 0:0 (2017), 1099636217714646. doi:10.1177/1099636217714646. Available online (https://doi.org/10.1177/1099636217714646).

[Yasuda and Yang 15] H Yasuda and Jinkyu Yang. "Reentrant Origami-Based Metamaterials with Negative Poisson's Ratio and Bistability." *Physical review letters* 114:18 (2015), 185502. Available online (http://faculty.washington.edu/jkyang/research/publication/2015_PRL_origami.pdf).

[Yasuda et al. 13] Hiromi Yasuda, Thu Yein, Tomohiro Tachi, Koryo Miura, and Minoru Taya. "Folding behaviour of Tachi–Miura polyhedron bellows." In *Proc. R. Soc. A*, 469, 469, p. 20130351. The Royal Society, 2013. Available online (http://rspa.royalsocietypublishing.org/content/469/2159/20130351.short).

[Yasuda et al. 16] H. Yasuda, C. Chong, E. G. Charalampidis, P. G. Kevrekidis, and J. Yang. "Formation of rarefaction waves in origami-based metamaterials." *Phys. Rev. E* 93 (2016), 043004. doi:10.1103/PhysRevE.93.043004. Available online (https://link.aps.org/doi/10.1103/PhysRevE.93.043004).

---

Sam Calisch

Center for Bits and Atoms, Massachusetts Institute of Technology, Cambridge, MA, U.S.A.
e-mail: sam.calisch@cba.mit.edu

Neil Gershenfeld

Center for Bits and Atoms, Massachusetts Institute of Technology, Cambridge, MA, U.S.A.
e-mail: neil.gershenfeld@cba.mit.edu

## Acknowledgement

This work was supported by the Center for Bits and Atoms research consortia. The authors would also like to thank David Lackner for help fabricating the foam test specimens.

# The Editors of Origami[7]

The papers in Origami[7] were reviewed by a Programme committee of 17 members headed up by Robert J Lang. The committee included; Robert Lang (Chair), Norma Boakes, Mark Bolitho, Chris Budd, Yan Chen, Mary Frecker, Simon Guest, Thomas Hull, Yves Klett, Jun Mitani, Jorge Pardo, Glaucio Paulino, Mark Schenk, Tomohiro Tachi, Ryuhei Uehara, Patsy Wang-Iverson and Zhong You.

**Robert J. Lang** is an independent origami artist and consulting, who has exhibited his origami artwork at museums around the world and has consulted for both companies and academia on the applications of origami to technological problems.
(robert@langorigami.com)

**Norma Boakes** is an Associate Professor of Education for the School of Education at Stockton University in New Jersey, USA. Her scholarly agenda includes extensive study of Origami as a tool for teaching math, science, and STEM/STEAM concepts at the K-12 and higher education level.
(Norma.Boakes@stockton.edu)

**Mark Bolitho** is part of the organising team for 7OSME. Mark works as an Origami Artist through his company, Creaselightning Ltd. Formerly Chairman of the British Origami Society, he has been a guest at various international Origami events in countries including Japan, Korea, China and Germany. For more information see the website; www.creaselightning.co.uk
(mark@creaselightning.co.uk)

**Chris Budd** OBE is Professor of Mathematics at the University of Bath and Gresham Professor of Geometry. He has worked on applications of origami to rock folding, aircraft wing design and structural mechanics. A passionate populariser of mathematics he frequently uses origami to show young people how creative mathematics can be, and the strong links between maths and art. (C.J.Budd@bath.ac.uk)

**Yan Chen** is a Professor of Engineering at Tianjin University, China, where she established and now heads the Motion Structure Laboratory. (yan_chen@tju.edu.cn)

**Mary Frecker** is a Professor of Mechanical Engineering and Biomedical Engineering and is the Associate Department Head for Graduate Programs in Mechanical & Nuclear Engineering at the Pennsylvania State University. She is a Fellow of the ASME. Research interests include optimal design of compliant mechanisms, adaptive structures, and origami-inspired devices. (mxf36@engr.psu.edu)

**Simon Guest** is Professor of Structural Mechanics at the University of Cambridge, where he has spent his entire academic career. His research work straddles the border between traditional structural mechanics, and the study of mechanisms. He has worked with origami concepts from his earliest research work on the development of novel deployable structures. (sdg@eng.cam.ac.uk)

**Thomas C. Hull** is an associate professor of mathematics at Western New England University in Springfield, MA, USA. He has been collecting and creating research on the mathematics of paper folding since 1990 and is considered a go-to expert in the area. He has written a book, Project Origami: Activities for Exploring Mathematics (2nd edition, 2013, CRC Press) on using origami to teach mathematics in university-level math classes. (thomas.hull@wne.edu)

**Yves Klett** had an early infatuation with paper planes, his love for things flying led him to study aerospace engineering in Stuttgart. Folding caught up with him there as well, this time for engineering purposes. He leads the Sandwich group at the Institute of Aircraft Design in Stuttgart and is co-founder of two folding-related companies. (yves.klett@gmail.com)

**Jun Mitani** is a professor of Information and Systems at University of Tsukuba. He was born in Shizuoka pref. in 1975, He received his Ph.D. in engineering from the University of Tokyo in 2004. He has been present post since April 2015. His research interests center on computer graphics. (jmitani@gmail.com)

**Jorge Pardo** is the Director of the Educational Museum of Origami, Zaragoza (EMOZ) in Spain. Formerly an Industrial Engineer he studied at Zaragoza University. His creative design is mainly modular origami models, the most famous model is "Flexiball" with 60 squares. He has been a guest at origami conventions in Spain, Italy, Peru, USA and Japan. (jorgepardo05@yahoo.es)

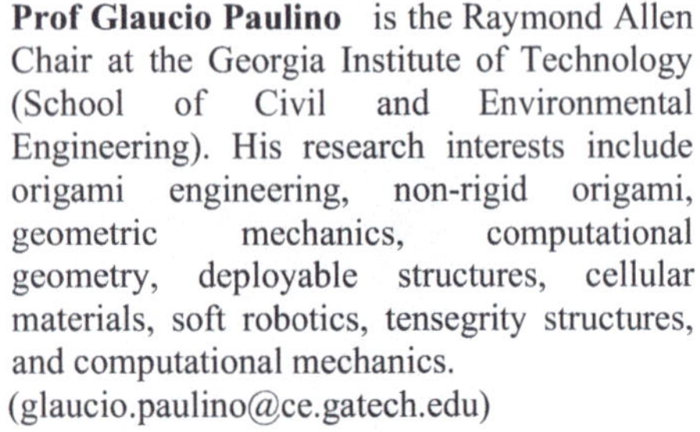

**Prof Glaucio Paulino** is the Raymond Allen Chair at the Georgia Institute of Technology (School of Civil and Environmental Engineering). His research interests include origami engineering, non-rigid origami, geometric mechanics, computational geometry, deployable structures, cellular materials, soft robotics, tensegrity structures, and computational mechanics.
(glaucio.paulino@ce.gatech.edu)

**Mark Schenk** is a Lecturer in Aerospace Engineering at the University of Bristol (UK), and works within the Bristol Composites Institute (ACCIS). He works on deployable, morphing and folding structures, and is developing novel experimental methods for nonlinear structures. He has worked as Research Associate at Surrey Space Centre, applying origami techniques to deployable structures for small satellites, and the Advanced Structures Group at the Cambridge University Engineering Department. His PhD was on Engineering Origami structures at Cambridge University, and his BSc/MSc degrees are from Delft University of Technology.
(m.schenk@bristol.ac.uk)

**Tomohiro Tachi** Tomohiro Tachi is an associate professor in Graphic and Computer Sciences at the University of Tokyo. He has been designing origami from 2002 and keeps exploring three-dimensional and kinematic origami through computation. His research interests include origami, structural morphology, computational design, and fabrication.
(tachi@idea.c.u-tokyo.ac.jp)

**Ryuhei Uehara** is a full professor in School of Information Science, Japan Advanced Institute of Science and Technology (JAIST). His research interests include computational complexity, algorithms and data structures, and graph algorithms. Especially, he is engrossed in computational origami, games and puzzles from the viewpoints of theoretical computer science.
(uehara@jaist.ac.jp)

**Patsy Wang-Iverson**, biochemist by training, is Vice President for Special Projects at The Gabriella & Paul Rosenbaum Foundation. She co-organized 5OSME and co-edited Origami^5 and Origami^6. She is managing editor of OrigamiUSA's online magazine, The Fold.
(pwangiverson@gmail.com)

**Zhong You** is leading the organising team for 7OSME. He is Professor of Engineering Science at Department of Engineering Science, University of Oxford. He works on deployable and origami structures, especially the kinematics of rigid origami and mechanical properties of origami structures.
(zhong.you@eng.ox.ac.uk)

# 7OSME Sponsors and Supporting Partners

Thanks to the following organisations who have provided support and sponsorship to 7OSME.

University of Oxford, Tianjin University, Creaselightning Ltd, British Origami Society, Daiwa Foundation, Great British Sasakawa Foundation, the Lubbock Grant, Institute of Mathematics, Tarquin, Origami USA, Gabriella & Paul Rosenbaum Foundation, Oxford Space Systems, and Yupo Europe GmbH.

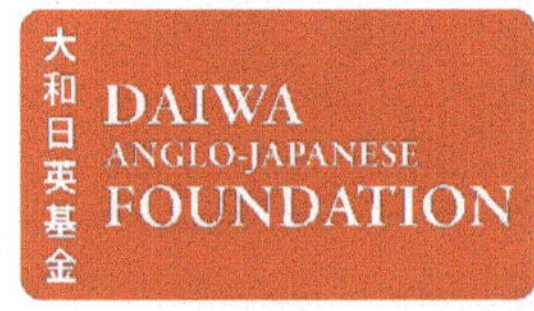

Gabriella & Paul Rosenbaum Foundation

# Acknowledgements

Many thanks are due to the team that worked on the papers in the Engineering volumes (3 and 4) of Origami[7].

Papers were managed by the following Programme committee members: Yan Chen, Mary Frecker, Simon Guest, Yves Klett, Robert J. Lang, Jun Mitani, Glaucio Paulino, Mark Schenk, Tomohiro Tachi and Zhong You.

This process was supported by reviews provided by Manan Arya, Emmanuel Baranger, Alex Bateman, John Bowers, Henri Buffart, Jianguo Cai, Sam Calisch, Sebastien Callens, Changqing Chen, Yao Chen, Wensu Chen, Kenneth Cheung, Rostislav Chudoba, Fehmi Cirak, Erica Crampton, Marcelo Dias, David Dureisseix, Yuki Endo, Andreas Falk, Hongbin Fang, Evgueni Filipov, Riccardo Foschi, Matthew Gardiner, Joseph Gattas, Amanda Ghassaei, Steve Grey, Marc Grzeschik, Hongwei Guo, Rachel Hanlon, Hong Hao, Sebastian Heimbs, Martin Hillebrandt, Susanne Hoffmann, Larry Howell, Hanqing Jiang, Florian Kovacs, Suyi Li, Sungjoon Lim, Ke Liu, Guoxing Lu, Jiayao Ma, Spencer Magleby, Rupert Maleczek, Chinthaka Mallikarachchi, Allan McRobie, Graham McShane, Peter Middendorf, Marc Miskin, Fabian Muhs, Juan Musto, Todd Nelson, Larissa Novelino, Johannes Overvelde, Bas Overvelde, Alexander Pagano, Jascha Paris, Edwin Peraza Hernandez, Phanisri Pratapa, Anup Pydah, Austin Reid, Dong Ruan, Jason S. Ku, Kazuya Saito, Fabrizio Scarpa, Simon Schleicher, Keith Seffen, Deanna Sessions, Jesse Silverberg, Rebecca Strzelec, Cynthia Sung, Hirotaka Suzuki, Tibor Tarnai, Sean Tolman, Javier Vila Moran, Andrew Viquerat, Paris von Lockette, Martin Walker, Hairui Wang, Yuka Watanabe, Xinmei Xiang, Jinkyu Yang, Yunfang Yang, Hiromi Yasuda, Amir Zadpoor, Yan Zhao and Xiang Zhou.

**Manan Arya** is a technologist and space origami engineer at NASA's Jet Propulsion Laboratory.

**Alex Bateman** is Head of Protein Sequence Resources at the European Bioinformatics Institute with an interest in Origami tesselations.
**John Bowers** works on discrete and computational geometry at James Madison University.

**Sam Calisch** is a PhD student and researcher at the Massachusetts Institute of Technology's Center for Bits and Atoms.
**Sebastien Callens** is a PhD student on biomechanics at the Delft University of Technology (S.J.P.Callens@tudelft.nl - Sebastien won't be on the formal review list, but has helped out Amir, listed above, and deserves recognition)
**Yao Chen** works on prestressed structures and origami structures at Southeast University, China.
**Dr. Wensu Chen** is ARC DECRA Fellow from Centre for Infrastructural Monitoring and Protection, Curtin University, Australia e-mail: wensu.chen@curtin.edu.au
**Fehmi Cirak** works on computational structural mechanics at the University of Cambridge.

**Erica Crampton** has an MSc in Mechanical Engineering from Brigham Young University and is currently working as a Mechanical Design Engineer.

**Evgueni Filipov** is an Assistant Professor at the University of Michigan. His research interests are in the mechanics of origami-inspired deployable and reconfigurable structures.

**Riccardo Foschi** is a PhD student University of Bologna (UNIBO) riccardo.foschi2@unibo.it

**Matthew Gardiner** is an artist most well known for his work with origami and robotics. **Joe Gattas** is a Senior Lecturer and leads the 'Folded Structures Lab' research group at the University of Queensland, Australia. ( j.gattas@uq.edu.au).

**Sebastian Heimbs** is Head of Airframe Dynamic Solutions at Airbus. **Susanne Hoffmann** – architect & origami enthusiast – aims to transfer origami principles to engineering applications in her work at the Chair of Structures and Structural Design at RWTH Aachen University. Dipl.-Ing. Susanne Hoffmann, Chair of Structures and Structural Design/RWTH Aachen University, research topics: transfer origami principles to engineering applications (21 words) **Larry L. Howell** is professor and associate dean of the College of Engineering at Brigham Young University (USA).

**Flórián Kovács** works on structural mechanics at the Budapest University of Technology and Economics. **Jason Ku,** lecturer in the Electrical Engineering and Computer Science Department at MIT (Cambridge, MA, USA), who is also a world-class origami designer and researcher in computational origami and mechanical origami designs.

**Rupert Maleczek** is an educated architect, investigating the relation between origami, computational design, structural performance and the physical fabrication process. (Rupert.Maleczek@uibk.ac.at). **Allan McRobie** works on structural mechanics at the University of Cambridge. **Graham McShane** works on the mechanics of materials at the University of Cambridge. **Fabian Muhs**, (Dipl.-Ing) Researcher – Sandwich Technologies, Institute of Aircraft Design, University of Stuttgart

**Keith Seffen** works on shape-changing structures at the University of Cambridge. **Tibor Tarnai** works on structural mechanics at the Budapest University of Technology and Economics.

**Andrew Viquerat** works on the design and analysis of lightweight deployable and flexible structures at the University of Surrey, UK.

**Amir Zadpoor** works on biomechanics at the Delft University of Technology.

# Engineering Contributors

**Roland Aigner**, Ars Electronica Linz GmbH & Co KG, Austria. (w) http://www.aec.at, (e) roland.aigner@aec.at

**Alex Avila**, Brigham Young University, United States, (e) tealex13@gmail.com

**Terri Bateman**, Brigham Young University, United States, (e) terri_bateman@byu.edu

**Brakhage Karl-Heinz**, RWTH Aachen University, Germany, (e) brakhage@igpm.rwth-aachen.de

**Henri Buffart**, RWTH Aachen University, Germany. (w) http://trako.arch.rwth-aachen.de, (e) buffart@trako.arch.rwth-aachen.de

**Dakota Burrow**, Brigham Young University, United States, (e) dburrow@byu.edu

**Jianguo Cai**, Southeast University, Key Laboratory of C & PC Structures of Ministry of Education, China, (e) j.cai@seu.edu.cn

**Sam Calisch**, Center for Bits and Atoms, MIT, United States, (e) calisch@MIT.EDU

**Wensu Chen**, Curtin University, Australia. (w) https://staffportal.curtin.edu.au/staff/profile/view/Wensu.Chen, (e) wensu.chen@curtin.edu.au

**Yan Chen**, Tianjin University, China. , (e) yan_chen@tju.edu.cn

**Pisak Chermprayong**, Imperial College London, United Kingdom, (e)

**Rostislav Chudoba**, RWTH Aachen University, Germany, (e) rostislav.chudoba@rwth-aachen.de

**Burkhard Corves**, Department of Mechanism Theory and Dynamics of Machines, RWTH Aachen University, Germany. (w) http://igm.rwth-aachen.de, (e) corves@igm.rwth-aachen.de

**Erica Crampton**, Brigham Young University, United States, (e) ebcrampton@gmail.com

**Wei Qi Cui**, The University of Queensland, Australia, (e) weiqi.cui@uq.net.au

**Simon Dehn**, Institute for Machine Elements and Systems Engineering, RWTH Aachen University, Germany, (e) Simon.Dehn@rwth-aachen.de

**Erik D. Demaine**, Massachusetts Institute of Technology, United States, (e) edemaine@mit.edu

**Xiang Deng**, University of Pennsylvania, United States, (e) dxiang@seas.upenn.edu

**Zongquan Deng**, Harbin Institute of Technology, China, (e) dengzq@hit.edu.cn

**Luis Diago**, Interlocus Inc, Japan. (w) http://www.i-locus.com/, (e) ldiago@i-locus.com

**Daniel Eatough**, University of Cambridge, United Kingdom, (e) dte22@cam.ac.uk

**Marc Emmanuelli**, Imperial College London, United Kingdom, (e)

**Jian Feng**, Southeast University, Key Laboratory of C & PC Structures of Ministry of Education, China, (e) fengjian@seu.edu.cn

**Dilum Fernando**, The University of Queensland, Australia, (e) dilum.fernando@uq.edu.au

**Riccardo Foschi**, Unibo, Italy. (w) https://www.unibo.it/sitoweb/riccardo.foschi2, (e) riccardo.foschi2@unibo.it

**Taro Fujikawa**, Tokyo Denki University, Japan, (e) fujikawa@fr.dendai.ac.jp

**Matthew Gardiner**, Ars Electronica Linz GmbH & Co KG, Austria. (w)
http://www.matthewgardiner.net, (e) Matthew.Gardiner@aec.at
**Joseph M. Gattas**, The University of Queensland, Australia. (w)
http://www.joegattas.com, (e) j.gattas@uq.edu.au
**Neil Gershenfeld**, Massachusetts Institute of Technology, United States, (e)
neil.gershenfeld@cba.mit.edu
**Timo Gfeller**, ETH Zurich, Switzerland, (e) tgfeller@student.ethz.ch
**Amanda Ghassaei**, Massachusetts Institute of Technology, United States, (e)
amandaghassaei@gmail.com
**Steven Grey**, Bristol Composites Institute (ACCIS), University of Bristol, United
Kingdom, (e) steven.grey@bristol.ac.uk
**Marc Grzeschik**, University Stuttgart, Institute of Aircraft Design, Germany. (w)
http://www.ifb.uni-stuttgart.de, (e) grzeschik@ifb.uni-stuttgart.de
**Hongwei Guo**, Harbin Institute of Technology, China, (e) guohw@hit.edu.cn

**Ichiro Hagiwara**, Meiji University, Japan. (w)
http://www.isc.meiji.ac.jp/~hagilab/index.html, (e) ihagi@meiji.ac.jp
**Rachel Hanlon**, Ars Electronica Linz GmbH & Co KG, Austria. (w)
http://www.rachelhanlon.com, (e) rachel.hanlon@aec.at
**Hong Hao**, Curtin University, Australia. (w)
https://staffportal.curtin.edu.au/staff/profile/view/Hong.Hao, (e) hong.hao@curtin.edu.au
**Qian He**, Hokkaido University, Japan, (e) heatherhe8@hotmail.com
**Michael T. Heitzmann**, The University of Queensland, Australia, (e)
m.heitzmann@uq.edu.au
**Janette Herron**, Brigham Young University, United States, (e) jviolin27@gmail.com
**Susanne Hoffmann**, Chair of Structures and Structural Design, RWTH Aachen University,
Germany. (w)http://trako.arch.rwth-aachen.de, (e) hoffmann@trako.arch.rwth-aachen.de
**Larry Howell**, Brigham Young University, United States. (w)
http://compliantmechanisms.byu.edu/, (e) lhowell@byu.edu
**Jiangping Huang**, College of mechanical and electrical engineering, Harbin Institute of
Technology, China, (e) 15B908065@hit.edu.cn
**Mathias Huesing**, RWTH Aachen University, Germany. (w) http://igm.rwth-aachen.de,
(e) huesing@igm.rwth-aachen.de

**Georg Jacobs**, Institute for Machine Elements and Systems Engineering, RWTH Aachen
University, Germany, (e) jacobs@ikt.rwth-aachen.de

**Yoshihiro Kawahara**, The University of Tokyo, Japan, (e) kawahara@akg.t.u-tokyo.ac.jp
**Yves Klett**, University Stuttgart, Institute of Aircraft Design, Germany. (w)
http://www.ifb.uni-stuttgart.de, (e) yves.klett@gmail.com
**Goran Konjevod**, organicorigami.com, United States. (w) http://organicorigami.com, (e)
goran.konjevod@gmail.com
**Mirko Kovac**, Imperial College London, United Kingdom, (e)
**Kaori Kuribayashi-Shigetomi**, Hokkaido University, Japan, (e) kaorik@ist.hokudai.ac.jp

**Arthur Lebée**, ENPC, France, (e) arthur.lebee@enpc.fr

**Ting Uei Lee**, The University of Queensland, Australi , (e) jeff.lee@uq.edu.au

**Junlan Li**, Tianjin University, China, (e) jlli@tju.edu.cn

**Meng Li**, Qian Xuesen Laboratory of Space Technology, China, (e) limeng@qxslab.cn

**Zhejian Li**, Curtin University, Australia, (e) zhejian.li@postgrad.curtin.edu.au

**Ke Liu**, Georgia Institute of Technology, United States, (e) kliu89@gatech.edu

**Rongqiang Liu**, Harbin Institute of Technology, China, (e) liurq@hit.edu.cn

**Xinying Lv**, School of Aeronautics & Astronautics, Shanghai Jiao Tong University, China, (e) xinying.lv@sjtu.edu.cn

**Jiayao Ma**, Tianjin University, China, (e) jiayao.ma@tju.edu.cn

**Spencer Magleby**, Brigham Young University, United States. (w) http://compliantmechanisms.byu.edu/, (e) magleby@byu.edu

**Rupert Maleczek**, University of Innsbruck; Institute of Design |i.sd - Structure and Design, Austria. (w) http://www.koge.at, (e) rupert.maleczek@uibk.ac.at

**R. Daniel Maynes**, Brigham Young University, United States, (e) maynes@byu.edu

**Judith Merz**, RWTH Aachen University, Germany. (w) http://igm.rwth-aachen.de, (e) merz@igmr.rwth-aachen.de

**Peter Middendorf**, Institut für Flugzeugbau / Universität Stuttgart, Germany. (w) http://www.ifb.uni-stuttgart.de, (e) middendorf@ifb.uni-stuttgart.de

**Jun Mitani**, University of Tsukuba, Japan, (e) mitani@npal.cs.tsukuba.ac.jp

**Yoshinobu Miyamoto**, Aichi Institute of Technology, Japan. (w) http://www.flickr.com/photos/31375127@N07/sets/, (e) yoshinobu.miyamoto@gmail.com

**Laurent Monasse**, INRIA, France, (e) laurent.monasse@inria.fr

**David Morgan**, Brigham Young University, United States, (e) dcmorgan@byu.edu

**Fabian Muhs**, Institute of Aircraft Design, Germany, (e) muhs@ifb.uni-stuttgart.de

**Juan Musto**, Chair of Structures and Structural Design, RWTH-Aachen, Germany. (w) https://trako.arch.rwth-aachen.de, (e) musto@trako.arch.rwth-aachen.de

**Haris Nadeem**, Imperial College London, United Kingdom, (e)

**Chie Nara**, Meiji university, Japan, (e) cnara@jeans.ocn.ne.jp

**Hussein Nassar**, University of Missouri, United States, (e) nassarh@missouri.edu

**Syed Nauroze**, Georgia Institute of Technology, United States, (e)

**Ryuma Niiyama**, The University of Tokyo, Japan, (e) niiyama@isi.imi.i.u-tokyo.ac.jp

**Larissa Novelino**, Georgia Institute of Technology, United States, (e) larissanovelino@hotmail.com

**Hideaki Ogawa**, Ars Electronica Linz GmbH & Co KG, Austria. (w) http://www.aec.at, (e) hideaki.ogawa@aec.at

**Takaharu Okajima**, Hokkaido University, Japan, (e) okajima@ist.hokudai.ac.jp

**Fuminori Okuya**, The University of Tokyo, Japan, (e) okuya23@akg.t.u-tokyo.ac.jp

**Jascha Paris**, Institute of Mechanism Theory, Machine Dynamics and Robotics, RWTH Aachen University, Germany. (w)http://igm.rwth-aachen.de, (e) Paris@igm.rwth-aachen.de

**Glaucio Paulino**, Georgia Institute of Technology, United States. (w) http://paulino.ce.gatech.edu, (e) glaucio.paulino@ce.gatech.edu

**James Pikul**, University of Pennsylvania, United States, (e) pikul@seas.upenn.edu

**Erwin Reitboeck**, Ars Electronica Linz GmbH & Co KG, Austria. (w) http://www.aec.at, (e) erwin.reitboeck@aec.at
**Esther Rivas-Adrover**, University of Cambridge, United Kingdom, (e) era.esther@gmail.com
**Julian Romero**, Meiji University, Japan, (e) jaromero_18@hotmail.com

**Kazuya Saito**, The University of Tokyo, Japan, (e) ksaito@akg.t.u-tokyo.ac.jp
**Pooya Sareh**, Imperial College London, United Kingdom, (e) p.sareh@imperial.ac.uk
**Fabrizio Scarpa**, Bristol Composites Institute (ACCIS), University of Bristol, United Kingdom, (e) f.scarpa@bristol.ac.uk
**Mark Schenk**, Bristol Composites Institute (ACCIS), University of Bristol, United Kingdom, (e) m.schenk@bristol.ac.uk
**Keith Seffen**, University of Cambridge, United Kingdom. (w) http://www.eng.cam.ac.uk/profiles/kas14, (e) kas14@cam.ac.uk
**Kendall Seymour**, Brigham Young University, United States, (e) seymourkendall@gmail.com
**Kristina Shea**, ETH Zurich, Switzerland, (e) kshea@ethz.ch
**Junichi Shinoda**, Interlocus Inc, Japan. (w) http://www.i-locus.com/, (e) jshinoda@i-locus.com
**Justus Siebrecht**, Chair and Institute for Engineering Design, RWTH Aachen University, Germany. (w)http://ikt.rwth-aachen.de/, (e) siebrecht@ikt.rwth-aachen.de
**Kathrin Spütz**, Institute for Machine Elements and Systems Engineering, RWTH Aachen University, Germany, (e) kathrin.spuetz@rwth-aachen.de
**Tino Stankovic**, ETH Zurich, Switzerland, (e) tinos@ethz.ch
**Gabriel Stern**, University of Innsbruck; Institute of Design |i.sd - Structure and Design, Austria, (e) Gabriel.Stern@student.uibk.ac.at
**Chad Stucki**, Brigham Young University, United States, (e) stuckic25@gmail.com
**Cynthia Sung**, University of Pennsylvania, United States, (e) crsung@seas.upenn.edu

**Tomohiro Tachi**, The University of Tokyo, Japan. (w) http://www.tsg.ne.jp/TT/, (e) tachi@g.ecc.u-tokyo.ac.jp
**Dewei Tang**, Harbin Institute of Technology, China, (e) dwtang@hit.edu.cn
**Yuzhen Tang**, Harbin Institute of Technology, China, (e) tangyuzhen92@163.com
**Emmanouil Tentzeris**, Georgia Institute of Technology, United States, (e)
**Kyler Tolman**, Brigham Young University, United States, (e) kylertolman@gmail.com
**Martin Trautz**, Chair of Structures and Structural Design, RWTH Aachen, Germany. (w) http://trako.arch.rwth-aachen.de, (e) trautz@trako.arch.rwth-aachen.de
**Naoya Tsuruta**, Tokyo University of Technology, Japan, (e) tsurutany@stf.teu.ac.jp

**Martin Walker**, University of Oxford, United Kingdom, (e) mgw39@cam.ac.uk
**Cheng Wang**, Tianjin University, China, (e) kwkck@tju.edu.cn
**Hairui Wang**, Dalian University of Technology, China, (e) whrlixue@126.com
**Minjie Wang**, Dalian University of Technology, China, (e) mjwang@dlut.edu.cn
**Naohiko Watanabe**, National Institute of technology, Gifu college, Japan, (e) watanabe@gifu-nct.ac.jp
**Yuka Watanabe**, University of Tsukuba, Japan, (e) yukakohno@hotmail.com
**Chantal Weigel**, Institute for Machine Elements and Systems Engineering, RWTH Aachen University, Germany. (w)http://ikt.rwth-aachen.de, (e) weigel@ikt.rwth-aachen.de

**Yohei Yamamoto**, Independent, Japan, (e) yohey.yamamort@gmail.com
**Xiaochen Yang**, Tianjin University, China, (e) xiaochen_yang@tju.edu.cn
**Yunfang Yang**, University of Oxford, United Kingdom, (e) yangyunfang12@gmail.com
**Zhong You**, University of Oxford, United Kingdom, (e) zhong.you@eng.ox.ac.uk
**Hang Yuan**, University of Pennsylvania, United States, (e) yuanhang@seas.upenn.edu

**Shixi Zang**, Tianjin University, China, (e) sx_zang@tju.edu.cn
**Qian Zhang**, Southeast University, Key Laboratory of C & PC Structures of Ministry of Education, China. , (e) 220150947@seu.edu.cn
**Danyang Zhao**, Dalian University of Technology, China, (e) zhaody@dlut.edu.cn
**Xiang Zhou**, School of Aeronautics & Astronautics, Shanghai Jiao Tong University, China, (e) xiangzhou@sjtu.edu.cn
**Luca Zimmermann**, ETH Zurich, Switzerland, (e) lucaz@ethz.ch

# Origami[7] Index

Printed and bound by CPI Group (UK) Ltd, Croydon, CR0 4YY

06/07/2026

02157566-0001